Computational Methods for Algebraic Spline Surfaces

Tor Dokken · Bert Jüttler

Computational Methods for Algebraic Spline Surfaces

ESF Exploratory Workshop

With 102 Figures

 Springer

Tor Dokken
SINTEF Applied Mathematics
P.O. Box 124 Blindern
N-0314 Oslo, Norway
e-mail: tor.dokken@sintef.no

Bert Jüttler
Johannes Kepler University
Institute of Applied Geometry
Altenberger Straße 69
A-4040 Linz, Austria
e-mail: bert.juettler@jku.at

Mathematics Subject Classification (2000): 14Pxx, 14Qxx, 53A05, 65D17, 68U05

ISBN 978-3-642-06233-9   e-ISBN 978-3-540-27157-4

Springer is a part of Springer Science+Business Media

springeronline.com

© Springer-Verlag Berlin Heidelberg 2010
Printed in Germany

Cover design: *design & production* GmbH, Heidelberg

Printed on acid-free paper 46/3142YL – 5 4 3 2 1 0

Preface

This volume contains revised papers that were presented at the international workshop entitled *Computational Methods for Algebraic Spline Surfaces* ("COMPASS"), which was held from September 29 to October 3, 2003, at Schloß Weinberg, Kefermarkt (Austria).

The workshop was mainly devoted to approximate algebraic geometry and its applications. The organizers wanted to emphasize the novel idea of approximate implicitization, that has strengthened the existing link between CAD / CAGD (Computer Aided Geometric Design) and classical algebraic geometry. The existing methods for exact implicitization (i.e., for conversion from the parametric to an implicit representation of a curve or surface) require exact arithmetic and are too slow and too expensive for industrial use. Thus the duality of an implicit representation and a parametric representation is only used for low degree algebraic surfaces such as planes, spheres, cylinders, cones and toroidal surfaces. On the other hand, this duality is a very useful tool for developing efficient algorithms. Approximate implicitization makes this duality available for *general* curves and surfaces.

The traditional exact implicitization of parametric surfaces produce global representations, which are exact everywhere. The surface patches used in CAD, however, are always defined within a small box only; they are obtained for a bounded parameter domain (typically a rectangle, or – in the case of "trimmed" surface patches – a subset of a rectangle). Consequently, a globally exact representation is not really needed in practice. Instead of a single exact high–degree implicit representation, the methods of approximate implicitization produce piecewise implicit surfaces of relatively low degree, which may cover the shape with any desired accuracy. This results in so–called *algebraic spline surfaces*, which can be expected to replace the exact implicit representation in many algorithms.

Compared to the traditional parametric representations, such as rational curves and surfaces (so–called Non-Uniform-Rational-B-Splines – NURBS), algebraic spline surfaces offer several computational advantages. For instance, by exploiting the duality between implicit and parametric representation, the intersection of two surfaces can easily be traced if one of the surfaces is given in implicit, and the other surface is given in parametric form. In this case, the problem can be reduced to a two-dimensional root-finding problem. In the case of two parametric surfaces, one has to solve a four–dimensional problem instead. As another advantage, the fitting of surfaces to scattered data, which is a fundamental tool for generating free-form geometry from prototypes, can be done without mapping the data into a plane – a process which often limits the flexibility and usefulness of the surface fitting techniques which are available today. We also foresee a number of other applications, e.g., in the computer game industry, virtual reality, medical imaging, and scientific computing.

The workshop, and the papers collected in this volume, was devoted both to the theoretical fundamentals and to the various computational aspects which arise in applications of approximate algebraic geometry. These applications are based on techniques developed in different branches of mathematics and computer science, including numerical analysis and scientific computing, algebraic geometry, applied geometry, and computer graphics. For instance, numerical methods are needed to efficiently generate implicit representations, and algebraic techniques are essential for detecting and analyzing singularities, which may help to solve practical problems arising in applied geometry and computer graphics.

Traditionally, these fields are represented by several fairly disjoint scientific communities, which traditionally do not communicate much. In order to stimulate the exchange of ideas, and to promote interdisciplinary research, the workshop brought together experts from the various fields involved.

The papers included in this volume provide an overview about the state-of-the-art in approximative implicitization and various related topics, including both the theoretical basis and the existing computational techniques. This can be expected to encourage and promote the use of approximate implicitization for solving geometric problems in computer-aided design. In some of the papers and in the panel discussion at COMPASS, which is also documented in this volume, the authors try to identify a number of problems (both theoretical and practical ones) which need to be addressed by the different research communities, in order to exploit the potential of implicit representations.

The editors are convinced that this volume will support the mutual exchange of ideas between the various research communities, promoting interdisciplinary research. The interactions between different mathematical disciplines such as approximation theory, classical algebraic geometry and computer aided geometric design will play an essential role for exploiting the new idea of approximate algebraic geometry.

The editors of this volume are indebted to the European Science Foundation (ESF) for providing generous financial support of the COMPASS event, which was organized as an ESF Exploratory workshop (EW 02/55). They would like to thank the staff at Springer–Verlag, Heidelberg, for the constructive cooperation during the preparation production of these proceedings. The time and the effort of the 37 referees, whose reports have greatly helped to improve the quality and the presentation of the material, is gratefully acknowledged. Last, but not least, they would like to thank Elmar Wurm and Martin Aigner for collecting the papers and preparing the final manuscript.

Oslo and Linz, May 2004

Tor Dokken
Bert Jüttler

Table of Contents

Approximate Parametrisation of Confidence Sets 1
Zbyněk Šír

Challenges in Surface-Surface Intersections 11
Vibeke Skytt

Computing the Topology of Three-Dimensional Algebraic Curves 27
G. Gatellier, A. Labrouzy, B. Mourrain, J.P. Técourt

Distance Properties of ϵ–Points on Algebraic Curves 45
Sonia Pérez-Díaz, Juana Sendra, J.Rafael Sendra

Distance Separation Measures Between Parametric Curves and Surfaces
Toward Intersection and Collision Detection Applications 63
Gershon Elber

Elementary Theory of Del Pezzo Surfaces 77
Josef Schicho

The Geometry of the Tangent Developable 95
Pål Hermunn Johansen

Numerical and Algebraic Properties of Bernstein Basis Resultant Matrices 107
Joab R. Winkler

Polynomial C^2 Spline Surfaces Guided by Rational Multisided Patches 119
Kęstutis Karčiauskas, Jörg Peters

A Recursive Taylor Method for Algebraic Curves and Surfaces 135
Huahao Shou, Ralph Martin, Guojin Wang, Adrian Bowyer, Irina Voiculescu

Self-Intersection Problems and Approximate Implicitization 155
Jan B. Thomassen

Singularities of Some Projective Rational Surfaces 171
Ragni Piene

On the Shape Effect of a Control Point: Experimenting with *NURBS* Surfaces .. 183
Panagiotis Kaklis, Spyridon Dellas

Third Order Invariants of Surfaces 193
Jens Gravesen

Universal Rational Parametrizations and Spline Curves on Toric Surfaces 213
Rimvydas Krasauskas, Margarita Kazakevičiūtė

Panel Discussion 233

Approximate Parametrisation of Confidence Sets[*]

Zbyněk Šír

Charles University, Sokolovská 83, Prague, Czech Republic,
sir@karlin.mff.cuni.cz

Abstract. In various geometrical applications, the analysis and the visualization of the error of calculated or constructed results is required. This error has very often character of a nontrivial multidimensional probability distribution. Such distributions can be represented in a geometrically interesting way by a system of so called confidence sets. In our paper we present a method for an approximate parametrisation of these sets. In sect. 1 we describe our motivation, which consists in the study of the errors of so called Passive Observation Systems (POS). In sect. 2 we give a result about the intersection of quadric surfaces of revolution, which is useful in the investigation of the POS. In sect. 3 we give a general method for an approximate parametrisation of the confidence sets via simultaneous Taylor expansion. This method, which can be applied in a wide range of geometrical situations, is demonstrated on a concrete example of the POS.

1 Motivation

Our research was motivated by concrete problem of the analysis and the visualization of the errors of so called Passive Observation Systems (POS).

1.1 Passive Observation Systems

The POS have been successfully constructed and produced in Czech Republic since the 1960's as an alternative to the classical radars. These systems, which do not transmit any signal (therefore passive), are based on the principle of the time difference. A pulse in the transmission of an object (a plane) is received at four (or more) observation sites. In practice any plane is forced to transmit some signals, at least in order to ensure its orientation. From the differences of the time of reception of the pulse the position of the object can be determined.

The POS have two main advantages comparing to the standard radars. As they do not transmit any signal they can not be itself detected and have very low energy consumption.

In addition the error of the POS has a different characteristic comparing to the classical radars. For this reason a simultaneous use of the POS and the classical radars can be very interesting. For more details about the principle of the POS and for the basic information about their precision see [1, Chapter 5].

---

[*] The author's research has been supported by the grant No. 201/03/D113 of the Czech Science Foundation.

1.2 Geometry of POS

The construction of POS creates many difficult problems on the level of the electrical engineering, but the underlying geometry is quite simple. Let a pulse transmitted by an object X be received at the sites A and A' respectively at times t_A and $t_{A'}$. Multiplying the difference $t_A - t_{A'}$ by the speed of the signal (typically the speed of light) we get the difference $d_{AA'}$ of distances from the object X to the sites A and A'. The object X must therefore lie on one of sheets of the two-sheet hyperboloid of revolution, which is determined by its foci A, A' and the measured difference of distances $d_{AA'}$. The sign of $d_{AA'}$ indicates which of the two sheets must be taken.

Repeating the same procedure for two other pairs of sites (B, B') and (C, C'), we get in all three hyperboloids on which the object X must lie and its position can be therefore determined as their intersection. The space coordinates $[x_1, x_2, x_3]$ of X are then computed from the measured distance differences $d_{AA'}$, $d_{BB'}$ and $d_{CC'}$.

The difference vector $[d_{AA'}, d_{BB'}, d_{CC'}]$ can be easily computed from $[x_1, x_2, x_3]$, and the corresponding mapping $F : [x_1, x_2, x_3] \rightarrow [d_{AA'}, d_{BB'}, d_{CC'}]$ can be explicitly expressed. If the sites A, A' have the space coordinates $[a_1, a_2, a_3]$ and $[a'_1, a'_2, a'_3]$ respectively, then for example

$$d_{AA'} = \sqrt{(x_1 - a_1)^2 + (x_2 - a_2)^2 + (x_3 - a_3)^2} - \sqrt{(x_1 - a'_1)^2 + (x_2 - a'_2)^2 + (x_3 - a'_3)^2}.$$

On the other hand the inversion mapping F^{-1} can not be in general expressed explicitly and the position of X must be computed from $[d_{AA'}, d_{BB'}, d_{CC'}]$ numerically as a solution of a system of algebraic equations of the total degree 8.

In practice a network of observation sites should be used. But the smallest operational system consists of four sites only. In this case one site $O = A' = B' = C'$ is considered as central one and the position of the object X is computed from the distance differences $[d_{AO}, d_{BO}, d_{CO}]$. In the sequel we will restrict ourselves to this simplest case. As we will show, in this case an explicit inversion formula for F^{-1} can be always given.

1.3 Measurement Error of the POS

Suppose, that a pulse is received at four observation sites O, A, B and C at times t_O, t_A, t_B and t_C. The error of the vector $[t_O, t_A, t_B, t_C]$ of independently measured times can be well modeled by a multivariate normal distribution, characterized by its mean value $[0, 0, 0, 0]$ and the variation-covariation matrix having on the diagonal the variations of the time errors at the four sites, which are not necessarily the same

$$\begin{bmatrix} \sigma_O{}^2 & 0 & 0 & 0 \\ 0 & \sigma_A{}^2 & 0 & 0 \\ 0 & 0 & \sigma_B{}^2 & 0 \\ 0 & 0 & 0 & \sigma_C{}^2 \end{bmatrix}. \tag{1}$$

The differences d_{AO}, d_{BO} and d_{CO} have no more independent errors, but the error of the vector $[d_{AO}, d_{BO}, d_{CO}]$ has still a normal distribution characterized by its mean value $[0, 0, 0]$ and the variation-covariation matrix

$$c^2 \begin{bmatrix} \sigma_A{}^2 + \sigma_O{}^2 & \sigma_O{}^2 & \sigma_O{}^2 \\ \sigma_O{}^2 & \sigma_B{}^2 + \sigma_O{}^2 & \sigma_O{}^2 \\ \sigma_O{}^2 & \sigma_O{}^2 & \sigma_C{}^2 + \sigma_O{}^2 \end{bmatrix},\tag{2}$$

where c is the speed of light. See [3] for the details about multivariate distributions and their characteristics.

If we compute the position $[x_1, x_2, x_3]$ using d_{AO}, d_{BO} and d_{CO} we transform the error distribution by the mapping F^{-1}. The transformed distribution will be no more normal. For this reason the mean value and the variation-covariation matrix are no more sufficient characteristics of this transformed error distribution.

In fact the analysis of such complex multivariate distributions is a difficult problem. This is due to the fact that the standard concepts used in in the case of one dimensional distributions, are insufficient for the description of the geometry of the multivariate distributions. We are convinced that the methods of the applied geometry would be very useful in the analysis of both theoretical distributions and experimental data. See [2] for one possible approach based on the concept of the data depth.

1.4   Confidence Sets

The confidence sets (called also tolerance regions) are perhaps geometrically the most interesting characteristics of probability distributions.

Definition 1. *For a given random variable U having the density function p_U and for a given probability $\alpha \in (0, 1]$ we define the confidence set $C_{U,\alpha}$ as a region for which*

$$\int_{x \in C_{U,\alpha}} p_U(x) = \alpha \tag{3}$$

In other words a confidence set is a region in which the random variable U lies with the probability α. In practice α is set quite high, for example 0.99, and thus a confidence set is simply a region in which the random variable lies with a reasonable certitude.

It is clear from the definition, that for a given probability $\alpha < 1$ there is in general more then one confidence set. There are natural additional properties which can be required of the confidence sets. First of all the confidence sets should be as small as possible in order to give good information about the probability density. For the same reason their boundaries should be the iso-lines (iso-surfaces) of the density function. In the case of multivariate normal distributions it is customary to use suitable ellipsoids as confidence sets. These ellipsoids satisfy both additional requirements (see for example [3, 45.9]).

The distribution of the error of the vector $[d_{AO}, d_{BO}, d_{CO}]$ can be described by a system of ellipsoids (confidence sets) depending on the probability α and on the values $[d_{AO}, d_{BO}, d_{CO}]$ (the error may in general depend on the value of $[d_{AO}, d_{BO}, d_{CO}]$). Transforming this system by F^{-1} we will get a new system of confidence sets describing the error of the position $[x_1, x_2, x_3]$. The boundaries of these new confidence sets will be iso-surfaces of the new density function.

2 Explicit Inversion Formula

The importance of an explicit formula for F^{-1} is obvious from the previous section. As X is obtained as intersection of three quadric surfaces, the resulting system of equations has degree 8. Therefore there is seemingly no possibility to obtain an explicit expression for F^{-1}. However for concrete examples, we were able to reduce the degree of the problem and even obtain a simple explicit formula. A deeper investigation of this fact has shown, that this simplification is due to the following interesting property.

2.1 Intersection of Quadric Surfaces of Revolution

Proposition 2. *Let S_1, S_2 be two quadric surfaces of revolution, each of which obtained by rotating a conic section around its main axis. (The only axis for a parabola and the axis passing through the foci for an ellipse or an hyperbola.) Suppose that S_1 and S_2 have a common focus. Then their intersection can be decomposed into curves of degree 2.*

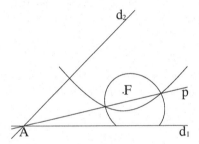

Proof. Let F be the common focus. Clearly the axes of S_1 and S_2 intersect in the point F and therefore they lie in a plane. The previous fig. represents this plane and its intersections with all mentioned objects.

We can characterize the surfaces S_1 and S_2 using the focus-directrix property of the generating conic sections. Obviously the surface S_1 is precisely the set of points in the space, having a constant ratio of distances to the focus F and a directrix plane d_1, perpendicular to the main axis: $S_1 = \{X, \frac{|XF|}{|Xd_1|} = r_1\}$ for some fixed ratio r_1. For $r_1 = 1$ we get a paraboloid, for $r_1 < 1$ an ellipsoid and for $r_1 > 1$ a two-sheet hyperboloid. In the same way the surface S_2 can be characterized as the set $S_2 = \{X, \frac{|XF|}{|Xd_2|} = r_2\}$ for some plane d_2 perpendicular to the axis of S_2 and for some fixed ratio r_2.

For the points of the intersection $X \in S_1 \cap S_2$ we thus get $\frac{|Xd_1|}{|Xd_2|} = \frac{r_2}{r_1}$. This equality characterizes all the points lying in two planes passing through the intersection $d_1 \cap d_2$. One of these planes is denoted p on the figure. As the intersection of a quadric surface with a plane is of degree 2, the intersection $S_1 \cap S_2$ must have a component of degree 2. As $S_1 \cap S_2$ is itself of degree 4, the proposition is proved. □

So the intersection of two hyperboloids, which is in general a curve of degree 4, will have components of degree 2 (conic sections) if the two hyperboloids share a focus.

Consequently the degree 8 system describing a general POS will be decomposed if two of hyperboloids have a common focus. If particular if the three hyperboloids have a common focus - the site O - the problem will be reduced twice and the resulting system can be decomposed to the degree 2 systems. In this case therefore an explicit inversion formula can be always obtained.

We will not describe this explicit formula in general, but we will study in detail one particular example. In this example the situation is simplified even more by the additional condition, that the four sites A, B, C, O are coplanar.

2.2 Example

Let us consider the POS in which the four sites lie in one plane and have the coordinates: $O = [0, 0, 0]$, $A = [30, 0, 0]$, $B = [-15, 26, 0]$ and $C = [-15, -26, 0]$. The mapping F is then expressed by formulae:

$$
\begin{aligned}
d_{AO} &= \sqrt{x_1^2 - 60x_1 + 900 + x_2^2 + x_3^2} - \sqrt{x_1^2 + x_2^2 + x_3^2} \\
d_{BO} &= \sqrt{x_1^2 + 30x_1 + 901 + x_2^2 - 52x_2 + x_3^2} - \sqrt{x_1^2 + x_2^2 + x_3^2} \\
d_{CO} &= \sqrt{x_1^2 + 30x_1 + 901 + x_2^2 + 52x_2 + x_3^2} - \sqrt{x_1^2 + x_2^2 + x_3^2}
\end{aligned}
\tag{4}
$$

We implicitise these equations and obtain implicit algebraic equations of the three hyperboloids H_{AO}, H_{BO} and H_{CO}. For example the implicit equation of H_{AO} is

$$
4d_{AO}^2 \left(x_1^2 + x_2^2 + x_3^2 \right) - \left(900 - 60x_1 - d_{AO}^2 \right)^2 = 0.
$$

Due to the Proposition 2 any two of these hyperboloids intersect in two conic sections. Because of the symmetry with regard to the plane $x_3 = 0$ these conics lie in the planes perpendicular to the plane $x_3 = 0$. Their projections to this plane will be therefore lines.

For the determination of the 8 intersections of the hyperboloids H_{AO}, H_{BO} and H_{CO} we first evaluate the resultant with respect to x_3 of the implicit equations of H_{AO} and H_{BO}. Because of the previous observations this resultant (of degree 4 in x_1, x_2) can be factorised in two linear factors (each of them of with multiplicity two) describing two stright lines p_1 and p_2. In a similar way from the equations of H_{AO} and H_{CO} we get two lines q_1 and q_2. As intersection of this two pairs of lines we get four points $X_{i,j} = p_i \cap q_j$, $i, j = 1..2$, each of them being projection of two symetrical intersections of the three hyperboloids. The signs of d_{AO}, d_{AO} and d_{AO} will indicate which of the four points $X_{i,j}$ must be taken. The last coordinate x_3^i can be calculated from the equation of any of the three hyperboloids.

Let us give the explicit formula of one of the 4 pairs of solutions of our example system (4):

$$
x_1 = \frac{d_{AO}(d_{BO}^2 + d_{CO}^2 - 1802) + (900 - d_{AO}^2)(d_{BO} + d_{CO})}{60(d_{AO} + d_{BO} + d_{CO})}
\tag{5}
$$

$$
x_2 = \frac{d_{AO}(d_{CO}^2 - d_{BO}^2) + (d_{AO}^2 - 2d_{BO}d_{CO} - 2702)(d_{BO} - d_{CO})}{104(d_{AO} + d_{BO} + d_{CO})}
\tag{6}
$$

$$
x_3 = \pm \frac{\sqrt{P_6(d_{AO}, d_{BO}, d_{CO})}}{d_{AO} + d_{BO} + d_{CO}}
\tag{7}
$$

where $P_6(d_{AO}, d_{BO}, d_{CO})$ is a polynomial of degree 6 in d_{AO}, d_{BO} and d_{CO}. The x_3 is usually supposed to be positiv, as the object (plane) is usually "over" the observation sites.

A similar explicit form of F^{-1} can be in general obtained for any POS having the four sites in a plane. In this case, the first two coordites x_1, x_2 can be expressed as rational functions in d_{AO}, d_{BO} and d_{CO}, but the expression of x_3 will involve a square root.

If the four sites are not coplanar, an explicit formula can be still obtained, but square roots will appear in the expressions of all coordinates.

For a general POS, based on three independent pairs of sites (A, A'), (B, B') and (C, C'), no closed expression of F^{-1} can be obtained.

3   Approximate Representation

The explicit inversion formula is not available for the POS in the general position. In some other cases the inversion formula can be too complicated. For this reason we will describe in this section a general method for the approximation of F^{-1}.

3.1   General Setting

Let us consider the following general setting. Suppose that $x = [x_1, \ldots, x_n]$ is a set of parameters which is transformed by a local diffeomorphism F to a second set of parameters $y = [y_1, \ldots, y_n]$:

$$F : [x_1, \ldots, x_n] \rightarrow [y_1, \ldots, y_n] \tag{8}$$

Suppose in addition that an algebraic implicitisation of F is available. We mean by this a system of algebraic equations

$$G(x, y) = 0 \tag{9}$$

which hold if and only if $y = F(x)$.

Next suppose that in the space of parameters y the system of confidence sets (for example a system of ellipsoids) is described. We want to obtain a description of the transformed system of the confidence sets in the space of the parameters x.

3.2   Implicit Representation

If the confidence sets in the space of parameters y are described implicitly we can obtain an implicit description in the space of parameters x in a straightforward way. Suppose, that the boundaries of the confidence sets in the space of parameters y are given by implicit equations

$$E_{\alpha, \bar{y}}(y) = 0 \tag{10}$$

depending algebraically on the measured value $\bar{y}$. Then substituting $y = F(x)$ and $\bar{y} = F(\bar{x})$ in this equations we get mplicit representations of the boundaries of the confidence sets in the space x depending on $\bar{x}$.

The drawbacks of this methods are obvious. As the transformation F is not necessarily rational, we obtain in general a complicated (non algebraic) implicit representation depending in a complicated way on $\bar{x}$.

3.3 Approximation by the Taylor Expansion

Another natural possibility is to approximate the inversion F^{-1} by its Taylor expansion in a suitable point $\overline{y}$:

$$F^{-1}(y) = F^{-1}(\overline{y}) + D_1 F_{\overline{y}}^{-1}(y-\overline{y}) + \frac{1}{2}D_2 F_{\overline{y}}^{-1}(y-\overline{y}) + \frac{1}{6}D_3 F_{\overline{y}}^{-1}(y-\overline{y}) + \dots \quad (11)$$

where $D_i F_{\overline{y}}^{-1}$ is the i-th total differential of F^{-1} at the point $\overline{y}$. See [4, par. 3.14] for the details about the multivariate Taylor expansion. The value of $\overline{x} = F^{-1}(\overline{y})$ can be calculated numerically from (9) and the operators $D_i F_{\overline{y}}^{-1}$ can be obtained by the implicit differentiation of (9), or from the known partial derivatives of F at the point $\overline{x}$. This approximation can be used for an approximate representation of the confidence sets in the space of parameters x. In particular if we have a parametrisation of the boundaries of the confidence sets in the space of parameters y, we can compose this parametrisation with the Taylor expansion and this way obtain an approximate parametrisation of the boundaries of the confidence sets in the space of parameters x.

The disadvantage of this approach is that the Taylor expansion can give a sufficiently good approximation in the proximity of the point $\overline{y}$ but will not be sufficient for more distant points.

3.4 Symbolic Computation of the Taylor Expansion

We propose a different approach, which consists in the symbolic computation of the Taylor expansion simultaneously in all points. If the mapping F^{-1} can not be expressed explicitly, there is no hope to get a general expression of the Taylor expansion depending on the point $\overline{y}$. On the other hand it is possible to get such general expression depending on the target point $\overline{x} = F^{-1}(\overline{y})$.

The total differentials $D_i F_{\overline{y}}^{-1}$ can be symbolically computed via partial differentiation of the equality

$$G(F^{-1}(y), y) = 0 \quad (12)$$

For example by taking all the partial derivatives of the first order $\frac{\partial}{\partial y_i}$ for $i = 1..n$, we obtain a system of n linear equations for n unknown partial derivatives $\frac{\partial F^{-1}}{\partial y_i}$. The coefficients of these equations are polynomials in y and $F^{-1}(y)$. This system can be symbolically solved and we get $\frac{\partial F^{-1}}{\partial y_i}$ in the form of a rational function of y and $F^{-1}(y)$. If we use in a similar way the higher partial derivatives of (12), we get the same king of expression for the higher partial derivatives. See [4, par. 4.5] for more details about the implicit differentiation.

Substituting these expressions into (11), we obtain the Taylor expansion having all the coefficients dependent rationally on $\overline{y}$ and $F^{-1}(\overline{y})$. In this expression we can simply substitute $F(\overline{x})$ for $\overline{y}$ and $\overline{x}$ for $F^{-1}(\overline{y})$ and we obtain the desired simultaneous Taylor expansion depending on $\overline{x}$.

3.5 Example

Let us demonstrate the described general procedure on the following example. Consider a two-dimensional version of the POS, which can be used for example for the location of ships on the surface of sea. In this case F^{-1} can be expressed explicitly, but using squre roots.

Suppose that we have three observation sites with coordinates $O = [0,0]$, $A = [30,0]$ and $B = [-26,15]$. For to be coherent with the general notation introduced in the paragraph 3.1, we will denote the distance differences d_{AO} and d_{BO} by y_1 and y_2. The mapping F is then given by:

$$\begin{aligned} y_1 &= \sqrt{x_1{}^2 - 60x_1 + 900 + x_2{}^2} - \sqrt{x_1{}^2 + y_1{}^2} \\ y_2 &= \sqrt{x_1{}^2 + 52x_1 + 901 + x_2{}^2} - 30x_2 - \sqrt{x_1{}^2 + x_2{}^2} \end{aligned} \tag{13}$$

Implicitisating these formulae we get a system of algebraic equations G:

$$\begin{aligned} &y_1{}^4 - 4\,y_1{}^2 x_1{}^2 - 4\,y_1{}^2 x_2{}^2 + 120\,y_1{}^2 x_1 - 1800\,y_1{}^2 + 3600\,x_1{}^2 - \\ &-108000\,x_1 + 810000 \\ &y_2{}^4 - 4\,y_2{}^2 x_1{}^2 - 4\,y_2{}^2 x_2{}^2 - 104\,y_2{}^2 x_1 + 60\,y_2{}^2 x_2 - 1802\,y_2{}^2 + \\ &+2704\,x_1{}^2 - 3120\,x_1 x_2 + 900\,x_2{}^2 + 93704\,x_1 - 54060\,x_2 + 811801 = 0 \end{aligned} \quad \begin{aligned} = 0 \\ \\ \tag{14} \end{aligned}$$

By implicit partial differentiation we were able, using the program Maple 8, to symbolically compute the partial derivatives of F^{-1} up to the degree 3. As the formulae become quickly very complicated, let us give just one example. The first component of the first partial derivative $\frac{\partial F^{-1}}{dy_1}$ at the point $F([x_1, x_2])$ is equal to

$$\begin{aligned} &(780\,x_1 + 13515 - 15\,y_2{}^2 - 450\,x_2 + 2\,y_2{}^2 x_2)\,y_1\,(2\,x_1{}^2 - 60\,x_1 + 900 - y_1{}^2 + 2\,x_2{}^2)/ \\ &(-780\,x_1{}^2 y_1{}^2 + 702000\,x_1{}^2 + 1633500\,x_1 - 1815\,x_1 y_1{}^2 - 182452500 + 202725\,y_1{}^2 + \\ &+15\,y_2{}^2 x_1 y_1{}^2 - 13500\,x_1 y_2{}^2 + 202500\,y_2{}^2 - 225\,y_2{}^2 y_1{}^2 - 902\,x_2 x_1 y_1{}^2 - 405000\,x_1 x_2 + \\ &+6075000\,x_2 - 30176\,x_2 y_1{}^2 + 1800\,y_2{}^2 x_2 x_1 - 27000\,y_2{}^2 x_2 + 56\,y_2{}^2 x_2 y_1{}^2 + 780\,x_2{}^2 y_1{}^2) \end{aligned} \tag{15}$$

Substituting (13) into this expression we get $\frac{\partial F^{-1}}{dy_1}$ depending on the target point $[x_1, x_2]$. Doing the same for all partial derivatives up to the degree 3, we get a general Taylor expansion of the third order, in all points, depending on $[x_1, x_2]$. This general expression can be now used for the simultaneous description of the system of the confidence sets representing the error of x.

Let us suppose, for simplicity, that the time measurement error is the same at the three sites and is independent on the measured values. For a given probability the system of the confidence sets representing the error of $[y_1, y_2]$ consists simply of the circles of the same radius. If we take a parametrisation of these circles and compose it with general Taylor expansion, we get the parametrisation of the system of confidence sets representing the error of $[x_1, x_2]$.

The fig. 1 shows the position of the three sites O (central), A (right) and B (left) and the 3rd order approximation of several confidence sets (scaled by 100) corresponding to the error of x at the points marked by small crosses. These confidence sets correspond to the probability $\alpha = 0.99$ in the case of a standard time measurement error.

The practical interpretation of this fig. is as follows: If an object is situated at a point marked by a small crosses, then the POS will with the probability 99% detect its

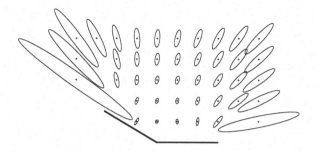

Fig. 1. System of the confidence sets representing the error of x, scaled by 100

position within the corresponding set. The obtained general description of the system of the confidence sets is clearly very usefull for the visualisation and the analysis of the precision of the POS and of its range of operation.

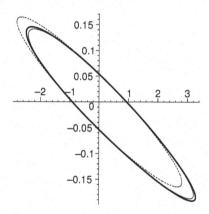

Fig. 2. Approximation of the confidence set at the point $x = [60, 2]$, using the Taylor expansion of order 1 (dotted line), 2 (thin solid line) and 3 (thick solid line). Note the different scaling of both axes

The fig. 2 shows more in detail the approximate parametrisation of the confidence set at the point $x = [60, 2]$, using the Taylor expansion of the 1st, the 2nd and the 3rd order. The first order approximation gives an ellipse. The third order approximation is indiscernible from the numerically computed confidence set.

4 Conclusion

The application of the described methods is not limited to the POS. The result presented in the paragraph 2.1 can be very useful in the construction of any devices using the quadric surfaces of revolution. This is for example the case of various observation systems based on the sum of distances, in which the ellipsoids occur.

The method described in the paragraph 3.4 can be applied in all situations satisfying the general setting 3.1. It is particularly interesting in the cases, in which we are interested by the analysis and the visualisation of the error depending not on the measured values y, but on the resulting values x. Let us mention for example the case of parallel robots, for which we want to know which positions can be reached with a prescribed precision.

References

1. Wiley, R.G.: Electronic Intelligence: The Interception of Radar Signals, Artech House 1985, 284 p.
2. Mizera, I.: On depth and deep points: A calculus, Ann. Stat. **30**(2002), pp 1681-1736
3. Johnson, N.L.; Kotz, S.; Balakrishnan, N.: Continuous multivariate distributions, Vol. 1: Models and Applications, New York, Wiley 2000, 752 p.
4. Walter, W.: Analysis 2, Grundwissen Mathematik Bd. 4, Springer-VErlag 1990, 396p.

Challenges in Surface-Surface Intersections

Vibeke Skytt

SINTEF, Norway

Abstract. Tangential and singular situations are still challenges in a system for surface-surface intersections. This paper presents several real world examples of hard intersection problems, and proposes methods on how to deal with them. In particular, solutions which use the possibility of representing a parametric surface as an algebraic surface through the use of approximate implicitization, are in focus. This allows us to transform an intersection between two parametric surfaces to the problem of finding zeroes of a function of two parameters.

1 Introduction

An important part of a CAD system is the ability to perform Boolean operations. These operations are again totally dependent on having access to a stable intersection functionality. Intersections between two surfaces are of particular interest, but also intersections involving curves are important, both in its own right and because they appear as parts of a surface-surface intersection.

The type of surfaces involved in a Boolean operation depends on the surfaces supported by the CAD system. Simple algebraic surfaces and parametric surfaces of type NURBS and rectangular Bezier surfaces have a most widespread use. This article will concentrate on intersections between two surfaces where either both surfaces are of type NURBS or where one NURBS surface is intersected with an algebraic surface.

Curve and surface tangency frequently arise in modern design. The preferred surface models often have smooth transitions between surfaces. The CAD system which produced a surface model will often be able to avoid intersections between for instance tangential surfaces. However, a geometry model is often shared between several systems using a standard geometry format. Then history information tends to get lost, and intersections that otherwise would be avoided are performed. A good intersection package should be able to handle tangential and singular situations in order to meet the needs of current computer aided design.

This article will focus on finding solutions to singular intersection cases and related problems and illustrate it through an extensive use of examples. The intersections are computed using a combination of recursive subdivision and marching methods. The intersection problem will be formulated in the next section with special emphasize on tangential intersections and tolerances. Next, we will give an overview of methods for surface intersections and of typical situations that provide challenges for the intersection. More emphasize is put on recursive subdivision in sect. 4. Section 5 discusses the tolerance aspect for intersections between one parametric and one algebraic surface while sect. 6 gives a short presentation of approximate implicitization. Finally, we will go deeper into some aspects of recursive subdivision and illlustrate it by examples of singular intersection problems.

2 Problem Formulation

Two parametric surfaces $F(u,v) : R^2 \rightarrow R^3$ and $G(s,t) : R^2 \rightarrow R^3$ that are mapping from planar parameter domains into 3 dimensional geometry space, are given. The surfaces intersect if $F(u,v) = G(s,t)$ for some quadruples (u,v,s,t). We want to compute all parameter values where the two surfaces intersect. In practice this problem formulation does not make sense. Due to the surface representation in the computer and to how the surfaces were constructed initially, a tolerance must be introduced. We want to find all parameter values of the two surfaces where the distance between the surfaces is less than a given tolerance, i.e. $|F(u,v) - G(s,t)| < \varepsilon$. The tolerance ε will depend on the accuracy with which the surfaces were constructed. This accuracy will again depend on the parent CAD system. Large tolerances will, in general, give rise to more complicated intersection problems than smaller ones. Computing intersections within a tolerance will imply that every intersection result will be an area. This area will normally be represented by the most significant point or curve within the area. For transversal intersections the difference between the exact and the tolerance depended formulation, is small, for tangential and singular cases the difference is large.

An intersection between one algebraic surface $G((x,y,z)^T) = 0$, and one parametric surface $F(u,v) : [a_1,a_2] \times [b_1,b_2] \rightarrow R^3$, is solved if we compute all parameter pairs (u,v) where the two surfaces are closer than the tolerance ε. More elaboration on this topic can be found in sect. 5.

Consider a curve-curve intersection as illustrated in fig. 1. Two curves intersect in isolated points or in coincidence intervals. The points can be of type

Transversal It is a clear unique intersection. A small perturbation of the given curves lead to small changes in the resulting intersection point.

Near transversal The two curves intersect transversally, but there is a small angle between the curve tangents in the intersection point. If the input curves are perturbed slightly, the topology of the intersection point will remain, but the position can change by a large distance.

Tangential The curves intersect in one point, but a small perturbation of one of the curves can change the position of the intersection point significantly. The intersection result easily change from tangential to near tangential or near transversal.

Near tangential The curves do not intersect, but lie closer than the tolerance in an area, or the curves intersect in two points, and the curves lie closer than the tolerance in an interval including the intersection points. A perturbation of a curve can change the configuration completely.

Intersections between other types of input objects can be classified correspondingly, the case of two parametric surfaces will be treated in some detail in sect. 9.

3 Surface-Surface Intersections

Similar to the curve case will transversal surface-surface intersections represent a stable and well defined problem. Tangential and singular situations are much more complicated and represent challenges in surface-surface intersection. However, also other

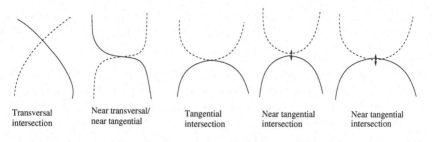

| Transversal intersection | Near transversal/ near tangential | Tangential intersection | Near tangential intersection | Near tangential intersection |

⏸ Size of tolerance

Fig. 1. Categories of curve-curve intersections

challenging situations exist and we will present a few of them in this section. Before going into the difficult situations we will shortly present some often used methods for computing intersections as the challenges will vary depending on the chosen method. More information on methods for surface-surface intersections can be found in e.g. [2], [9], [10], [16], [17] and [19].

Lattice evaluation A curve or point mesh is generated from the surfaces. The curve mesh of each surface is intersected with the other surface to detect intersection tracks. A restricted initial mesh can lead to loss of small intersection loops or isolated points or to wrong connectivity.

Marching methods From one point on the intersection curve, we find a new point on the curve. The direction of the curve in the given point is computed from local information and a guess on the step length for the marching is made. It is important to ensure that there exist start points for every intersection curve. Start points can be found by performing intersections between boundary curves or other curves in one surface and the other surface. A search for closed inner intersection loops can be performed to avoid loosing intersection tracks. Care must be taken to avoid jumping between different intersection branches during marching, and singular and near singular situations can create problems. Marching approaches are discussed in [1], [3], [11] and [8].

Recursive subdivision For the intersection of two NURBS surfaces recursive subdivision is an often used approach. Recursive methods are based on the idea that the intersection problem gets simpler if we perform subdivision and intersect subsurfaces. However, if care is not taken the sub-problems can get more complicated than the original problem. Subdivision can be performed until sub-pieces do not intersect or the sub-pieces are flat enough to represent intersections between them as linear curve pieces. At each recursion step, intersection points at the boundaries of the sub surfaces are computed. If the number of recursion steps is limited, intersection results can get lost. Otherwise the process can be very resource demanding. A combination of recursive subdivsion with some marching method speed up the intersection. A more detail view on this combined method is presented in sect. 4. See also [7].

The remaining part of this section mainly concentrates on challenging situations for subdivision methods although the same situations tend to create problems also for other methods.

Singular and near singular intersections are challenging situations. Several intersection curves meet or do almost coincide in a singular branch point. Typically, there is an area around the branch point where the two surfaces lie closer than a tolerance. This problem is a main topic for this paper and described in sect. 9. Tangential and near tangential intersection curves are also the result of a singular intersection problem and a small part of the much more complicated problem of partial coincidences. This problem treated in sect. 10.

One issue in a recursive subdivision method is to decide whether two surfaces or sub surfaces may intersect at all. This can be difficult if the surfaces are curved, nearly parallel and lie close together. The impliciation is that the recursion can be very deep. This problem is discussed with some detail in sect. 7.

Intersection between constant parameter curves in one surface and the other surface is a part of both the recursive subdivision and the lattice evaluation method for surface-surface intersection. A transversal intersection on the level of surface-surface intersection may lead to tangential intersections at the curve-surface level. The intersections can take a long time and the output can be inaccurate. Figure 1 illustrates that the position of an intersection point can be badly defined. The situation occurs if some intersection curve is tangential to a surface boundary. It can also occur for constant parameter curves internal to the surface, but it is often possible to avoid it by a wise choice of constant parameter curves to intersect. The implication for a subdivision method is that the position of subdivision must be carefully selected. A uniform approach will not always give a good result.

Constant parameter curves in the two surfaces can be nearly parallel and at the same time almost parallel to an intersection curve. This is a situation that frequently occur in design and that is hard to avoid even if a good subdivision strategy is applied. It will, similar to the situation described above, create unstable lower order intersection problems.

Degenerate surfaces can create problems if an intersection curve passes close to the degeneracy. Intersections close to a degenerate edge in a surface are unstable and especially the iterations involved in a surface-surface intersection are sensitive to degeneracies. Thus, special treatment of these parts of the surfaces is recommended.

Not well-behaved surfaces, for instance surfaces with highly non-isometric parameterization or very close, distinct knots, are critical. There is a risk of numerical instabilities. Iterations, especially of Newton type, can be unstable when the surface parametrization is far from isometric.

Some of the situations described above will be treated in more detail in later sections. The surface intersection challenges will be illustrated through the use of real world examples appearing as problem cases in Boolean operations.

4 Recursive Subdivision

We will in the remaining parts of this paper focus on computation of intersections, and in particular singular ones, using a combination of recursive subdivision and marching. A short description of a recursive procedure is outlined below. Here we assume that two parametric surfaces are given, but a similar procedure apply for a problem involving one parametric and one implicit surface. The examples to be presented is computed according to this procedure.

- Make sure that all intersections on the surface boundaries are found. This leads to intersection problems of a reduced number of parameter directions, i.e. curve-surface intersections. These intersection problems are solved following a similar approach to the one described here. Intersections at the surface boundaries and curve endpoints are computed before any intersections in the inner of the objects.
- Check if any intersections are possible. This is normally done by performing a box test. More sophisticated interception methods can also be used, see sect. 7.
- Check if there is any possibilities for a closed inner intersection loops. This is denoted a simple case test, see sect. 8. For curve-curve intersections and surface-curve intersections, a simple case implies that not more than one intersection point between the geometry objects can exist.
- If a simple case situation is reached, all intersection points found at the boundaries are connected into tracks. These tracks will later be refined or marched out to achieve good approximations to the true intersection curves. In the case of curve-curve or surface-curve intersections, we iterate to an intersection point.
- Otherwise, check for total coincidence.
- If no further subdivision is possible, define the most consistent intersection result for this situation. The situation should be avoided through a good subdivision strategy since its occurence imply bad performance and possibly inaccurate results. It is included as a security net.
- If there are still possibilities for intersections and no simple case, subdivide the current geometry objects to simplify the problem. The decision on how to subdivide is very crucial, and the performance of the intersection procedure is very dependent on this decision.
- Treat sub-problems. It is no limits on the number of recursions.
- Clean up in intersection results.

The result of this procedure is the complete topology of the intersection result. All intersection points at surface boundaries are exactly computed and so are the singular intersection points. The correctness of the result can be guaranteed, but there might be a performance problem for complicated intersection cases. We will later look at details in the recursive procedure that can be improved in order to cope with for instance singular situations, but first we will have a look at intersections involving an implicitely defined surface.

5   Intersections Between one Parametric and one Algebraic Surface

So far the considerations on surface-surface intersections have mainly been focused on intersections between two parametric surfaces. In many aspects the methods for handling intersections between one parametric and one algebraic surface are similar. The expression for the parametric surface is put into the implicit equation resulting in a function. The task is then to compute all zeroes of this functions. This may be handled correspondingly to the case of two parametric surface, but due to the reduced number of parameter directions, it is normally less complicated. Alternatively, special methods for computing the zeroes of a function can be applied, see [17] and references therin. However, a special emphasize has to be put on the tolerance concept.

One parametric surface $F(u, v) : R^2 \to R^3$ and one algebraic surface $H(x, y, z) = 0$ are defined. We want to compute all intersections between these two surfaces relative to a given tolerance ε. This problem simplifies to the problem of computing all zeroes of a function of two parameters

$$f(u, v) = H(F_x(u, v), F_y(u, v), F_z(u, v))$$

with respect to a tolerance δ, i.e. we want to find parameters (u^*, v^*) such that $f(u^*, v^*) < \delta$. We will now shortly outline the relation between the tolerances ε and δ.

In an intersection between one parametric and one implicitly defined surface, we can define $a(u, v)$ which is the direction in which the error in geometry space is measured. The expression $H(F(u, v) - \varepsilon a(u, v))$ is zero where the distance between the two surfaces is exactly ϵ with respect to the direction of measurement. We require $|a(u, v)| \equiv 1$. Performing Taylor expansion of this expression, we get

$$H(F(u, v) - \varepsilon a(u, v)) \approx H(F(u, v)) - \epsilon \nabla H(F(u, v)) \cdot a(u, v) \qquad (1)$$

plus higher order terms in ε. We now set

$$\delta = |H(F(u, v))| \approx \varepsilon |\nabla H(F(u, v)) \cdot a(u, v)| \leq \varepsilon |\nabla H(F(u, v))| \qquad (2)$$

We choose the direction of error measurement to always be equal to the gradient of the implicitly defined surface H in the point closest to the current point $F(u, v)$. Then we can choose the tolerance for the transformed problem to be

$$\delta = \delta(u, v) = \varepsilon |\nabla H(F(u, v))| \qquad (3)$$

The higher order terms in ε will be neglect-able if ε is sufficiently small. However, if the gradient of the implicit surface varies in the area of interest, the tolerance δ is non-constant. This is a problem only for tangential intersections. Transversal intersections is less sensitive to tolerances, but tangential intersections might get lost using a too small tolerance. For planes, spheres and cylinder, δ will be constant.

Figure 2 illustrates a situation where the tolerance ε is quite large. A very small B-spline surface is intersected with a cylinder, and using the tolerance δ as outlined above, the result is total coincidence. However, at least one corner of the B-spline surface

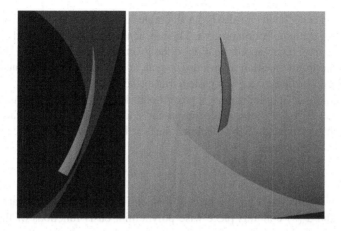

Fig. 2. Intersection between a small B-spline surface and a cylinder

is more distant from the cylinder than ε. To see what happens consider a circle with radius r and center in origo. We want to find the conditions for coincidence between a parametric curve in the plane $f(t) = (x(t), y(t))$ and this circle with respect to the tolerance ε. A point (x, y) on the curve is closer than the tolerance to the circle if

$$r - \varepsilon < \sqrt{x^2 + y^2} < r + \varepsilon$$

Elaborating this expression, we get

$$r^2 - 2r\varepsilon + \varepsilon^2 < x^2 + y^2 < r^2 + 2r\varepsilon + \varepsilon^2$$
$$|x^2 + y^2 - r^2 - \varepsilon^2| < 2r\varepsilon$$

Thus, higher order terms of ε enters the expression in a way that translates the coincidence interval for the function H created by inserting the expression for the B-spline curve into the equation of the circle. The same situation occurs when a B-spline surface is inserted into a cylinder equation. Note that the positions of the exact zeroes of H are not changed. For small tolerance values the term ε^2 is neglectable, but in this particular example the introduction of this term resulted in the intersection result shown in the second picture of fig. 2. Only two of the boundaries of the B-spline surface are found to intersect the cylinder. Parts of the inner of the surface are closer to the cylinder than the tolerance although this is not represented in the intersection result.

6   Approximate Implicitization

In many cases we prefer to deal with a problem of two parameter directions instead of a problem with four directions. The algebraic surface which is intersected with a parametric surface can be a conic surface which is supported by the CAD system using the intersection code, but it can also be an algebraic surface generated from a parametric one to simplify the intersection process. This surface will, in order to avoid a very high polynomial degree, be an approximation to the given parametric surface. The concept of approximate implicitization allows us to create a dual surface represented as a low degree algebraic surface, to a given parametric surface.

The approximate implicitization approach is based on combining an unknown algebraic surface $H(x) = 0$ of total degree m and a known parametric surface $F(u, v)$. The unknown coefficients of the algebraic surface is organized in the vector b. This combination can be written

$$0 = H(F(u, v)) = (Db)T(s)$$

The entries in the matrix D are products of the coefficients of the coordinate functions of $F(u, v)$, and $T(s)$ contains piece-wise polynomial basis functions of the surface. In our case $T(s)$ contains basis functions that are a partition of unity, and then the smaller singular values of D identifies coefficient vectors of H that can be used for approximating $F(u, v)$. If the algebraic surfaces are represented by a Bernstein basis in a barycentric coordinate system the approach has nice numerical properties. The topic was first introduced in in [4] and later in [5], see also [6]. See [22] for exact implicitization.

7   Interception Testing

Interception testing is a fundamental tool in recursive subdivision. The purpose of interception testing is to check whether two geometry objects may intersect, or whether a function may have any zeroes. If no intersection is possible the current sub problem is solved. The classical method for interception is box testing. Coordinate boxes surrounding the two objects are created. Intersection between these boxes is easily checked, and if the boxes do not intersect, the objects cannot intersect either. We can also make rotated boxes, i.e. the boxes are rotated according to some properties of the combination of geometry objects in order to improve the interception. A standard rotation by letting the box corners be cut off whenever possible also improves the interception at a low cost. Peters and Wu present some new methods for interception in [18]. For a function, a box is similar to an interval. The intersection tolerance must be taken into account while doing box testing to avoid loosing near tangential intersections. The first picture in fig. 3 shows a surface-surface intersection where interception by coordinate boxes has very good efficiency.

Another approach is to place a plane or a cylinder or some other simple standard geometry between the geometry objects, i.e. we want to define a splitting geometry that separates the two objects. If these objects do not intersect the standard geometry, they do not intersect each other either. However, intersection cases occur where none of these tools are effective for interception, see the second surface configuration in fig. 3.

For near parallel, non-planar, parametric, sculptured surfaces, the box testing is not effective. The recursion continues very deeply and the performanc is not acceptable. The same is true for nearly parallel curves or a curve lying very close to a surface. We want the problem in some sense to become flat, and do so by generalizing the concept

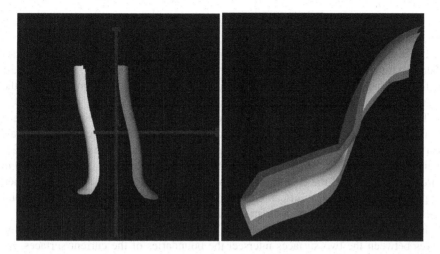

Fig. 3. Surface configurations where there are no intersections between the surfaces

of splitting geometry. The object representing the splitting geometry can be any kind of algebraic surface, not only a standard conic. Moreover, instead of placing the splitting geometry between the two initial curves or surfaces, it is defined to coincide with one of the objects. If the other object does not intersect the splitting geometry, it does not intersect the object being coincident to the splitting geometry either.

Assume that two parametric surfaces $F(u, v)$ and $G(s, t)$ are given. Assume also that a good algebraic approximation, H, to G exist, see sect. 6, i.e. $H(G(s, t)) \approx 0$. Then we can define the function $f(u, v) = H(F(u, v))$. If this function does not have any zeroes with respect to the tolerance δ outlined in sect. 5, then the surfaces F and G do not intersect. The problem illustrated in the second picture of fig. 3 is easily solved by this approach. However, δ is, as we have seen, not a constant. It is a function in u and v and it is dependent on the gradient of the algebraic surface H. Thus, H does not only need to be a good approximation to $G(s, t)$ in the area where G exist, it must also posses a gradient that is close to constant in this area.

8   Simple case Situations

The aim is to know when we have enough information to define the topology of an intersection problem in a recursive subdivision setting or whether a start point to all intersection tracks are found in a marching approach. The action is to check for a simple case situations.

For the problem of finding all zeroes in a function of one parameter or to compute intersections between two curves or between a curve and a surface, it is a simple case situations if there cannot be more than one zero or one intersection point in the current problem. In surface-surface intersection or the problem of computing zeroes of a function of two parameters, there is a simple case where there is no closed intersection loops in the surfaces or function. In a recursive subdivision method, one goal is to reach a simple case situation.

In the function case, we look for monotonicity. If one parameter direction exists where the function is monotone, the situation is simple, and we can stop subdividing. If the function have two parameters, this parameter direction can be a combination of the two main parameter directions.

Two surfaces cannot intersect in a closed inner loop if their corresponding surface normals do not overlap. This can be checked by bounding all possible surface normals by a cone and check whether these cones overlap. The loop detection problem is extensively treated in [20], [21], [13], [14] and [24]. Similarly, two curves cannot intersect in more than one point if the cones surrounding the curve tangents do not overlap. Then the possible intersection point can be found by iteration. Iterating for a closest point gives a more stable result than a direct iteration for the intersection point.

In surface-surface intersection a simple case situation implies that all intersection curves between the two surfaces intersect the boundaries of the current surfaces. Intersection points at the boundaries are already found at this stage, and they can be connected into tracks. The direction of the intersection curve in a point is an important tool in order to make correct connections, and this direction can normally be found by computing the cross product of the two surfaces in this point. In singular situations a more complex computation is required, see [12]

A monotonicity test for a function of two parameters is a stronger tool than a check for overlap between two cones bounding surface normals. The assumption is that we allow for monotonicity in other parameter directions than the two standard ones. Since an intersection between two parametric surfaces can be transformed to the problem of computing zeroes of a function through the use of approximate implicitization, the monotonicity test tool can be available also for intersections between two parametric surfaces. Figure 7 shows an intersection problem where four intersection curves meet in a singular point. Two of the curves lie in the same quadrant of the parameter domain for both surfaces. Thus, a simple case situation will never be reached for recursive subdivision alone independent on the recursion depth. However, if only one parameter domain exists and a corresponding function has two zero curves in the same quadrant of this domain, there might still be some parameter direction where the function is monotone.

9   Singular Surface-Surface Intersections

Tangential and singular situations frequently occur in design, and even if the knowledge of e.g. a tangential surface-surface intersection prevents the designing system to perform an intersection between the related surfaces, a solid- or surface model often has a second life outside of the originating CAD system. Then it is no guard against performing such an intersection. We have to accept that tangential and singular surface-surface intersections exist, and should be handled in the best possible way.

Not only do singular situations exist, they are seldom very accurately described. We see two intersection curves that almost meet in a singular point, and where the surfaces in this point are closer than the given tolerance. We see two intersection curves

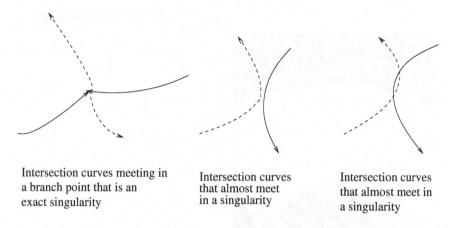

Intersection curves meeting in
a branch point that is an
exact singularity

Intersection curves
that almost meet
in a singularity

Intersection curves
that almost meet in
a singularity

Fig. 4. Categories of branch points

that intersect in two points where the whole area is in intersection. Cases exist where surfaces almost touch in one curve, or intersect in two extremely close curves. Figure 4 illustrates some configurations. The direction of the intersection curves are shown by arrows.

Given two parametric surfaces $F(u, v)$ and $G(s, t)$, the parameter quadruple $(\tilde{u}, \tilde{v}, \tilde{s}, \tilde{t})$ represent a singular intersection point if

$$F_u(\tilde{u}, \tilde{v}) \cdot (F(\tilde{u}, \tilde{v}) - G(\tilde{s}, \tilde{t})) = 0$$
$$F_v(\tilde{u}, \tilde{v}) \cdot (F(\tilde{u}, \tilde{v}) - G(\tilde{s}, \tilde{t})) = 0$$
$$G_s(\tilde{s}, \tilde{t}) \cdot (F(\tilde{u}, \tilde{v}) - G(\tilde{s}, \tilde{t})) = 0$$
$$G_t(\tilde{s}, \tilde{t}) \cdot (F(\tilde{u}, \tilde{v}) - G(\tilde{s}, \tilde{t})) = 0$$

and $|F(\tilde{u}, \tilde{v}) - G(\tilde{s}, \tilde{t})| < \varepsilon$. Singular intersection points can be of several types:

Isolated singularity The point does not belong to any intersection curves. If the intersection point exists only within a tolerance, we have to take care not to loose the point.

Tangential intersection point The point belongs to a tangential intersection curve. If the point is exact, there is a unique tangential intersection curve passing through this point. Normally, it will be possible to compute the direction of this curve in the point using second derivative information from the surface, but not the orientation. Tracing of tangential intersections is treated in [12] and differential geometry of the curves in [23]. If the point is a near singularity, there will be two intersection curves very close to each other. The directions of the two curves are very similar, but the two curves have opposite orientation. The area between the curves does also belong to the intersection.

Branch point Several intersection curves meet in the point. If the point is exact, it is normally possible to compute the directions of the various curves in the point.

Figure 5 shows a near singular situation. The two surfaces intersect in two intersection curves that lie very close in an area. Subdivision in the singular point between the

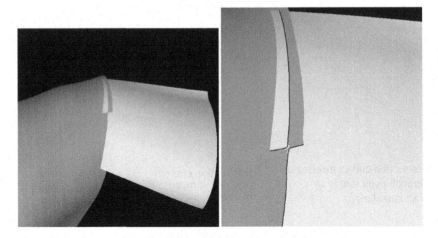

Fig. 5. Two intersection curves do almost meet in a singular point

intersection curves leads to sub problems where a simple case situation can be recognized. The singular point is found by iteration.

Two surfaces that intersect in a near singularity very often have almost parallel constant parameter curves in the singular intersection point. To reach a simple case situation, it is crucial to place singular intersection points at corners of the sub surfaces. If subdivision along these guide lines is to be performed for both surfaces, we construct very tangential curve-curve and surface-curve intersections.

Figure 7 shows two surfaces that intersect. Using a very small tolerance, the intersection curves are separate. For the tolerance given by the CAD system, the two curves meet in a singular intersection point, but the accuracy in this point is not good. Moreover, the different intersection branches can not be separated by subdivision since two of the intersection curves lie in the same quarter of the surface. The strategy is to subdive in the singularity. Then each sub problem is transformed to a problem of computing zeroes of a function by the means of approximate implicitization. We can then use the monotonicity approach in simple case checking for each sub problem, and simple case situations are reached. In this case, subdivision in the singularity leads to unstable curve-surface intersections. Thus, finding a clean representation of the intersection results in the area around the singular point is a challenge.

10   Partial Coincidences and Tangential Intersections

Also partial coincidences are singular situations. Figure 8 shows a partial coincidence between two B-spline surfaces. The surfaces are made by loft through a set of planar profiles where some of the profiles are the same for both surfaces. This approach creates two surfaces that lie very close in a large area, but they differ more than the given tolerance. There is a lot of singular branch points, and tangential and nearly tangential

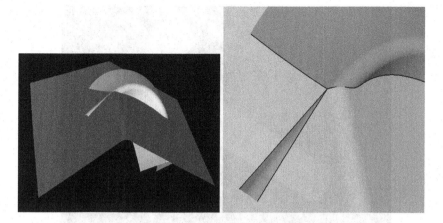

Fig. 6. Two intersection curves do almost meet in a singular point

Fig. 7. Two intersection curves intersect in a singular point within the tolerance

intersection curves in this area. Computing these intersections are very effort demanding and error proof and do not give a useful result for the application. Other methods for construction is in this case required to reduce the demand on the surface intersection functionality.

Figure 9 shows an intersection problem where a partial coincidence area is well defined. The two surfaces touch exactly in the area. The boundaries of the partial coincidence area follow constant parameter curves in the surfaces. In this case the coincidence in the inner of the area is verified by sampling, but also methods that use approximate implicitization and transform the problem to a function, can be used.

Also surfaces that intersect in tangential intersection curves provide a partial coincidence situation that is possible to handle. Two surfaces touch along a curve. Due to the most common construction methods this tangential intersection curve will be a constant parameter curve in one of the surfaces when both surfaces are parametric. The true intersection with respect to the tolerance is an area around the intersection curve. We represent this area by the curve itself.

Near tangential intersection curves are more complex. One surface is attached to a curve in another surface to create a smooth transition by approximate methods, for instance when creating a blend between two surfaces. The blend surface approximately intersects the two initial surfaces along their boundaries, and the intersection curves are nearly tangential. In reality there are a lot of minor intersections in an area close to the boundaries. The application is typically not interested in these intersections although the entire areas where they occur are not partial coincidence areas. The information that a near tangential intersection curve is found along a blend boundary is more constructive. Figure 10 shows an intersection between a blend surface and one of its mothers.

Figure 11 shows two surfaces intersecting tangentially along the boundary of one of the surfaces. Moreover, there are two intersection curves joining the tangential inter-

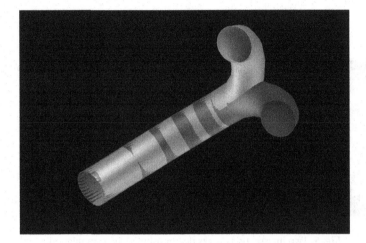

Fig. 8. Partial coincidence

section curve in branch points that are singularities if high order. These branch points are hard to find by iteration. We follow the trend of the transversal intersection curves toward the singular points to find appropriate positions for the branch points.

11   Conclusion

In general, finding transversal intersections is a stable and well defined problem. Computing tangential intersections, on the other hand, is highly unstable. The intersection appears and disappears depending on the size of the tolerance, the classification can change from one or more points to intervals or partial coincidence surfaces or the other way around, and the positioning the intersection results can be changed significantly by a small change in the input geometry.

We want to represent this unstable intersection result in a relatively stable way. Thus, we seek to represent the singularites and classify them according to their type: singular branch point, tangential intersection curve, near tangential intersection curve, and so on. The main properties of the current intersection are reported while details in the critical area can be found by a post process if required.

The choices of subdivision directions and positions in recursive subdivision are actively used to achieve sub problems where good interception methods or the recognition of a simple case can solve the problem. Moreover, we want to avoid creating complicated intersection problems of a reduced number of parameter directions by choosing a wrong position for subdivision. The subdision strategy is a very important part of a recursive subdivision method.

It is a challenge to be able to separate intersection problems that can be handled by a wise choise of tools, from the problems where a solution is not realistic, see fig. 8. The type of problems where we put our effort should be real life problems where we have a good chance to find a solution.

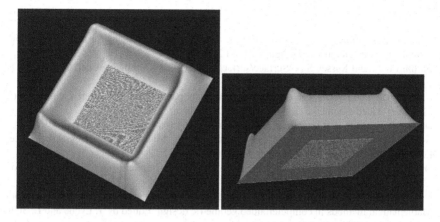

Fig. 9. A well defined partial coincidence between two surfaces

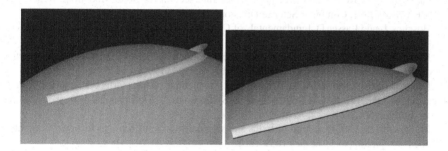

Fig. 10. Intersection between a blend surface and its mother

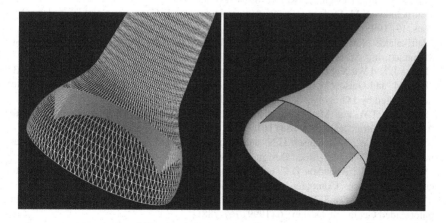

Fig. 11. Tangential intersection with singularities

References

1. Bajaj C.L., Hoffmann C.M., Hopcroft J.E., and Lynch R.E. Tracing surface intersections. Computer Aided Geometric Design 5 (1988) 285-307
2. Barnhill R.E. and Kersey S.N. Surface/surface intersection. Computer Aided Geometric Design 20 (1986) 18-36.
3. Barnhill R.E. and Kersey S.N. A marching method for parametric surface/surface intersection. Computer Aided Geometry Design 7 (1990) 257-280
4. Dokken T. Aspects of Intersection Algorithms and Approximation, Doctor thesis. 1997.
5. Dokken T. Approximate Implicitization. Mathematical Methods for Curves and Surfaces. Edited by T. Lyche and L. Schumaker (Oslo 2000) 81-102
6. Dokken T. and Thomassen J.B. Overview of Approximate Implicitization. To appear in Contemporary Mathematics (CONM) book series by AMS
7. Dokken T. Skytt V. and Ytrehus A-M. Recursive Subdivision and Iteration in Intersections. Mathematical methods in computer aided geometric design. Edited by T. Lyche and L. Schumaker. (Oslo 1989) 207-214
8. Grandine T.A. and Klein F.W. A new approach to the surface intersection problem. Computer Aided Geometric Design 14 (1997) 111-134.
9. Hohmeyer M.E. Robust and Efficient Surface Intersection for Solid Modelling. Report No UCB/CSK 92/681, Computer Science Division, University of California (1992)
10. Hoschek J. and Lasser D. Fundamentals of Computer Aided Geometric Design. A.K. Peters (1993)
11. Kriezis G.A., Patrikalakis N.M. and Wolter F-E. Topolgical and differential-equation methods for surface intersections. Computer-Aided Design 24 (1992) 41-55
12. Luo R.C., Ma Y. and McAllister D.F. Tracing tangential surface-surface intersections. Proceedings of the Third ACM Solid Modeling Symposium. C. Hoffman and J. Rossignac, editors. (1995) 255-262
13. Ma Y. and Lee Y-S. Detection of loops and singularities of surface intersections. Computer Aided Design 30, No 14 (1998) 1059-1067
14. Ma Y. and Luo R.C. Topological metod for loop detection of surface intersection problems. Computer-Aided Design 27 (1995) 811-820
15. Markot R.P. and Magedson R.L. Solutions of tangential surface and curve intersections. Computer Aided Design 21, No 7 (1989) 421-429
16. Patrikalakis N.M. Surface-to-surface intersections. IEEE Computer Graphics and Applications, 13 (1993) 89-95
17. Patrikalakis N.M. and Maekawa T. Shape Interrogation for Computer Aided Design and Manufacturing. Springer 2002.
18. Peters J. and Wu X. SLEVEs for planar spline curves. Submitted to Elsvier Science (2003)
19. Pratt M. and Geisow A. Surface/Surface Intersection Problems. IMA Mathematics of Surfaces. Edited by J Gregory, Clarendon Press (1986) 117-142
20. Sederberg T.W. and Meyers R.F. Loop detection in surface patch intersection. Computer Aided Geometric Design 5 (1988) 161-171.
21. Sederberg T.W., Christiansen H.N. and Katz S. Improved test for closed loops in surface intersections. Computer-Aided Design 21 (1989) 505-508
22. Sederberg T.W., Anderson D.C. and Goldman R.N. Implicit Representation of parametric curves and surfaces. Computer Vision, Graphics and Image Processing 29 (1984) 72-84
23. Ye X. and Maekawa T. Differential geometry of intersection curves of two surfaces. Computer Aided Geometric Design 16 (1999) 767-788
24. Zudel A.K. Surface-Surface Intersection: Loop Destruction Using Bezier Clipping and Pyramidal Bounds. Thesis for Doctor of Philosophy, Brigham Young University (1994)

Computing the Topology of Three-Dimensional Algebraic Curves*

G. Gatellier, A. Labrouzy, B. Mourrain, and J.P. Técourt

GALAAD, INRIA
BP 93, 06902 Sophia Antipolis

Abstract. In this paper, we present a new method for computing the topology of curves defined as the intersection of two implicit surfaces. The main ingredients are projection tools, based on resultant constructions and 0-dimensional polynomial system solvers. We describe a lifting method for points on the projection of the curve on a plane, even in the case of multiple preimages on the 3D curve. Reducing the problem to the comparison of coordinates of so-called critical points, we propose an approach which combines control and efficiency. An emphasis in this work is put on the experimental validation of this new method. Examples treated with the tools of the library AXEL[1] (Algebraic Software-Components for gEometric modeLing) are showing the potential of such techniques.

1 Introduction

Numerical modeling plays an increasingly role in fields at the border between data processing and mathematics. This is the case for example in CAD (Computer-aided design, where the objects of a scene or a piece to be built are represented by parameterized curves or surfaces such as NURBS), robotics (problem of the parallel robot, or vision), or molecular biology (rebuilding of a molecule starting from the matrix of the distances between its atoms obtained by NMR). A fundamental operation in this context is the intersection of geometric models, which leads to algebraic questions.

In this paper, we focus on the problem of computing the topology of the intersection of two algebraic surfaces. Such a question is critical in many solid geometry operations, involved in the digital modeling or construction process of shapes. In the case of two parameterized surfaces, in order to reduce to such a situation, we may compute the implicit equation of one of the rational surfaces [4]. This reduces the problem of intersection to the case of an implicit and a parameterized representation, which boils down, by substitution, to the case of a curve defined by an implicit equation in the plane of parameters. Our main concern will be the case of implicit curves, either in the plane or defined by two polynomial equations, in a 3-dimensional space.

This intersection problem received a lot of attention in the past literature. See for instance [9, 19, 16]. Different techniques (subdivision, lattice evaluation, marching methods) have been experimented [18, 10, 1, 13, 12, 19], but they suffer from the problem of certifying the topology of the result.

---

* This work is partially supported by GAIA II IST-2001-34919 European project
[1] http://www-sop.inria.fr/galaad/logiciels/axel/

In this paper, we present a new method to compute the topology of an algebraic curve in 3D, based on an extension of the 2D approach [13], [11]. Our objective is to devise a certified and output-sensitive method, in order to combine control and efficiency. We show that it reduces to the comparison of coordinates of points of intersections of two curves or three surfaces. This task can be fulfilled by using exact methods, such as the one described in [3], [8], which reduces to comparison of roots of univariate problems. Our approach is combining symbolic and numeric techniques, in order to filter the numerical computation. We present preliminary experiments in the library AXEL[2] (Algebraic Software-Components for gEometric modeLing), devoted to algebraic tools for geometric modeling. We are interested in the efficiency and also in the numerical behavior and stability of the method. The experiments are made using the package SYNAPS[3] [6] (SYmbolic and Numeric APplicationS), which provides a set of polynomial solvers. We apply in particular solvers, based on algebraic manipulations [20], or resultant constructions. This leads to eigenvalue computations, which are performed by LAPACK subroutines [2]. For more details on the polynomial solving algorithms, we refer to [8].

The main objective being the description of the curve of intersection of two implicit surfaces, the method that we present yields "only" the topology of such a curve, that is a graph of 3D points, connected by segments, with the same topology as the algebraic curve. Producing a good geometric approximation of the curve, which is the next step of a complete method, will not be considered here. It consists in applying marching techniques on the regular branches of the curve.

In the next section, we will describe quickly the algebraic ingredients that we need. The algorithm will be detailed in section 3 and some implementation topics and experimentations are presented in the last section.

2   Algebraic Tools

In this section, we introduce different algebraic tools that will be used later.

2.1   Resultant and Projection

For any polynomials $p_1, \ldots, p_k \in \mathbb{R}[z]$, we denoted by $\langle p_1, \ldots, p_k \rangle$ the vector space spanned by these polynomials. Let $\mathbb{R}[z]_d$ be the space of polynomials of degree less or equal to d, with basis $\{1, \ldots, z^d\}$.

Let us consider two univariate polynomials P and Q of $\mathbb{R}[z]$.

$$P = \sum_{k=0}^{p} a_k z^k, \quad Q = \sum_{k=0}^{q} b_k z^k.$$

We need first the following definition:

---
[2] http://www-sop.inria.fr/galaad/logiciels/axel/
[3] http://www-sop.inria.fr/galaad/logiciels/synaps/

Definition 1. *The Sylvester matrix of P and Q is the matrix of the application*

$$\begin{array}{ccc} \mathbb{R}[z]_{q-1} \oplus \mathbb{R}[z]_{p-1} & \longrightarrow & \mathbb{R}[z]_{p+q-1} \\ (u, v) & & P\,u + Q\,v \end{array}$$

of the form

$$\mathrm{Syl}(P,Q) = \begin{pmatrix} a_p & & & b_q & & \\ a_{p-1} & \ddots & a_p & b_{q-1} & \ddots & b_q \\ \vdots & & a_{p-1} & \vdots & & b_{q-1} \\ a_0 & & \vdots & b_0 & & \vdots \\ & \ddots & a_0 & & \ddots & b_0 \end{pmatrix}$$

Definition 2. *Determinantal polynomial. Let $\mathcal{M}$ be a matrix $k \times l$ with $l \leq k$. We define the determinantal polynomial of $\mathcal{M}$:*

$$detpol(\mathcal{M}) = \det(\mathcal{M}_k)\, z^{k-l} + \cdots + \det(\mathcal{M}_l)$$

where $\mathcal{M}_j$ denotes the submatrix of $\mathcal{M}$ consisting of the $l-1$ first rows of $\mathcal{M}$ followed by the jth.

Definition 3. *The polynomial subresultant of order i associated to P and Q is:*

$$S_i = detpol(z^{q-i-1}\,P, \ldots, P, z^{p-i-1}\,Q, \ldots, Q) = \sum_{k=0}^{l} S_{i,k} z^k$$

See [21]. These polynomials can be computed efficiently by Sturm-Habicht sequences [3]. Notice that S_0 is the determinant of the Sylvester matrix (i.e the resultant) of P and Q.

We will use the following result:

Proposition 4. *The last polynomial S_k associated to P and Q with $S_{k,k} \neq 0$ is the greatest common divisor of P and Q.*

Another important property is:

Proposition 5. *The corank of the Sylvester matrix associated to P and Q is the degree of $gcd(P, Q)$.*

Proof. Let $D = gcd(P, Q)$ and $\delta = \deg(D)$. Then we have

$$\mathrm{corank}(\mathrm{Syl}(P,Q)) = p + q - \dim\langle P, z\,P, \ldots, z^{q-1}\,P, \ldots, Q, z\,Q, \ldots, z^{p-1}Q\rangle.$$

As $\dim\langle P, z\,P, \ldots, z^{q-1}\,P, \ldots, Q, z\,Q, \ldots, z^{p-1}Q\rangle = \dim\langle D, z\,D, \ldots, z^{p+q-1-\delta}\,D\rangle$, we deduce that $\mathrm{corank}(\mathrm{Syl}(P,Q)) = \delta$.

2.2 Solving 0-Dimensional Systems

In this section, we are interested in solving polynomial systems of dimension 0 i.e that admit a finite number of (complex) solutions. Different approaches exist to solve such systems [8]. We focus on the algebraic approach that transforms the resolution problem into linear algebra problems.

Here are some notations: $R = \mathbb{R}[x, y, z]$, $f_1 = 0, \ldots, f_m = 0$ with $f_i \in R$, the equations we want to solve, $I = (f_1, \ldots, f_m)$ is the ideal generated by these polynomials, $\mathcal{A} = R/I$ the quotient algebra. We denote by

$$\mathcal{V}(f_1, \ldots, f_m) = \{(x, y, z) \in \mathbb{C}^3, f_i(x, y, z) = 0, \ i = 1, \ldots, m\},$$

the variety of $\mathbb{C}^3$ defined by the equations $f_i(x, y, z) = 0$.

   We deduce from the structure of the quotient algebra $\mathcal{A}$, the solutions $\mathcal{V}(I)$, from the following theorem:

Theorem 6. *Assume that* $\mathcal{V}(I) = \{\xi_1, \ldots, \xi_d\}$. *We have:*

   – *Let* $a \in \mathcal{A}$. *The eigenvalues of the operator* M_a *(and* M_a^t*) are* $a(\xi_1), \ldots, a(\xi_d)$.
   – *The common eigenvectors of* $(M_a^t)_{a \in \mathcal{A}}$ *are (up to a scalar)* $1_{\xi_1}, \ldots, 1_{\xi_d}$ *where* 1_{ξ_i} *is the linear form* $1_{\xi_i} : p \longrightarrow p(\xi_i)$.

This theorem reduces the resolution to a linear algebra problem [8] if we are able to work in $\mathcal{A}$. In order to turn this theorem into an effective method, we have to construct the matrices of multiplication in $\mathcal{A}$. For this purpose, we compute so called normal forms. One way is to use Groebner bases but for numerical stability, we prefer to use general normal forms [17], [20].

We summarize the main stages of the resolution process, in the following algorithm:

Algorithm 7. — **Solving 0-dimensional system.** *Input:* $I = (f_1, \ldots, f_m)$.

   – *Compute a basis of $\mathcal{A}$ and polynomials which yield a normal form reduction modulo I.*
   – *Deduce the matrices of multiplication by x, y, z in the basis of $\mathcal{A}$.*
   – *Compute simultaneous eigenvectors of M_x^t, M_y^t, M_z^t and the corresponding eigenvalues[8].*

Output: $\mathcal{V}(I) = \{\xi_i(\text{with multiplicity}), i = 1, \ldots, \dim \mathcal{A}\}$.

3 Topology of Algebraic Curves

By definition, a three dimensional algebraic curve $\mathcal{C}_\mathbb{C} = \mathcal{V}(f_1, \ldots, f_m)$ $(f_i \in \mathbb{R}[x, y, z])$ is an algebraic variety of dimension 1 in $\mathbb{C}^3$. We denote by $I(\mathcal{C}_C) \subset \mathbb{R}[x, y, z]$, the ideal of the curve $\mathcal{C}_C$ (that is the set of polynomials which vanish on $\mathcal{C}_C$) and by $g_1, \ldots, g_s \in \mathbb{R}[x, y, z]$ a set of generators: $I(\mathcal{C}_C) = (g_1, \ldots, g_s)$. By Hilbert's Nullstellensatz [5, 15], we have $I(\mathcal{V}(f_1, \ldots, f_k)) = \sqrt{I} \subset \mathbb{R}[x, y, z]$. It can be proved [7], [15], that 3 polynomials $g_1, g_2, g_3 \in \mathbb{R}[x, y, z]$ are enough to generate $I(\mathcal{C}_C)$.

   For simplicity reasons, we will consider here that the curve is described as the intersection of two surfaces $P_1(x, y, z) = 0$, $P_2(x, y, z) = 0$, with $P_1, P_2 \in \mathbb{R}[x, y, z]$.

We assume that the gcd of P_1 and P_2 in $\mathbb{R}[x, y, z]$ is 1, so that $\mathcal{V}(P_1, P_2) = \mathcal{C}_\mathbb{C}$ is of dimension 1, and all its irreducible components are of dimension 1. We are interested in describing the topology of the real part

$$\mathcal{C}_\mathbb{R} = \{(x, y, z) \in \mathbb{R}^3, \ P_1(x, y, z) = 0, \ P_2(x, y, z) = 0\},$$

that we will denote hereafter by $\mathcal{C}$.

In this paper, we assume that $I(\mathcal{C}) = (P_1, P_2)$ or equivalently that (P_1, P_2) *is a reduced ideal*: $(P_1, P_2) = \sqrt{(P_1, P_2)}$.

We will not consider examples such as $P_1 = x^2 + y^2 - 1, P_2 = x^2 + y^2 + z^2 - 1$, where $(P_1, P_2) = (x^2 + y^2 - 1, z^2)$ and $I(\mathcal{C}) = (x^2 + y^2 - 1, z)$, so that the curve $\mathcal{C}$ is defined "twice" by the equations $P_1 = 0, P_2 = 0$ (the two surfaces intersect tangently along $\mathcal{C}$). Such a property can be tested by projecting into a generic direction and testing if the equation computed from the resultant of P_1, P_2, is squarefree, or by more general methods such as computing the radical of (P_1, P_2) [14].

The general idea behind the algorithm that we are going to describe is as follows: we use a sweeping plane in a given direction (say parallel to the (y, z) plane) to detect the critical positions where *something* happen. We also compute the positions where something happen in projection on the (x, y) and (x, z) plane. Then, we connect the points of the curve of $\mathcal{C}$ on these critical planes. This yields a graph of points, connected by segments, with the same topology as the curve $\mathcal{C}$.

3.1   Critical Points and Generic Position

In this section, we precise what we mean by the points where *something* happen. These points will be called hereafter critical points.

Definition 8. *Let $I(\mathcal{C}) = (g_1, g_2, \ldots, g_s)$ and let M be the $s \times 3$ Jacobian matrix with rows $\partial_x g_i, \partial_y g_i, \partial_z g_i$.*

- *A point $p \in \mathcal{C}$ is* regular *(or smooth) if the rank of M evaluated at p is 2.*
- *A point $p \in \mathcal{C}$ which is not regular is called* singular.
- *A point $p = (\alpha, \beta, \gamma) \in \mathcal{C}$ is x-critical (or critical for the projection on the x-axis) if the curve $\mathcal{C}$ is tangent at this point to a plane parallel to the (y, z)-plane i.e the multiplicity of intersection of the plane with $I(\mathcal{C})$ at p is greater or equal to 2. The corresponding α is called a x-critical value.*

A similar definition applies for the orthogonal projection onto the y and z axis or onto any line in space. Notice that a singular point is critical for any direction of projection.

If $I(\mathcal{C}) = (P_1, P_2)$, then the x-critical points are the solutions of the system

$$P_1(x, y, z) = 0, \ P_2(x, y, z) = 0, \ (\partial_y P_1 \partial_z P_2 - \partial_y P_2 \partial_z P_1)(x, y, z) = 0. \quad (1)$$

In the case of a planar curve defined by $P(x, y) = z = 0$, with $P(x, y)$ squarefree so that $I(\mathcal{C}) = (P(x, y), z)$, this yields the following definitions: a point (α, β)

- is *singular* if $P(\alpha, \beta) = \partial_x P(\alpha, \beta) = \partial_y P(\alpha, \beta) = 0$.
- is *x-critical* if $P(\alpha, \beta) = \partial_y P(\alpha, \beta) = 0$.

This allows us to describe the genericity condition that we require for the curve $\mathcal{C}$, in order to be able to apply the algorithm:

Definition 9. *Let*

$$N_x(\alpha) = \#\{(\beta, \gamma) \in \mathbb{R}^2 \ st. \ (\alpha, \beta, \gamma) \ is \ a \ x-critical \ point \ of \ \mathcal{C}\}.$$

We say that $\mathcal{C}$ is in a generic position for the x-direction, if

- *$\forall \alpha \in \mathbb{R}$, $N_x(\alpha) \leqslant 1$, and*
- *there is no asymptotic direction of $\mathcal{C}$ parallel to the (y, z)-plane.*

We will show that by a random change of variables, the curve can be put in a generic position. In practice, instead of changing the variables, we may choose a random direction for the sweeping plane.

3.2 The Projected Curves

The algorithm that we are going to describe, uses the singular points of the projection of $\mathcal{C}$ onto the (x, y) and (x, z)-planes. We denote by $\mathcal{C}'$ (resp. $\mathcal{C}''$) the projection of the curve $\mathcal{C}$ onto the (x, y) (resp. (x, z))-plane. The equation of the curve $\mathcal{C}'$ is obtained as follows. We decompose the polynomials P_1, P_2 in terms of the variable z:

$$P_1(x, y, z) = a_{d_1}(x, y)z^{d_1} + \dots + a_0(x, y)$$
$$P_2(x, y, z) = b_{d_2}(x, y)z^{d_2} + \dots + b_0(x, y)$$

with $a_{d_1}(x, y) \neq 0$ and $b_{d_2}(x, y) \neq 0$. Then, the resultant polynomial

$$G(x, y) = \mathrm{Res}_z(P_1, P_2)$$

vanishes on the projection of the curve $\mathcal{C}$ on the plane (x, y). Conversely, by the resultant theorem [8], $G(x, y) = 0$ defines exactly the projection $\mathcal{C}'$ of the curve $\mathcal{C}$ if $a_{d_1}(x, y)$ and $b_{d_2}(x, y)$ do not vanish simultaneously on a component of dimension 1 of $\mathcal{C}'$, that is, if the gcd $c(x, y)$ of $a_{d_1}(x, y)$ and $b_{d_2}(x, y)$ in $\mathbb{R}[x, y]$ is 1. If it's not the case, G is a non-trivial multiple of the implicit equation of $\mathcal{C}'$. Such a situation can be avoided, by a linear change of variables. Nevertheless, since the critical points of the curve defined by $G(x, y) = 0$ contains the critical points of $\mathcal{C}'$, we will see hereafter that this change of variable is not necessary.

Notice, that $G(x, y)$ is not necessarily a squarefree polynomial. Consider for instance the case $P_1 = x^2 + y^2 - 1$, $P_2 = x^2 + y^2 + z^2 - 2$, where $g(x, y) = (x^2 + y^2 - 1)^2$. In this case, there are generically two (complex) points of $\mathcal{C}$ above a point of $\mathcal{C}'$.

We can easily compute the gcd of $G(x, y)$ and $\partial_y G(x, y)$ (using proposition 4), in order to get the squarefree part $g(x, y) = G(x, y)/gcd(G(x, y), \partial_y G(x, y))$ of $G(x, y)$.

Similarly, for the projection $\mathcal{C}''$ of $\mathcal{C}$ on the (x, z)-plane, we compute

$$H(x, z) = \mathrm{Res}_y(P_1, P_2)$$

and its square-free part $h(x, z)$ from the gcd of $H(x, z)$ and $\partial_z H(x, z)$. The equation $h(x, z) = 0$ defines a curve which is exactly $\mathcal{C}''$, if the gcd of the leading components of P_1, P_2 in y is 1. Its set of singular points contains those of $\mathcal{C}''$.

In order to analyse locally the projection of the curve $\mathcal{C}$, we recall the following definition:

Definition 10. *[22] Let X be an algebraic subset of $\mathbb{R}^n$ and let p be a point of X. The tangent cone at p to X is the set of points u in $\mathbb{R}^n$ such that there exist a sequence of points x_k of X converging to p and a sequence of real numbers t_k such that $\lim_{k \to +\infty} t_k(x_k - p) = u$.*

Notice, that at a smooth point of $\mathcal{C}$, the tangent cone is a line.

Proposition 11. *Let $p' = (\alpha, \beta)$ be a x-critical point of $\mathcal{C}'$, which is not singular. Then α is the x-coordinate of a x-critical point of $\mathcal{C}$.*

Proof. Let V be the set of points $p \in \mathcal{C}$, which project onto p'. From the previous definition, we directly deduce that the projection of the tangent cone at $p \in \mathcal{C}$ is contained in the tangent cone of the projection of p. Thus the tangent cone of $\mathcal{C}'$ at $p' = (\alpha, \beta)$ contains the projection of the tangent cones of the points $p \in V$. Since p' is regular, its tangent cone is a line parallel to the y direction. Therefore, the tangent cones of the points $p \in V$ are in the plane $x - \alpha = 0$, parallel to the plane (y, z). This implies that the intersection of $\mathcal{C}$ with the plane $x - \alpha = 0$ contains a point of multiplicity ≥ 2, that is a x-critical point. In other words, α is the x-coordinate of a x-critical point of $\mathcal{C}$. □

3.3 Lifting a Point of $\mathcal{C}'$

The problem we want to tackle here is the following: Assume we are given two surfaces defined by two implicit equations $P_1 = 0$ and $P_2 = 0$. Let us consider the projection of the curve of intersection of the two surfaces on the (x, y)-plane. Starting from a point (x_0, y_0) of the projected curve, how can we find the z-coordinate of the point(s) above (x_0, y_0) ?

We note $P(z) = P_1(x_0, y_0, z)$, $Q(z) = P_2(x_0, y_0, z)$ and $p = \deg(P)$, $q = \deg(Q)$. Consider the Sylvester submatrix $\mathrm{Syl}_1(x_0, y_0)$ of the application

$$\begin{array}{ccc} \mathbb{R}[z]_{q-2} \oplus \mathbb{R}[z]_{p-2} & \longrightarrow & \mathbb{R}[z]_{p+q-2} \\ (u, v) & \mapsto & P\,u + Q\,v \end{array}$$

If ξ is a common root of P and Q then $(1, \xi, \ldots, \xi^{p+q-2})$ is in the kernel of the transpose of $\mathrm{Syl}_1(x_0, y_0)$. If we assume that $\mathrm{Syl}_1(x_0, y_0)$ is of maximal rank, and if Δ_i denotes the minor of $\mathrm{Syl}_1(x_0, y_0)$ obtained by removing the row i, then the (nonzero) vector $[\Delta_1, -\Delta_2, \ldots, (-1)^{p+q-1} \Delta_{p+q-1}]$ is in the kernel of the transpose of $\mathrm{Syl}_1(x_0, y_0)$. Thus $(1, \xi, \ldots, \xi^{p+q-2})$ and $[\Delta_1, -\Delta_2, \ldots, (-1)^{p+q-1} \Delta_{p+q-1}]$ are linearly dependent. We deduce that $\xi = -\frac{\Delta_{p+q-1}}{\Delta_{p+q-2}} = -\frac{S_{1,0}(x_0, y_0)}{S_{1,1}(x_0, y_0)}$.

This method allows us to lift a point on $\mathcal{C}$, if there is only one point above (x_0, y_0), but it can be generalized when there are several points above. This generalization is closely related to the subresultant construction of univariate polynomials [21]. Here we want to exploit linear algebra tools from a numerical perspective. The aim is to make the matrix of multiplication by z in the quotient algebra $\mathbb{R}[z]/(P_1(x_0, y_0, z), P_2(x_0, y_0, z))$ appear, in order to compute its eigenvalues which yields z-coordinate of the points above (x_0, y_0) [8].

We proceed as follows: Given a point (x_0, y_0) of the projected curve $\mathcal{C}'$, we construct the Sylvester matrix associated to $P(z), Q(z)$. By construction, the columns of this matrix are $P, z\,P, \ldots, z^{q-1}\,P, Q, z\,Q, \ldots, z^{p-1}\,Q$, using $1, z, \ldots, z^{p+q-1}$ as a basis. Assume that the kernel of the transposed Sylvester matrix $\mathrm{Syl}(x_0, y_0)$ has dimension d and is generated by $\Lambda_1, \ldots, \Lambda_d$.

By transposition, we can interpret the Λ_i $(i = 1 \ldots d)$ as linear forms over $\mathbb{K}_{p+q-1}[z]$ vanishing on $P, z\,P, \ldots, z^{q-1}\,P, Q, z\,Q, \ldots, z^{p-1}\,Q$. We can extend the Λ_i over $\mathbb{R}[z]$, considering that these forms vanish over all the ideal generated by P and Q. So they can be considered as elements of the dual of $\mathcal{A} = \mathbb{R}[z]/(P(z), Q(z))$. As the linear forms Λ_i are independent, they also form a basis of this dual space. The coefficients of Λ_i in the dual basis $(1^*, \ldots, (z^{d-1})^*)$ of the monomial basis $\{1, z, \ldots, z^{d-1}\}$ of $\mathcal{A}$ are $[\Lambda_i(1), \Lambda_i(z), \ldots, \Lambda_i(z^{d-1})]$. By definition of the transposed operator, for any $a \in \mathcal{A}$, $M^t(\Lambda_i)(a) = \Lambda_i(M_z(a)) = \Lambda_i(z\,a)$. Thus we have the relation:

$$
\begin{pmatrix} \Lambda_1(z) & \cdots & \Lambda_d(z) \\ \vdots & & \vdots \\ \Lambda_1(z^d) & \cdots & \Lambda_d(z^d) \end{pmatrix} = M_z^t \begin{pmatrix} \Lambda_1(1) & \cdots & \Lambda_d(1) \\ \vdots & & \vdots \\ \Lambda_1(z^{d-1}) & \cdots & \Lambda_d(z^{d-1}) \end{pmatrix}
$$

where M_z is the operator of multiplication by z in $\mathbb{R}[z]/(P(z), Q(z))$. As $d = \dim \ker(\mathrm{Syl}(x_0, y_0)) = \dim \mathcal{A}$, and as $(1, z, ..., z^{d-1})$ form a basis of the quotient space, the matrix

$$
\begin{pmatrix} \Lambda_1(1) & \cdots & \Lambda_d(1) \\ \vdots & & \vdots \\ \Lambda_1(z^{d-1}) & \cdots & \Lambda_d(z^{d-1}) \end{pmatrix}
$$

is invertible. We deduce that computing the generalized eigenvalues of the previous matrices yields the eigenvalues of the operator M_z of multiplication by z in $\mathcal{A}$, that is the z-coordinate of the points above (x_0, y_0).

We summarize the algorithm here:

Algorithm 12. — Lifting the projection.

- *Compute the Sylvester matrix $S = \mathrm{Syl}(x_0, y_0)$.*
- *Compute a basis $\Lambda_1, \ldots, \Lambda_d$ of the kernel of S^t.*
- *Extract the submatrix A_0 of the coordinates of $\Lambda_1, \ldots, \Lambda_d$ corresponding to the evaluations in $1, \ldots, z^{d-1}$.*
- *Extract the submatrix A_1 of the coordinates of $\Lambda_1, \ldots, \Lambda_d$ corresponding to the evaluations in $z, \ldots, z^d$.*
- *Compute the generalized eigenvalues of A_1 and A_0 and output the corresponding z-coordinates of the point above (x_0, y_0).*

The last step can be replaced by the computation of $\det(A_1 - z\,A_0)$ and an univariate root finding step.

3.4 Computing Points of $\mathcal{C}$ at Critical Values

In this section, we are going to describe how we check the genericity condition and how we compute a finite set of points, which will allow us to deduce the topology of $\mathcal{C}$.

First, we check that there is no asymptotic direction parallel to the (y, z) -plane, by testing if the curve $\mathcal{C}$ has a point at infinity in the plane $x = 0$. This is done by checking if the system

$$P_1^\top(0, y, z) = P_2^\top(0, y, z) = 0$$

has a non-trivial solution, where $P^\top$ is the homogeneous component of highest degree of a polynomial P. It reduces to computing the projective resultant of these two homogeneous polynomials. Since the number of asymptotic directions of $\mathcal{C}$ is finite, by a generic linear change of variables, we can avoid the cases where $\mathcal{C}$ has an asymptotic direction parallel to the (y, z) plane.

Next, we compute the x-critical points of $\mathcal{C}$ by solving the system (1), using algorithm 7. This computation allows us to check that the system is zero-dimensional and that the x-coordinate of the real solutions are distinct. If this is not the case, we perform a generic change of coordinates.

The cases for which we have to do a change of coordinates are those where a component of $\mathcal{C}$ is in a plane parallel to (y, z) or where a plane parallel to (y, z) is tangent to $\mathcal{C}$ in two distinct points. Such cases are avoided by a generic change of coordinates.

We denote by $\Sigma_0 = \{\sigma_1^0, \ldots, \sigma_{k_0}^0\}$ the x-coordinates of the x-critical points: $\sigma_1^0 < \cdots < \sigma_{k_0}^0$.

Next, we compute the singular points of $\mathcal{C}'$ as (a subset of) the real solutions of the system

$$g(x, y) = 0, \partial_x g(x, y) = 0, \partial_y g(x, y) = 0, \tag{2}$$

and of $\mathcal{C}''$, as (a subset of) the real solutions of

$$h(x, z) = 0, \partial_x h(x, z) = 0, \partial_z h(x, z) = 0. \tag{3}$$

We denote by $\Sigma_1 = \{\sigma_1^1, \ldots, \sigma_{k_1}^1\}$ the x-coordinates of these singular points: $\sigma_1^1 < \cdots < \sigma_{k_1}^1$.

Let us denote by $\Sigma = \Sigma_0 \cup \Sigma_1 = \{\sigma_1, \ldots, \sigma_l\}$ (with $\sigma_1 < \cdots < \sigma_l$) the sequence of all the x-coordinate computed so far.

An important property of the projected curves $\mathcal{C}'$ and $\mathcal{C}''$, that will be used in the algorithm, is the following:

Proposition 13. *The arcs of the curve $\mathcal{C}'$ (resp. $\mathcal{C}''$) above $]\sigma_i, \sigma_{i+1}[$ do not intersect.*

Proof. By definition, the arcs of $\mathcal{C}'$ above $]\sigma_i, \sigma_{i+1}[$ can only intersect at the x-critical points of $\mathcal{C}'$. Let σ be the x-coordinate of such a point. According to proposition 11, σ is either

- the x-coordinate of a x-critical point of $\mathcal{C}$ ($\in \Sigma_0$),
- or the x-coordinate of a singular point of $\mathcal{C}'$ ($\in \Sigma_1$).

Thus, $\sigma \in \Sigma$ and $\sigma \notin]\sigma_i, \sigma_{i+1}[$, which implies that the arcs of $\mathcal{C}'$ above $]\sigma_i, \sigma_{i+1}[$ do not intersect. The same proof applies for $\mathcal{C}''$.

3.5   Connecting the Branches

The approach that we are going to describe now for the branch connection, can be seen as an extension of the approach of [13], [11] to the three-dimensional case.

The previous step yields a sequence of strictly increasing values

$$\Sigma = \{\sigma_1, \ldots, \sigma_l\},$$

such that above $]\sigma_i, \sigma_{i+1}[$, the branches of $\mathcal{C}$ are smooth and the arcs of $\mathcal{C}', \mathcal{C}''$ do not intersect. We will use this property to connect the points of $\mathcal{C}$ above the values σ_i. Notice that proposition 13 is still true if we refine the sequence $\sigma_1, \ldots, \sigma_l$. In particular, it is valid if we consider the x-coordinates of the singular points of a curve, defined by a multiple of the equation of $\mathcal{C}'$ (resp. $\mathcal{C}''$). It is also valid, if we insert new values in between these critical values: $\delta_0 < \sigma_1 < \mu_1 < \cdots < \sigma_l < \delta_1$, where $\mu_i := \frac{\sigma_i + \sigma_{i+1}}{2}$ for $i = 0, \ldots, l - 1$, and δ_0, δ_1 are any value such that $]\delta_0, \delta_1[$ contains Σ. We denote by

$$\alpha_0 < \cdots < \alpha_m$$

this new refined sequence of values and by L_i, the set of points on $\mathcal{C}$ above α_i, for $i = 0, \ldots, m$. These points are computed, either

- by substituting $x = \alpha_i$ and solving the 2-dimensional system $P_1(\alpha_i, y, z) = 0$, $P_2(\alpha_i, y, z) = 0$.
- or by computing the points of $\mathcal{C}'$ above α_i and by lifting them to $\mathcal{C}$ (algorithm 12).

This construction implies the following lemma, which is used in the next theorem, in order to describe how the computed points have to be connected:

Lemma 14. *Two distinct points of a regular section of $\mathcal{C}$ with the same y-coordinate (resp. z-coordinate) are connected to two points of the next section, with the same y-coordinate (resp. z-coordinate) or to a critical point.*

Proof. We denote by L the regular section at $x = \alpha$ of $\mathcal{C}$ and by L' the next section, at $x = \alpha'$. Let $p = (\alpha, \beta, \gamma) \in L, q = (\alpha, \beta, \delta) \in L$ with $\gamma \neq \delta$. They are connected by $\mathcal{C}$ respectively to $p' = (\alpha', \beta', \gamma'), q' = (\alpha', \epsilon', \delta') \in L'$. Assume that $\beta' \neq \epsilon'$. Then there are two arcs of the projection $\mathcal{C}'$, connecting (α, β) to (α', β') and to (α', ϵ'), above $[\alpha, \alpha']$. This implies that there exists a point $r \in \mathcal{C}'$ with $x(r) \in [\alpha, \alpha'[$ belonging to 3 branches. Such a point cannot be regular, in contradiction with the fact that $\mathcal{C}'$ is smooth above $[\alpha, \alpha'[$. Exchanging the role of y and z, we get the same property for the z-coordinates. ∎

Theorem 15. *Under the genericity condition of definition 9, the curve $\mathcal{C}$ can connect the points L_i to the points L_{i+1}, only in one way.*

Proof. By construction, for any pair (α_i, α_{i+1}), at least one of the two values is not in Σ. Let us assume that $\alpha_i \notin \Sigma$ and $\alpha_{i+1} \in \Sigma$ (the treatment of the other possibility being symmetric). To simplify the notations, let $L = L_i \subset \mathcal{C}$ and $L' = L_{i+1} \subset \mathcal{C}$.

By the genericity assumption, L' contains at most one x-critical point c of $\mathcal{C}$. Since $\alpha_i \notin \Sigma$, each point in L is regular. Moreover, by construction, the arcs of $\mathcal{C}$ above

$]\alpha_i, \alpha_{i+1}[$ have no x-critical point. Since there is no asymptotic direction of $\mathcal{C}$ in the (y, z)-direction, by the implicit function theorem, a (regular) point $p \in L$ is connected by an arc of $\mathcal{C}$, to a point of L'. Conversely, each regular point of L' is connected by an arc of $\mathcal{C}$ to a single point of L, which implies that $|L| \geq |L'|$.

We are going to prove by induction on $|L'|$ that there is a unique way to connect the points of L to the points of L', if the arcs of the (x, y) and (x, z) projections of $\mathcal{C}$ do not intersect above $]\alpha_i, \alpha_{i+1}[$.

If $|L'| = 1$, the curve connects any point of L to the unique element of L'.

Let us assume that the induction hypothesis is true for the cases where $|L'| < |L_{i+1}|$. Let q' be the greatest point of L', for the lexicographic order with $x > y > z$ and let V' be the set of points of L' which y-coordinate is $y(q')$. Let s be the cardinal of V'.

Assume first that $c \notin V'$. This implies that the points of V' are regular. We denote by V the set of s greatest points of L for the lexicographic order with $x > y > z$.

We denote by p the point of V with the greatest y-coordinate among those with greatest z-coordinate. We are going to proof that $\mathcal{C}$ has to connect $p \in L$ and $q' \in L'$.

Assume the converse, so that we have $p \in L$ connected to $p' \in L'$ and $q \in L$ connected to $q' \in L'$, with $p \neq q$, $p' \neq q'$. We consider the following possible cases:

1. $q \in V, p' \in V'$. Then we have $z(p') < z(q')$ (since q' is the greatest point of V' for the lexicographic ordering). From lemma 14, we deduce that $z(p) \neq z(q)$ and as p has the greatest z-coordinate of V, we have $z(q) < z(p)$. This implies that the projection on the (x, z)-plane of the arcs of $\mathcal{C}$ connecting p to p' and q to q' intersect. It contradicts the hypothesis that $\mathcal{C}''$ is smooth above $]\alpha_i, \alpha_{i+1}[$.

2. $q \notin V, p' \in V'$. Since every (regular) point of V' is connected to a single point in L and $|V| = |V'|$, there exists a point r of V connected to a point $r' \notin V'$. By definition of V', we have $y(r') < y(q')$, thus by lemma 14, we have $y(r) \neq y(q)$ and as the y-coordinate of the points not in V are smaller that those in V, we have $y(r) > y(q)$. This implies that the projection on the (x, y)-plane of the arcs of $\mathcal{C}$ connecting r to r' and q to q' intersect. This contradicts the hypothesis that $\mathcal{C}'$ is smooth above $]\alpha_i, \alpha_{i+1}[$.

3. $q \in V, p' \notin V'$. Then there exists a point r' of V' connected to a point $r \notin V$. By definition of V', we have $y(r') > y(p')$, thus by lemma 14 we deduce $y(r) \neq y(p)$. As $r \notin V$ and as the y-coordinate of the points not in V are smaller that those in V, we have $y(r) < y(p)$. It leads to another contradiction, since $\mathcal{C}'$ is smooth above $]\alpha_i, \alpha_{i+1}[$.

4. $q \notin V, p' \notin V'$. As $p' \notin V'$, we have $y(p') < y(q')$, thus by lemma 14, $y(p) \neq y(q)$. As $q \notin V$ and as the y-coordinate of the points not in V are smaller that those in V, we have $y(q) < y(p)$. It leads to another contradiction, since $\mathcal{C}'$ is smooth above $]\alpha_i, \alpha_{i+1}[$.

In all these cases, we obtain a contradiction. Thus p and q' have to be connected by an arc of $\mathcal{C}$, above $[\alpha_i, \alpha_{i+1}]$. Removing these points respectively from L and L', we apply the induction hypothesis to proof the result.

Suppose now that c is in V' but not in the set W' of points with the same y-coordinate as the lowest point of L'. We apply the same proof, replacing greatest by smallest in the previous constructions.

The last case, which remains to be treated, is the case where $L' = V' = W'$ is the set of points with the same y-coordinate as the x-critical point c. We define p as the point of L with greatest y-coordinate among those with greatest z-coordinate and q' as the point with greatest z-coordinate.

If $q' \neq c$, then p has to be connected to q'. Otherwise $p \in L$ is connected to $p' \in L'$ and $q \in L$ connected to $q' \in L'$, with $p \neq q$, $p' \neq q'$. Then we have $z(p') < z(q')$. By lemma 14 we deduce $z(p) \neq z(q)$ and as p has the greatest z-coordinate: $z(p) > z(q)$, which contradicts the fact that C'' is smooth above $]\alpha_i, \alpha_{i+1}[$. Thus the curve C connects p to q'. Removing these points respectively from L and L', we apply the induction hypothesis to conclude.

If $q' = c$ and the point with the smallest z-coordinate in L' is not c, we apply the same construction, replacing greatest by smallest.

The remaining case is when $|L'| = 1$, which has already been treated.

To summarize, the connection of the branches from one plane section of C to the next one, is performed as follows:

Algorithm 16. — Connecting the branches.
If there is no x-critical point in L_i and possibly a x-critical point c of C in L_{i+1}, do the following:

1. *Decompose L_{i+1} into the subsets $V'_1, \ldots, V'_k$ of the points with the same y-coordinate, listed by increasing y. Let $s_j = |V'_j|$.*
2. *Compute the index j_0 such that $c \in V'_{j_0}$. Decompose L_i into the subsets $V_1, \ldots, V_k$ in the following way:*
   - *For $j > j_0$, V_j is the set of s_j greatest points for the lexicographic order with $x > y > z$, among $L_i - \cup_{l>j} V_l$.*
   - *For $j < j_0$, V_j is the set of s_j smallest points for the lexicographic order, among $L_i - \cup_{l<j} V_l$.*
   - *V_{j_0} is the remaining set of points $L_i - \cup_{l \neq j_0} V_l$.*
3. *For $j \neq j_0$ connect the points of V_j to the points of V'_j, according to there z-coordinates, by segments.*
4. *For $j = j_0$, let A'_{j_0} (resp. B'_{j_0}) be the set of regular points of V'_{j_0}, with z-coordinate $< z(c)$ (resp. $> z(c)$).*
   - *Connect the $|A'_{j_0}|$ points of smallest z-coordinates in V_{j_0} to the points in A'_{j_0}, according to their z-coordinate, by segments.*
   - *Connect the $|B'_{j_0}|$ points of greatest z-coordinates in V_{j_0} to the points in B'_{j_0}, according to their z-coordinate, by segments.*
   - *Connect the remaining points in V_{j_0} to c, by segments.*

If there is a x-critical point of C in L_i, exchange the role of L_i and L_{i+1} in the previous steps.

Proposition 17. *Assume that we are in a generic position. Then the topology of the curve above the segment $[\alpha_i, \alpha_{i+1}]$ is the same as the set of segments produced by the algorithm 16.*

Proof. Since we are in a generic position, by theorem 15, the algorithm 16 produces the only way the arcs of the curve $\mathcal{C}$ above $[\alpha_i, \alpha_{i+1}]$ connect the points of L_i to the points of L_{i+1}. Since the algorithm only involves the coordinates of the end points, the projection of these segments on the (x, y) and (x, z) planes either coincide or do not intersect above $]\alpha_i, \alpha_{i+1}[$. Consequently, the curve $\mathcal{C}$ above $[\alpha_i, \alpha_{i+1}]$ is homeomorph to the set of segments joining the corresponding end points.

3.6   The Algorithm

We summarize the complete algorithm below:

Algorithm 18. — Representation of the curve $\mathcal{C}$ defined by $P_1(x, y, z) = P_2(x, y, z) = 0$.
Input: polynomials $P_1(x, y, z), P_2(x, y, z)$.

- *Compute the x-critical points of $\mathcal{C}$ and their x-coordinates $\Sigma := \{\sigma_1^0, \ldots, \sigma_k^0\}$ with $\sigma_1^0 < \cdots < \sigma_k^0$.*
- *Check the generic position; If the curve is not in a generic position, apply a random change of variables and restart from the first step.*
- *Compute the square-free part $g(x, y)$ of $\mathrm{Res}_z(P_1, P_2)$.*
- *Compute the square-free part $h(x, z)$ of $\mathrm{Res}_y(P_1, P_2)$.*
- *Compute the singular points of the curves $g(x, y) = 0$ and $h(x, z) = 0$ and insert their x-coordinate in Σ.*
- *Compute the μ_i, δ_0, δ_1 and the ordered sequence $\alpha_1 < \cdots < \alpha_l$. Above each α_i for $i = 1, \ldots, l$, compute the set of points L_i on the curve $\mathcal{C}$.*
- *For each $i = 0, \ldots, l - 1$, connect the points L_i to those of L_{i+1} by algorithm 16.*

Output: the graph of 3D points connected by segments, with the same topology as the curve $\mathcal{C}$.

Remark 19. This algorithm can be easily adapted to the computation of the topology of $\mathcal{C}$ in a box (resp. bounded domain), by considering the points on the border of the box (resp. domain) as x-critical points.

Remark 20. By a generic change of variables, the set of x-coordinates of the x-critical points of $\mathcal{C}'$ will contain those of $\mathcal{C}$ and the resolution of the system (1) can be replaced by the computation of the x-critical points of $\mathcal{C}'$ and by a lifting operation on $\mathcal{C}$. This allows us to treat unreduced curves, such that $I(\mathcal{C}) \neq (P_1, P_2)$, by using only the squarefree part of $G(x, y)$ and $H(x, z)$. However the verification, a posteriori, of the correctness of the result is more delicate.

4   Implementation and Experiments

The previous algorithm has been implemented in the C++ library called AXEL[4] (Algebraic Software-Components for gEometric modeLing), where classes for implicit curves and surfaces are available:

---

[4] http://www-sop.inria.fr/galaad/logiciels/axel/

```
namespace implicit
{
  template<class C, class R> curve2d<C,R=MPol<C> >;
  template<class C, class R>
      curve3d<C,R=shared_object<std::vector<MPol<C> > > >;
  template<class C, class R> surface<C,R=MPol<C> >;
}
```

where C is the type of coefficient and R is the internal representation used to store the object. In the case of a planar curve curve2d, the default value is a bivariate polynomial MPol<C> from the SYNAPS[5] library. For a 3D curve, the default value is a vector of multivariate polynomials. For a 3D surface, the default value is a multivariate polynomial MPol<C> from the SYNAPS library.

Since the algorithm depends heavily on the algebraic solver used to recover the critical points of $\mathcal{C}$, we parameterize the implementation as

```
template <class M>
class Projection
{
  ...
  template <class G, class Surface>
  void topology(G & graph, const Surface & s0, const
                                        Surface & s1);

  template <class G, class Curve3d>
  void topology(G & graph, const Curve3d & c0);
  ...
}
```

where the parameter M is the type of method used to solve the 0-dimensional systems. The class Projection represents the type of method that we used, to compute the topology of the curve.

Our tests are based on solvers provided by the SYNAPS library, such as Newmac (see [20]). Here is an illustration of the way it can be used:

```
  ...
  typedef double coef;
  MPol<coef> P=..., Q= ...;
  vector<MPol<coef> > v; v.push_back(P); v.push_back(Q);
  implicit::curve3d c(v);

  affine::point_graph<coef> g;

  Projection<Newmac<coef> > Method;
  Method.topology(g,c);
  ...
```

---

[5] http://www-sop.inria.fr/galaad/logiciels/synaps/

Other examples which are currently under test are:

```
Projection<Newmac<QQ> >().topology(g,c);
Projection<Sylvester<double> >().topology(g,c);
Projection<Subdivision<double> >().topology(g,c);
```

Notice that the solutions given from equations (1) are computed numerically, and we have to sort the lists of critical values to check the generic position of the curve and the points in the lists L_i. For this purpose, we introduce two thresholds ϵ_x and ϵ_y which are the precision on the x-coordinate and the y-coordinate. We have $\epsilon_y \gg \epsilon_x$ because the computation of the y-coordinate has already noisy input (the x-coordinate with a precision ϵ_x). Concretely, this means that given a curve $\mathcal{C}$ we are able to compute correctly the topology, if two critical points are separated at least by $2\epsilon_x$ on the x-coordinate, and two points on $\mathcal{C}$ with the same x-coordinate are separated by ϵ_y at least.

In the cases where we are not able to distinguish within the precision ϵ_x, ϵ_y, two strategies can be applied. Either we use an exact method for representing the solution of the corresponding polynomial system, assuming exact input. Or we consider input polynomial with approximate coefficients and we identify the x-points which are within the prescribed precision. This is what has been experimented.

We run a non-optimized implementation of the algorithm, on cases case where the resultant does not define twice or more the projected curves. The solver that we use is the one of Ph. Trébuchet [20], giving the biggest precision for the smallest $\epsilon_x, \epsilon_y, \epsilon_z$ (about 10^{-6}), compared with the resultant solvers. Further experiments are required to analyse the behavior of such solvers and to compare them correctly, in the context of the topology computation problem. The experimentations have been performed on a Pentium 2Ghz workstation.

4.1 Examples of Planar Curves

The algorithm that we describe, can be easily specialized to the case of planar curves. This is illustrated by fig. 1. The topology of the curve is represented by a graph of 2D points. The time needed to compute this graph of points is given in seconds.

4.2 Examples of 3D Curves

Figure 2 presents some experiments with 3D curves defined by two polynomials, showing the set of segments describing the topology of the curve and the time needed to compute them (in seconds).

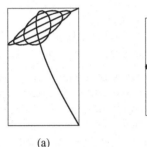 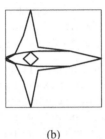 

(a)   (b)   (c)

a) $P = -y^8 + x^7 - 7x^6y + 21x^5y^2 - 35x^4y^3 + 35x^3y^4 - 21x^2y^5 + 7xy^6 - y^7 + 8y^6 - 7x^5 + 35x^4y - 70x^3y^2 + 70x^2y^3 - 35xy^4 + 7y^5 - 20y^4 + 14x^3 - 42x^2y + 42xy^2 - 14y^3 + 16y^2 - 7x + 7y - 2$

  time: 0.77s

b) $P = 35.9x^6 + 2589.4x^4y^2 + 46728x^2y^4 + 1296y^6 + 217x^5 + 15588x^3y^2 + 2.7994e + 05xy^4 - 2303.9x^4 - 72774x^2y^2 + 3.7066e + 05y^4 - 15583x^3 - 5.5969e + 05xy^2 + 26044x^2 - 7.4529e + 05y^2 + 2.7976e + 05x + 3.7333e + 05$

  time: 0.16s

c) $P = -8y^7 - 7x^6 + 42x^5y - 105x^4y^2 + 140x^3y^3 - 105x^2y^4 + 42xy^5 - 7y^6 + 48y^5 + 35x^4 - 140x^3y + 210x^2y^2 - 140xy^3 + 35y^4 - 80y^3 - 42x^2 + 84xy - 42y^2 + 32y + 7$

  time: 0.28s

Fig. 1. Examples of planar curves

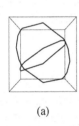 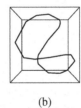 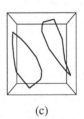 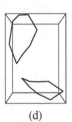

(a)   (b)   (c)   (d)

a) $P = 0.85934x^2 + 0.259387xy + 0.880419y^2 + 0.524937xz - 0.484008yz + 0.510242z^2 - 1$

  $Q = 0.95309x^2 + 0.303149xy + 0.510242y^2 - 0.200075xz + 0.64647yz + 0.786669z^2 - 1$

  time: 0.17s

b) $P = -0.125x^2 - 0.0583493xy + 0.493569y^2 + 0.966682xz - 1.5073yz - 0.368569z^2 - 0.865971x - 0.433067y - 0.250095z$

  $Q = x^2 + y^2 + z^2 - 2$

  time: 0.15s

c) $P = 2x^2 + y^2 + z^2 - 4$

  $Q = x^2 + 2xy + y^2 - 2yz - 2z^2 + 2zx$

  time: 0.13s

d) $P = x^4 + y^4 + 2x^2y^2 + 2x^2 + 2y^2 - x - y - z$

  $Q = x^4 + 2x^2y^2 + y^4 + 3x^2y - y^3 + z^2$

  time: 1.21s

Fig. 2. Examples of space curves

References

1. K. Abdel-Malek and H.-J. Yeh. On the determination of starting points for parametric surface intersections. *Computer-Aided Design*, 28:21–35, 1997.
2. E. Anderson, Z. Bai, C. Bischof, J. Demmel, J. Dongarra, J. Du Croz, A. Greenbaum, S. Hammarling an A. McKenney, S. Ostrouchov, and D. Sorensen. *LAPACK Users' Guide*. SIAM, Philadelphia, 1992. http://www.netlib.org/lapack/.
3. S. Basu, R. Pollack, and M.-F. Roy. *Algorithms in Real ALgebraic Geometry*. Springer-Verlag, Berlin, 2003. ISBN 3-540-00973-6.
4. L. Busé, M. Elkadi, and B. Mourrain. Using projection operators in computer aided geometric design. In *Topics in Algebraic Geometry and Geometric Modeling,*, pages 321–342. Contemporary Mathematics, 2003.
5. D. Cox, J. Little, and D. O'Shea. *Ideals, Varieties, and Algorithms: An Introduction to Computational Algebraic Geometry and Commutative Algebra*. Undergraduate Texts in Mathematics. Springer Verlag, New York, 1992.
6. G. Dos Reis, B. Mourrain, R. Rouillier, and Ph. Trébuchet. An environment for symbolic and numeric computation. In *Proc. of the International Conference on Mathematical Software 2002*, World Scientific, pages 239–249, 2002.
7. D. Eisenbud. *Commutative Algebra with a view toward Algebraic Geometry*, volume 150 of *Graduate Texts in Math*. Berlin, Springer-Verlag, 1994.
8. M. Elkadi and B. Mourrain. *Introduction à la résolution des systèmes d'équations algébriques*, 2003. Notes de cours, Univ. de Nice (310 p.). Soumis pour publication dans la srie mathmatiques appliques (SMAI).
9. G. Farin. An ssi bibliography. In *Geometry Processing for Design and Manufacturing*, pages 205–207. SIAM, Philadelphia, 1992.
10. T. Garrity and J. Warren. Geometric continuity. *Comp. Aided Geom. Design*, 8:51–65, 1991.
11. Laureano González-Vega and Ioana Necula. Efficient topology determination of implicitly defined algebraic plane curves. *Comput. Aided Geom. Design*, 19(9):719–743, 2002.
12. T. A. Grandine. Applications of contouring. *SIAM Review*, 42:297–316, 2000.
13. T. A. Grandine and F. W. Klein. A new approach to the surface intersection problem. *Computer Aided Geometric Design*, 14:111–134, 1997.
14. Gert-Martin Greuel and Gerhard Pfister. *A **Singular** introduction to commutative algebra*. Springer-Verlag, Berlin, 2002. With contributions by Olaf Bachmann, Christoph Lossen and Hans Schönemann, With 1 CD-ROM (Windows, Macintosh, and UNIX).
15. J. Harris. *Algebraic Geometry, a first course*, volume 133 of *Graduate Texts in Math*. New-York, Springer-Verlag, 1992.
16. S. Krishnan and D. Manocha. An efficient intersection algorithm based on lower dimensional formulation. *ACM Transactions on Computer Graphics*, 16:74–106, 1997.
17. B. Mourrain. A new criterion for normal form algorithms. In M. Fossorier, H. Imai, Shu Lin, and A. Poli, editors, *Proc. AAECC*, vol. 1719 of *LNCS*, pages 430–443. Springer, Berlin, 1999.
18. J. Owen and A. Rockwood. Intersection of general implicit surfaces. In *Geometric Modeling: Algorithms and New Trends*, pages 335–345. SIAM, Philadelphia, 1987.
19. M. P. Patrikalakis and T. Maekawa. *Shape Interrogation for Computer Aided Design and Manufacturing*. Springer Verlag, 2002.
20. Ph. Trébuchet. *Vers une résolution stable et rapide des équations algébriques*. PhD thesis, Université Pierre et Marie Curie, 2002.
21. J. von zur Gathen and J. Gerhard. *Modern computer algebra*. Cambridge University Press, New York, 1999.
22. Hassler Whitney. *Complex analytic varieties*. Addison-Wesley Publishing Co., Reading, Mass.-London-Don Mills, Ont., 1972.

Distance Properties of ϵ–Points on Algebraic Curves

Sonia Pérez-Díaz[1], Juana Sendra[2], and J.Rafael Sendra[1]*

[1] Dpto de Matemáticas, Universidad de Alcalá, E-28871 Madrid, Spain,
sonia.perez@uah.es, rafael.sendra@uah.es,
[2] Dpto de Matemáticas, Universidad Carlos III, E-28911 Madrid, Spain,
jsendra@math.uc3m.es

Abstract. This paper deals with some mathematical objects that the authors have named ϵ–points (see [8]), and that appear in the problem of parametrizing approximately algebraic curves. This type of points are used as based points of the linear systems of curves that appear in the parametrization algorithms, and they play an important role in the error analysis. In this paper, we focus on the general study of distance properties of ϵ–points on algebraic plane curves, and we show that if $P^\star$ is an ϵ–point on a plane curve $\mathcal{C}$ of proper degree d, then there exists an exact point P on $\mathcal{C}$ such that its distance to $P^\star$ is at most $\sqrt{\epsilon}$ if $P^\star$ is simple, and $\mathcal{O}(\sqrt{\epsilon}^{1/2d})$ if $P^\star$ is of multiplicity $r > 1$. Furthermore, we see how these results particularize to the univariate case giving bounds that fit properly with the classical results in numerical analysis.

1 Introduction

An important step, and usually a hard step to deal with, in the development and analysis of approximate algorithms consists in estimating how "close" the input and the output of the algorithm are. If one is working with algebraic objects, for instance with polynomial gcd's (see [3]), this question may be approached by measuring relative errors of polynomials. However, when one is working with geometric entities, like for instance algebraic curves or algebraic surfaces, this approach might not be sufficient. It may happen that, even though the implicit equations of the input and output are close like polynomials, the algebraic varieties that they define (i.e. their set of zeroes) are far when seen as point of the usual Euclidean space. For example, we consider a tolerance $\epsilon = 0.005$, and three circles $\mathcal{C}_1$, $\mathcal{C}_2$ and $\mathcal{C}_3$ of equations

$$f_1(x, y) = 1 + y + 0.008x^2 + 0.007y^2,$$

$$f_2(x, y) = 1.004994 + y + 0.012994x^2 + 0.011994y^2,$$

$$f_3(x, y) = 1.00501 + y + 0.008x^2 + 0.007y^2.$$

* This work is partially supported by BMF2002-04402-C02-01 (*Curvas y Superficies: Fundamentos, Algoritmos y Aplicaciones*), Acción Integrada Hispano-Austriaca HU2001-0002 (*Computer Aided Geometric Design by Symbolic-Numerical Methods*), and GAIA II (IST-2002-35512).

The relative error of the polynomials are

$$\frac{\|f_1 - f_2\|}{\|f_2\|} = 0.004969183896 < \epsilon,$$

$$\frac{\|f_1 - f_3\|}{\|f_3\|} = 0.0049850250 < \epsilon.$$

However, when plotting the circles (see Fig. 1) one realizes that C_1, C_2 are not close, but C_1, C_3 are close.

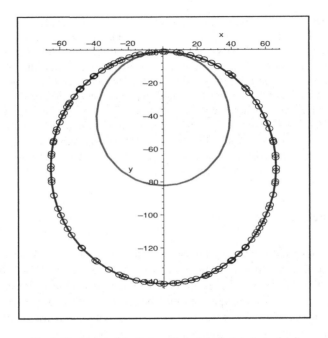

Fig. 1. C_1 (big circle), C_2 (small circle), C_3 (circle in dots)

This phenomenon can be controlled, for instance, by requiring that each geometric entity lies in the offset region of the other at some small distance (see [8, 9] for further details). In the previous example, one clearly sees that even though the relative errors are smaller than the tolerance, the circles C_1, C_2 do not satisfy the offset property while C_1, C_3 do.

A powerful technique to handle this difficulty, i.e. guaranteeing that an algebraic curve (or an algebraic surface) is within the offset region of another, is the use of ϵ-points (see Definition 1), and more precisely, distance properties of this type of points. For instance, applying Corollary 17, Corollary 20, and Corollary 21, in this paper, one may directly deduce the offset property by means of the existence of ϵ–points.

The notion of ϵ–point of an algebraic variety (in practical applications, a curve or a surface over the field of the real numbers) is quite intuitive. It essentially consists in a point such that when substituted in the implicit equations of the variety gives values of small modulus. Observe that the notion of ϵ-point generalizes the concept of "exact" point, since every exact point on a variety is an ϵ–point, for any $\epsilon > 0$. Note that, in the previous example, almost all exact points on $\mathcal{C}_1$ are ϵ–points on $\mathcal{C}_3$, but this is not true for $\mathcal{C}_1$ and $\mathcal{C}_2$.

Theoretical properties and algorithmic questions of this type of mathematical objects have been studied by several authors for the univariate case. For instance, bound analysis of roots of univariate polynomials can be found in [2, 6, 7], formulae for separating small roots of univariate polynomials are given in [10], the problem of constructing univariate polynomials with exact roots at some specific ϵ–roots (see Definition 6 for the notion of ϵ–root) is analyzed in [5], condition numbers of ϵ–roots are studied in [11], etc.

Nevertheless, the corresponding problems for the bivariate case have not been studied so extensively, and only some contributions can be found in the literature (see e.g. [4, 8, 9]). In this paper, we focus on the problem of analyzing distance properties of ϵ–points for the bivariate case; i.e. ϵ–points of algebraic curves. More precisely, we show that if $P^\star$ is an ϵ–point on a plane curve $\mathcal{C}$ of proper degree d, then there exists an exact point P on $\mathcal{C}$ such that its distance to $P^\star$ is at most $\sqrt{\epsilon}$ if $P^\star$ is simple, and $\mathcal{O}(\sqrt{\epsilon}^{1/2d})$ if $P^\star$ is of multiplicity $r > 1$. As a consequence, we give theoretical results to guarantee that an algebraic curve lies in the offset region of another at some small distance. Furthermore, we see how these results particularize to the univariate case giving bounds that fit properly with the classical results in numerical analysis.

This paper is structured as follows. In sect. 2 we introduce the basic notions and we state some preliminary properties of ϵ–points and ϵ–roots. Section 3 is devoted to the analysis of distance properties of ϵ–points for the univariate case; i.e. for ϵ–roots. Section 4 focusses on the corresponding analysis of distance properties for the bivariate case.

2 Basic Notions and Preliminary Results

In this section, we recall the basic notions on ϵ–points, we introduce the concept of proper degree of an algebraic plane curve, and we establish some basic properties of these type of objects. In addition, we see how these definitions may be specialized to the case of univariate polynomials.

Throughout this paper, we fix a tolerance $0 < \epsilon < 1$ and we will use the polynomial ∞–norm; i.e if $p(x, y) = \sum_{i,j \in I} c_{i,j} x^i y^j \in \mathbb{C}[x, y]$ then $\|p(x, y)\|$ is defined as $\max\{|c_{i,j}| \ / \ i, j \in I\}$, where for $c \in \mathbb{C}$, $|c|$ denotes its modulus. In particular if

$p(x, y)$ is a constant coefficient $\|p(x, y)\|$ will denote its modulus. Similarly one introduces $\|p(x)\| \in \mathbb{C}[x]$. We also will use the Euclidean distance $\| \cdot \|_2$ in the usual unitary $\mathbb{C}^n$. To be more precisely, if $p = (a_1, \ldots, a_n) \in \mathbb{C}^n$ then $\|p\|_2$ is defined as $\sqrt{|a_1|^2 + \cdots + |a_n|^2}$.

We start with the notion of ϵ–point on an algebraic affine plane curve. As we have mentioned in the introduction, the notion of ϵ–point is quite intuitive and, for the curve case, it essentially consists in points such that when substituted in the implicit equation of the curve one gets values of small modulus. Nevertheless, for our purposes and taking into account that we will be working in the frame of algebraic geometry, we are also interested in introducing the additional notion of ϵ–singularity. This is also quite intuitive, and it basically consists in asking the ϵ–point to be also an ϵ–point of some partial derivatives. More precisely, one has the following definition.

Definition 1. *We say that $P^\star = (a^\star, b^\star) \in \mathbb{C}^2$ is an ϵ–(affine) point of an algebraic plane curve $\mathcal{C}$ defined over $\mathbb{C}$ by a polynomial $f(x, y) \in \mathbb{C}[x, y]$ of degree d, if there exists $r \in \mathbb{N}$, $1 \leq r \leq d$, such that*

1. for $0 \leq i + j \leq r - 1$, it holds that

$$\left| \frac{\partial^{i+j} f}{\partial^i x \partial^j y}(P^\star) \right| < \epsilon \cdot \|f(x, y)\|,$$

2. there exist $i_0, j_0 \in \mathbb{N}$ with $i_0 + j_0 = r$ such that

$$\left| \frac{\partial^r f}{\partial^{i_0} x \partial^{j_0} y}(P^\star) \right| \geq \epsilon \cdot \|f(x, y)\|.$$

If $r = 1$ we say that $P^\star$ is an ϵ–(affine) simple point. Otherwise, we say that $P^\star$ is an ϵ–(affine) singularity of multiplicity r.

Observe that the bounds in the conditions (1) and (2), in Definition 1, involve the tolerance and the given polynomial norm. This fact guarantees that the notion of ϵ–point of an algebraic curve $\mathcal{C}$ does not depend on the choice of the implicit equation. In addition, note that condition (2) in Definition 1 requires that some partial derivative specialized at the ϵ–point is, in modulus, greater than the tolerance. This fact guarantees the exact value of the multiplicity of the ϵ–singularity. Nevertheless, for the study of some of the distance properties developed in the subsequent sections, we need to be more restricted on this condition, and we impose in (2) a bigger bound that leads to the notion of ϵ–strong point. More precisely, we consider the following definition.

Definition 2. *We say that $P^\star = (a^\star, b^\star) \in \mathbb{C}^2$ is an ϵ–(affine) strong point of an algebraic plane curve $\mathcal{C}$ defined over $\mathbb{C}$ by a polynomial $f(x, y) \in \mathbb{C}[x, y]$ of degree d, if there exists $r \in \mathbb{N}$, $1 \leq r \leq d$, such that*

1. for $0 \leq i + j \leq r - 1$, it holds that

$$\left| \frac{\partial^{i+j} f}{\partial^i x \partial^j y}(P^\star) \right| < \epsilon \cdot \|f(x, y)\|,$$

2. *there exist* $i_0, j_0 \in \mathbb{N}$ *with* $i_0 + j_0 = r$ *such that*

$$\left| \frac{\partial^r f}{\partial^{i_0} x \partial^{j_0} y}(P^\star) \right| \geq \sqrt{\epsilon} \cdot \| f(x,y) \|.$$

If $r = 1$ *we say that* $P^\star$ *is an* ε–*(affine) strong simple point. Otherwise, we say that* $P^\star$ *is an* ε–*(affine) strong singularity of multiplicity* r.

Note that from the above definition it follows that any ε–strong singularity of multiplicity r is indeed an ε–singularity of multiplicity r.

In the analysis of distance properties of ε–points of positive multiplicity, i.e ε–singularities, we will have to use the degree of the algebraic curve. However, since we are working under a fixed tolerance, it may happen that the leading terms of the defining polynomial are "superfluos", and therefore the degree of the curve is not properly counted. In order to control this phenomenon, we give the following definition where the notion of proper degree is introduced.

Definition 3. *We say that a polynomial* $f(x,y) \in \mathbb{C}[x,y]$ *has **proper degree** d if there exist* $i_p, j_p \in \mathbb{N}$, *with* $i_p + j_p = d$, *such that*

$$\frac{\left| \frac{\partial^d f}{\partial^{i_p} x \partial^{j_p} y} \right|}{i_p! j_p!} \geq \sqrt{\epsilon} \cdot \| f(x,y) \|.$$

*We say that a plane algebraic curve has **proper degree** d if its defining polynomial has proper degree d.*

Note that the defining polynomial of a plane algebraic curve is unique up to multiplication by elements of the ground field, in our case $\mathbb{C}$. However, in Definition 3 the norm of $f(x,y)$ appears and therefore the notion of proper degree of a plane curve is well defined.

In addition, we observe that the definition of proper degree of a polynomial implies that there exists a coefficient of the homogeneous form of maximum degree of the polynomial $f(x,y)$, such that its modulus is bigger than $\sqrt{\epsilon} \cdot \|f\|$. Therefore, the notion does not depend on whether the polynomial is represented around any point of $\mathbb{C}^2$ as a Taylor expansion.

In the following, we state some properties on ε–singularities of curves with proper degree.

Proposition 4. *Let* $f(x,y) \in \mathbb{C}[x,y]$ *be a polynomial of proper degree* $d > 0$, *and let* $P^\star \in \mathbb{C}^2$ *be an* ε–*singularity of multiplicity* r *of* $f(x,y)$. *Then, there exists* $s \in \mathbb{Z}$ *with* $r \leq s \leq d$, *such that* $P^\star$ *is an* $\sqrt{\epsilon}$–*singularity of multiplicity* s *of* $f(x,y)$

Proof. Note that since $P^\star$ is an ϵ–singularity of multiplicity r of $f(x, y)$, then for $0 \leq i + j \leq r - 1$, it holds that

$$\left| \frac{\partial^{i+j} f}{\partial^i x \partial^j y}(P^\star) \right| < \epsilon \cdot \|f\| \leq \sqrt{\epsilon} \cdot \|f\|.$$

Now, if there exist $i_0, j_0 \in \mathbb{N}$ with $i_0 + j_0 = r$ satisfying that

$$\left| \frac{\partial^r f}{\partial^{i_0} x \partial^{j_0} y}(P^\star) \right| \geq \sqrt{\epsilon} \cdot \|f\|,$$

then $s = r$, and the statement follows. Otherwise, one reasons inductively: since $f(x, y)$ has proper degree $d > 0$, then there exist $i_p, j_p \in \mathbb{N}$, with $i_p + j_p = d$, such that

$$\frac{\left| \frac{\partial^d f}{\partial^{i_p} x \partial^{j_p} y} \right|}{i_p! j_p!} \geq \sqrt{\epsilon} \cdot \|f\|,$$

which implies that

$$\left| \frac{\partial^d f}{\partial^{i_p} x \partial^{j_p} y} \right| \geq \sqrt{\epsilon} \cdot i_p! j_p! \cdot \|f\| \geq \sqrt{\epsilon} \cdot \|f\|,$$

and therefore the result follows. □

Proposition 5. *Let $f(x, y) \in \mathbb{C}[x, y]$ be a polynomial of proper degree $d > 0$, and let $P^\star \in \mathbb{C}^2$ be an ϵ–singularity of multiplicity r of $f(x, y)$. Then, there exists $s \in \mathbb{Z}$ with $r \leq s \leq d$, and there exist non-negative integers i_0, j_0 such that $s = i_0 + j_0$, and*

$$\left| \frac{\partial^s f}{\partial^{i_0} x \partial^{j_0} y}(P^\star) \right| \geq \sqrt{\epsilon} \cdot i_0! j_0! \cdot \|f(x, y)\|.$$

Proof. Let us assume that for every $i, j \in \mathbb{N}$ with $0 \leq i + j \leq d$, it holds that

$$\left| \frac{\partial^{i+j} f}{\partial^i x \partial^j y}(P^\star) \right| < \sqrt{\epsilon} \cdot i! j! \cdot \|f\|.$$

Then, in particular for $i_p, j_p \in \mathbb{N}$, with $i_p + j_p = d$, we have that

$$\frac{\left| \frac{\partial^d f}{\partial^{i_p} x \partial^{j_p} y} \right|}{i_p! j_p!} \geq \sqrt{\epsilon} \cdot \|f\|,$$

which is impossible because we are assuming that $f(x, y)$ has a proper degree d. Furthermore, note that since $P^\star$ is an ϵ–singularity of multiplicity r, then for $0 \leq i + j \leq r - 1$, it holds that

$$\left| \frac{\partial^{i+j} f}{\partial^i x \partial^j y}(P^\star) \right| < \epsilon \cdot \|f\| \leq \sqrt{\epsilon} \cdot i! j! \cdot \|f\|.$$

Thus, we deduce that $r \leq s \leq d$. □

These notions can be straightforward specialized to the case of univariate polynomials in terms of roots. Since we will consider a different treatment to the univariate case and to the bivariate case, we give the formal definitions for univariate polynomials, More precisely, one has the following concepts.

Definition 6. *We say that $a^\star \in \mathbb{C}$ is an ϵ–root of multiplicity r of a polynomial $h(x) \in \mathbb{C}[x]$, if for $0 \le i \le r - 1$, it holds that*

$$\left| \frac{\partial^i h}{\partial^i x}(a^\star) \right| < \epsilon \cdot \|h(x)\|,$$

and

$$\left| \frac{\partial^r h}{\partial^r x}(a^\star) \right| \ge \epsilon \cdot \|h(x)\|.$$

If $r = 1$ we say that $a^\star$ is an ϵ–simple root of $h(x)$.

Definition 7. *We say that $a^\star \in \mathbb{C}$ is an ϵ–strong root of multiplicity r of a polynomial $h(x) \in \mathbb{C}[x]$, if for $0 \le i \le r - 1$, it holds that*

$$\left| \frac{\partial^i h}{\partial^i x}(a^\star) \right| < \epsilon \cdot \|h(x)\|,$$

and

$$\left| \frac{\partial^r h}{\partial^r x}(a^\star) \right| \ge \sqrt{\epsilon} \cdot \|h(x)\|.$$

If $r = 1$ we say that $a^\star$ is an ϵ–simple strong root of $h(x)$.

Note that, as in the bivariate case, the above definitions imply that any ϵ–strong root of multiplicity r is an ϵ–root of multiplicity r. In addition, the notion of proper degree of a univariate polynomial can be defined as follows.

Definition 8. *We say that a polynomial $h(x) = a_d x^d + \cdots + a_0 \in \mathbb{C}[x]$, where $a_d \ne 0$, has proper degree d if*

$$|a_d| \ge \sqrt{\epsilon} \cdot \|h(x)\|.$$

Finally, in Propositions 9 and 10, we see how Propositions 4 and 5 can be stated for the univariate case. Their proofs are very similar and we omit them.

Proposition 9. *Let $h(x) \in \mathbb{C}[x]$ be a polynomial of proper degree $d > 0$, and let $a^\star \in \mathbb{C}$ be an ϵ–root of multiplicity r of $h(x)$. Then, there exists $s \in \mathbb{Z}$ with $r \le s \le d$ such that $a^\star$ is an $\sqrt{\epsilon}$–root of multiplicity s of $h(x)$.* □

Proposition 10. *Let $h(x) \in \mathbb{C}[x]$ be a polynomial of proper degree $d > 0$, and let $a^\star \in \mathbb{C}$ be an ϵ–root of multiplicity r of $h(x)$. Then, there exists $s \in \mathbb{Z}$ with $r \le s \le d$ such that*

$$\left| \frac{\partial^s h}{\partial^s x}(a^\star) \right| \ge \sqrt{\epsilon} \cdot s! \cdot \|h(x)\|.$$

□

To finish this section, we observe that ϵ–points on algebraic curves can be computed; see for instance sect. 2 in [8]. A similar reasoning can be done for ϵ–roots.

3 ϵ–Roots

In this section, we study distance properties of ϵ–roots of univariate polynomials over the complex numbers. We start recalling the following well known results on the sensitivity of roots of perturbed polynomials.

Lemma 11. Let $h^\star(x), p(x), h(x) \in \mathbb{C}[x]$, such that

$$h(x) = h^\star(x) + \epsilon \cdot p(x),$$

and $h(x)$ is not a constant. Then, if $a^\star \in \mathbb{C}$ is a root of multiplicity r of $h^\star(x)$, it holds that there exists a root $a \in \mathbb{C}$ of $h(x)$ such that

$$|a - a^\star| \leq \left(\frac{r! \cdot \epsilon \cdot |p(a^\star)|}{|\frac{\partial^r h^\star}{\partial^r x}(a^\star)|} \right)^{1/r}.$$

Proof. See sect. 5.8, pp. 303 in [2].   □

Lemma 11 focusses on distance properties of roots of a polynomial and its perturbation. Now, we see how the same analysis can be done for the case of ϵ–roots of a polynomial. We first study how close strong ϵ–roots and exact roots of a univariate polynomial are.

Theorem 12. Let $h(x) \in \mathbb{C}[x]$, and let $a^\star$ be an ϵ–strong root of multiplicity r of $h(x)$. Then, there exists a root $a \in \mathbb{C}$ of $h(x)$ such that

$$|a - a^\star| \leq \left(r! \cdot \sqrt{\epsilon} \right)^{\frac{1}{r}}.$$

Proof. Let d be the exact degree of $h(x)$. We consider the polynomial

$$g(x) = h(x + a^\star) = \sum_{i=0}^{d} \frac{\frac{\partial^i h}{\partial^i x}(a^\star)}{i!} x^i = \sum_{i=0}^{r-1} \frac{\frac{\partial^i h}{\partial^i x}(a^\star)}{i!} x^i + \sum_{i=r}^{d} \frac{\frac{\partial^i h}{\partial^i x}(a^\star)}{i!} x^i.$$

Let

$$g^\star(x) = \sum_{i=r}^{d} \frac{\frac{\partial^i h}{\partial^i x}(a^\star)}{i!} x^i,$$

and

$$q(x) = \sum_{i=0}^{r-1} \frac{\frac{\partial^i h}{\partial^i x}(a^\star)}{i!} x^i.$$

Note that since $a^\star \in \mathbb{C}$ is an ϵ–strong root of multiplicity r of $h(x)$ we have that

$$\left| \frac{\partial^i h}{\partial^i x}(a^\star) \right| < \epsilon \cdot \|h\|, \quad \text{for} \quad i = 0, \ldots, r-1,$$

and then we may write $q(x) = \epsilon \cdot p(x)$, where

$$p(x) = \sum_{i=0}^{r-1} b_i x^i,$$

and

$$|b_i| = \frac{\left|\frac{\partial^i h}{\partial^i x}(a^\star)\right|}{i! \cdot \epsilon} < \|h\|, \quad \text{for} \quad i = 0, \ldots, r-1.$$

Thus, we get that

$$g(x) = g^\star(x) + \epsilon \cdot p(x).$$

In these conditions, since

$$g^\star(x) = \sum_{i=r}^{d} \frac{\frac{\partial^i h}{\partial^i x}(a^\star)}{i!} x^i,$$

we have that $\frac{\partial^i g^\star}{\partial^i x}(0) = 0$ for $i = 0, \ldots, r-1$, and $\left|\frac{\partial^r g^\star}{\partial^r x}(0)\right| = \left|\frac{\partial^r h}{\partial^r x}(a^\star)\right| \geq \sqrt{\epsilon} \cdot \|h\| > 0$. Thus, we deduce that 0 is an exact root of multiplicity r of the polynomial $g^\star(x)$. Now, by Lemma 11, we deduce that there exists a root $x_0 \in \mathbb{C}$ of $g(x)$ such that

$$|x_0| \leq \left(\frac{r! \cdot \epsilon \cdot |p(0)|}{\left|\frac{\partial^r g^\star}{\partial^r x}(0)\right|}\right)^{1/r}.$$

Now, observe that since $a^\star \in \mathbb{C}$ is an ϵ–strong root of multiplicity r of $h(x)$ we have that

$$|p(0)| = |b_0| = \frac{|h(a^\star)|}{\epsilon} < \frac{\epsilon \cdot \|h\|}{\epsilon} = \|h\|,$$

and

$$\left|\frac{\partial^r g^\star}{\partial^r x}(0)\right| = \left|\frac{\partial^r h}{\partial^r x}(a^\star)\right| \geq \sqrt{\epsilon} \cdot \|h\|.$$

Thus,

$$|x_0| \leq \left(\frac{r! \cdot \epsilon \cdot \|h\|}{\sqrt{\epsilon} \cdot \|h\|}\right)^{1/r} = (r! \cdot \sqrt{\epsilon})^{1/r}.$$

Finally, since $x_0 \in \mathbb{C}$ is a root of $g(x)$, then $a = x_0 + a^\star \in \mathbb{C}$ is a root of $h(x)$, and

$$|a - a^\star| = |x_0| \leq (r! \cdot \sqrt{\epsilon})^{1/r}.$$

□

From Theorem 12 one has the following corollary.

Corollary 13. *Let $h(x) \in \mathbb{C}[x]$, and let $a^\star$ be an ϵ–strong simple root of $h(x)$. Then, there exists a root $a \in \mathbb{C}$ of $h(x)$ such that*

$$|a - a^\star| \leq \sqrt{\epsilon}.$$

□

As one may see there is a non-surprising different behavior between simple and singular ϵ–point. The bound for the simple case is quite sharp. However, the bound for the singular case introduces the factorial of the multiplicity. These bounds are given for the case of strong ϵ–roots. In the next theorem we see that, if one changes the hypothesis of being strong by asking proper degree, one gets a new bound where the factorial is not involved.

Theorem 14. *Let $h(x) \in \mathbb{C}[x]$ be a polynomial of proper degree $d > 0$, and let $a^\star$ be an ϵ-root of multiplicity r of $h(x)$. Then, there exists $s \in \mathbb{Z}$ such that $r \le s \le d$, and there exist a root $a \in \mathbb{C}$ of $h(x)$ such that:*

$$|a - a^\star| \le (\sqrt{\epsilon})^{\frac{1}{s}}.$$

Proof. First, since $a^\star$ is an ϵ-root of multiplicity r of $h(x)$, and since $h(x)$ has proper degree $d > 0$, by Proposition 10, one deduces that there exist $s \in \mathbb{Z}$ with $r \le s \le d$, such that

$$\left| \frac{\partial^s h}{\partial^s x}(a^\star) \right| \ge \sqrt{\epsilon} \cdot s! \cdot \|h\|.$$

Let s be the minimal integer satisfying this property.
We consider the polynomial

$$g(x) = h(x + a^\star) = \sum_{i=0}^{d} \frac{\frac{\partial^i h}{\partial^i x}(a^\star)}{i!} x^i = \sum_{i=0}^{s-1} \frac{\frac{\partial^i h}{\partial^i x}(a^\star)}{i!} x^i + \sum_{i=s}^{d} \frac{\frac{\partial^i h}{\partial^i x}(a^\star)}{i!} x^i.$$

Let

$$g^\star(x) = \sum_{i=s}^{d} \frac{\frac{\partial^i h}{\partial^i x}(a^\star)}{i!} x^i,$$

and

$$q(x) = \sum_{i=0}^{s-1} \frac{\frac{\partial^i h}{\partial^i x}(a^\star)}{i!} x^i.$$

Note that 0 is an exact root of multiplicity s of the polynomial $g^\star(x)$ since

$$\frac{\partial^i g^\star}{\partial^i x}(0) = 0, \quad \text{for} \quad i = 1, \ldots, s-1,$$

and

$$\left| \frac{\partial^s g^\star}{\partial^s x}(0) \right| = \left| \frac{\partial^s h}{\partial^s x}(a^\star) \right| \ge \sqrt{\epsilon} \cdot s! \cdot \|h\| > 0.$$

Furthermore, since s the minimal integer satisfying that

$$\left| \frac{\partial^s h}{\partial^s x}(a^\star) \right| \ge \sqrt{\epsilon} \cdot s! \cdot \|h\|,$$

one has that

$$\left| \frac{\partial^i h}{\partial^i x}(a^\star) \right| < \sqrt{\epsilon} \cdot i! \cdot \|h\|, \quad \text{for} \quad i = 0, \ldots, s-1.$$

In this conditions, since

$$q(x) = \sum_{i=0}^{s-1} \frac{\frac{\partial^i h}{\partial^i x}(a^\star)}{i!} x^i,$$

we may write $q(x) = \sqrt{\epsilon} \cdot p(x)$, where

$$p(x) = \sum_{i=0}^{s-1} b_i x^i,$$

and

$$|b_i| = \frac{\left|\frac{\partial^i h}{\partial^i x}(a^\star)\right|}{i! \cdot \sqrt{\epsilon}} < \|h\|, \quad \text{for} \quad i = 0, \dots, s-1.$$

In particular observe that

$$|p(0)| = |b_0| = \frac{|h(a^\star)|}{\sqrt{\epsilon}} < \frac{\epsilon \cdot \|h\|}{\sqrt{\epsilon}} = \sqrt{\epsilon} \cdot \|h\|.$$

Now, by Lemma 11 we deduce that there exists a root $x_0 \in \mathbb{C}$ of $g(x)$ such that

$$|x_0| \leq \left(\frac{s! \cdot \sqrt{\epsilon} \cdot |p(0)|}{\left|\frac{\partial^s g^\star}{\partial^s x}(0)\right|} \right)^{\frac{1}{s}} = \left(\frac{s! \cdot \sqrt{\epsilon} \cdot |p(0)|}{\left|\frac{\partial^s h}{\partial^s x}(a^\star)\right|} \right)^{\frac{1}{s}} \leq \left(\frac{s! \cdot \epsilon \cdot \|h\|}{\sqrt{\epsilon} \cdot s! \cdot \|h\|} \right)^{\frac{1}{s}} = \sqrt{\epsilon}^{\frac{1}{s}}.$$

Finally, since $x_0 \in \mathbb{C}$ is a root of $g(x)$, then $a = x_0 + a^\star \in \mathbb{C}$ is a root of $h(x)$, and

$$|a - a^\star| = |x_0| \leq \sqrt{\epsilon}^{\frac{1}{s}} \leq \sqrt{\epsilon}^{\frac{1}{d}}.$$

$\square$

From Theorem 14, one has the following corollary.

Corollary 15. *Let $h(x) \in \mathbb{C}[x]$ be a polynomial of proper degree $d > 0$, and let $a^\star$ be an ϵ–root of multiplicity r of $h(x)$. Then, there exists a root $a \in \mathbb{C}$ of $h(x)$ such that:*

$$|a - a^\star| \leq (\sqrt{\epsilon})^{\frac{1}{d}}.$$

$\square$

4 ϵ–Points on Curves

In this section, we focus on the analysis of the bivariate case, i.e we study distance properties of ϵ–points, and we see how the results in the previous section can be generalized. From these distance properties we deduce results on the offset behavior of the curve.

For this purpose, first we give a particular treatment to the case of ϵ–simple points, where we impose the condition of being strong. Secondly, we deal with the general case, i.e. ϵ–points of any multiplicity and non necessarily strong, but assuming proper degree.

Theorem 16. *Let C be a plane algebraic curve over $\mathbb{C}$, and let $P^\star \in \mathbb{C}^2$ be an ϵ–strong simple point of C. Then, there exists an exact point $P \in \mathbb{C}^2$ of C such that*

$$\|P^\star - P\|_2 \leq \sqrt{\epsilon}.$$

Proof. Let $f(x, y) \in \mathbb{C}[x, y]$ be the defining polynomial of $\mathcal{C}$, and let $f(x, y)$ be expressed as

$$f(x, y) = \sum_{i+j=0}^{d} a_{i,j}(x - a^\star)^i (y - b^\star)^j,$$

where $P^\star = (a^\star, b^\star)$. We consider the univariate polynomial $h(t) = f(t, b^\star)$. Note that

$$h(t) = \sum_{i=0}^{d} a_{i,0}(t - a^\star)^i \in \mathbb{C}[x].$$

Now, observe that $|h(a^\star)| = |a_{0,0}| = |f(P^\star)| < \epsilon \cdot \|f\|$. Moreover, since $P^\star$ is an ϵ–strong simple point of $\mathcal{C}$, it holds that there exist $i_0, j_0 \in \mathbb{N}$ with $i_0 + j_0 = 1$ such that

$$\left| \frac{\partial f}{\partial^{i_0} x \partial^{j_0} y}(P^\star) \right| \geq \sqrt{\epsilon} \cdot \|f\|.$$

Let us assume w.l.o.g that $i_0 = 1$ and $j_0 = 0$. Then,

$$\left| \frac{\partial h}{\partial t}(a^\star) \right| = |a_{1,0}| = \left| \frac{\partial f}{\partial x}(P^\star) \right| \geq \sqrt{\epsilon} \cdot \|f\|.$$

Therefore $a^\star$ is an ϵ–strong simple root of $h(t)$. Hence, by Corollary 13, one deduces that there exists a root $a \in \mathbb{C}$ of $h(t)$ such that

$$|a - a^\star| \leq \sqrt{\epsilon}.$$

Then, if $P = (a, b^\star) \in \mathbb{C}^2$ it holds that $f(P) = f(a, b^\star) = h(a) = 0$; i.e P is an exact point of $\mathcal{C}$, and

$$\|P^\star - P\|_2 = |a - a^\star| \leq \sqrt{\epsilon}.$$

$\square$

The next corollary shows how two curves are locally related, in terms of their offsets (see [1] for definition of offsets), when ϵ–simple points appear.

Corollary 17. *Let $\mathcal{C}$ and $\mathcal{C}^\star$ be two plane algebraic curves over $\mathbb{C}$ defined by the polynomials $f(x, y), f^\star(x, y) \in \mathbb{C}[x, y]$, respectively. Let $P \in \mathbb{C}^2$ be an exact point of $\mathcal{C}$ that is an ϵ–strong simple point of $\mathcal{C}^\star$. Then, in the neighborhood of the point P, the curve $\mathcal{C}$ is contained in the offset region of $\mathcal{C}^\star$ at distance at most $2\sqrt{\epsilon}$.*

Proof. Since $P = (a, b)$ is an ϵ–strong simple point of $\mathcal{C}^\star$, by Theorem 16, one deduces that there exists an exact point $P^\star = (a^\star, b^\star) \in \mathbb{C}^2$ of $\mathcal{C}^\star$ such that $\|P^\star - P\|_2 \leq \sqrt{\epsilon}$. In this situation, we consider the tangent line to $\mathcal{C}^\star$ at $P^\star$; i.e $T^\star(x, y) = n_x(x - a^\star) + n_y(y - b^\star)$, where (n_x, n_y) is the unit normal vector to $\mathcal{C}^\star$ at $P^\star$. Then, we bound the value $\|T^\star(P)\|_2$:

$$\|T^\star(P)\|_2 \leq |n_x| \cdot |a - a^\star| + |n_y| \cdot |b - b^\star| \leq \|P^\star - P\|_2(|n_x| + |n_y|) \leq 2\sqrt{\epsilon}.$$

Therefore, reasoning as in Subsection 2.2 of [4] one deduces that, in the neighborhood of the point P, the curve $\mathcal{C}$ is contained in the offset region of $\mathcal{C}^\star$ at distance at most $2\sqrt{\epsilon}$.

$\square$

Once ϵ–simple points have been analyzed, we study the distance properties of ϵ–singularities. As we have mentioned above, we will assume that curves are given with proper degree.

Theorem 18. *Let C be a plane algebraic curve over $\mathbb{C}$ of proper degree $d > 0$, and let $P^\star \in \mathbb{C}^2$ be an ϵ–singularity of multiplicity r of C. Then, there exists $s \in \mathbb{Z}$ such that $r \leq s \leq d$, and there exists an exact point $P \in \mathbb{C}^2$ of C such that*

$$\|P^\star - P\|_2 \leq \sqrt{2}\sqrt{\epsilon}^{\frac{1}{2s}}.$$

Proof. Let $f(x,y) \in \mathbb{C}[x,y]$ be the defining polynomial of C, and let $f(x,y)$ be expressed as

$$f(x,y) = \sum_{i+j=0}^{d} a_{i,j}(x - a^\star)^i(y - b^\star)^j,$$

where $P^\star = (a^\star, b^\star) \in \mathbb{C}^2$. Note that since $P^\star \in \mathbb{C}^2$ is an ϵ–singularity of multiplicity r of C, by Proposition 5, we deduce that there exists $s \in \mathbb{Z}$ with $r \leq s \leq d$, and there exist non-negative integers i_1, j_1 such that $s = i_1 + j_1$, and

$$\left| \frac{\partial^s f}{\partial^{i_1} x \partial^{j_1} y}(P^\star) \right| \geq \sqrt{\epsilon} \cdot i_1! j_1! \cdot \|f\|.$$

Let $s = i_1 + j_1$ be the minimal integer satisfying this property.
Now we distinguish two different cases depending on the values of i_1 and j_1.

1. First, we deal with the case $j_1 = 0$ or $i_1 = 0$. For instance, let $j_1 = 0$. Note that this implies that

$$|a_{i_1,0}| = |a_{s,0}| = \frac{\left| \frac{\partial^s f}{\partial^s x}(P^\star) \right|}{s!} \geq \sqrt{\epsilon} \cdot \|f\|.$$

In these conditions, we prove that there exists a point $P \in \mathbb{C}^2$ on C such that $\|P^\star - P\|_2 \leq \sqrt{\epsilon}^{\frac{1}{s}}$. Indeed; we consider the univariate polynomial $h(t) = f(t + a^\star, b^\star)$. Note that

$$h(t) = \sum_{i=0}^{d} a_{i,0}t^i = \sum_{i=s}^{d} a_{i,0}t^i + \sum_{i=0}^{s-1} a_{i,0}t^i.$$

Let

$$h^\star(t) = \sum_{i=s}^{d} a_{i,0}t^i,$$

and

$$q(t) = \sum_{i=0}^{s-1} a_{i,0}t^i.$$

Note that 0 is an exact root of multiplicity s of the polynomial $h^\star(t)$, since

$$\frac{\partial^i h^\star}{\partial^i t}(0) = 0, \quad \text{for} \quad i = 0, \ldots, s-1,$$

and

$$\left| \frac{\partial^s h^\star}{\partial^s t}(0) \right| = \left| \frac{\partial^s f}{\partial^s x}(P^\star) \right| \geq \sqrt{\epsilon} \cdot s! \cdot \|f\| > 0.$$

Furthermore, since s is the minimal integer satisfying that

$$\left| \frac{\partial^s h}{\partial^s t}(0) \right| = \left| \frac{\partial^s f}{\partial^s x}(P^\star) \right| = |a_{s,0}|s! \geq \sqrt{\epsilon} \cdot s! \cdot \|f\|,$$

one deduces that for $0 \leq i \leq s-1$, it holds that

$$\frac{\left| \frac{\partial^i h}{\partial^i t}(0) \right|}{i!} = |a_{i,0}| = \frac{\left| \frac{\partial^i f}{\partial^i x}(P^\star) \right|}{i!} < \sqrt{\epsilon} \cdot \|f\|.$$

In this conditions, since

$$q(t) = \sum_{i=0}^{s-1} a_{i,0} t^i,$$

we may write

$$q(t) = \sqrt{\epsilon} \cdot p(t),$$

where

$$p(t) = \sum_{i=0}^{s-1} b_i t^i,$$

and

$$|b_i| = \frac{\left| \frac{\partial^i h}{\partial^i t}(0) \right|}{i! \cdot \sqrt{\epsilon}} < \|f\|, \quad \text{for} \quad i = 0, \ldots, s-1.$$

In particular, observe that

$$|p(0)| = |b_0| = \frac{|f(P^\star)|}{\sqrt{\epsilon}} < \frac{\epsilon \cdot \|f\|}{\sqrt{\epsilon}} = \sqrt{\epsilon} \cdot \|f\|.$$

Now, by Lemma 11 one deduces that there exists a root $t_0 \in \mathbb{C}$ of $h(t)$ such that

$$|t_0| \leq \left(\frac{s! \cdot \epsilon \cdot |p(0)|}{\left| \frac{\partial^s h^\star}{\partial^s t}(0) \right|} \right)^{1/s} =$$

$$\left(\frac{s! \cdot \epsilon \cdot \|f\|}{\left| \frac{\partial^s h^\star}{\partial^s t}(0) \right|} \right)^{1/s} \leq \left(\frac{s! \cdot \epsilon \cdot \|f\|}{\sqrt{\epsilon} \cdot s! \cdot \|f\|} \right)^{1/s} = \left(\frac{\epsilon}{\sqrt{\epsilon}} \right)^{1/s} = \sqrt{\epsilon}^{\frac{1}{s}}.$$

Therefore,

$$P = (t_0 + a^\star, b^\star) \in \mathbb{C}^2$$

is a point of the curve $\mathcal{C}$, and

$$\|P^\star - P\|_2 = |t_0| \leq \sqrt{\epsilon}^{\frac{1}{s}}.$$

2. Now, we assume that $1 \leq i_1 \leq j_1 \leq s - 1 \leq d - 1$, and let

$$i_1 = \min_{1 \leq i \leq s-1} \left\{ i \in \mathbb{N} \quad \text{such that} \quad \left| \frac{\partial^s f}{\partial^i x \partial^{s-i} y}(P^\star) \right| \geq \sqrt{\epsilon} \cdot i!(s-i)! \cdot \|f\| \right\}.$$

In these conditions, we prove that there exists a point $P \in \mathbb{C}^2$ on $\mathcal{C}$ such that

$$\|P^\star - P\|_2 \leq \sqrt{2}\sqrt{\epsilon^{\frac{1}{2s}}}.$$

Indeed; we consider the univariate polynomial $h(t) = f(t^{1+d} + a^\star, t^d + b^\star)$. Note that

$$h(t) = \sum_{i+j=0}^{d} a_{i,j} t^{jd+i(1+d)} = \sum_{i+j=s}^{d} a_{i,j} t^{jd+i(1+d)} + \sum_{i+j=0}^{s-1} a_{i,j} t^{jd+i(1+d)}.$$

Let

$$h^\star(t) = \sum_{i+j=s}^{d} a_{i,j} t^{jd+i(1+d)},$$

and

$$q(t) = \sum_{i+j=0}^{s-1} a_{i,j} t^{jd+i(1+d)}.$$

Since s is the minimal integer satisfying that

$$\frac{\left| \frac{\partial^s f}{\partial^{i_1} x \partial^{j_1} y}(P^\star) \right|}{i_1! j_1!} = |a_{i_1,j_1}| \geq \sqrt{\epsilon} \cdot \|f\|,$$

one has that for $0 \leq i + j \leq s - 1$, it holds that

$$|a_{i,j}| = \frac{\left| \frac{\partial^{i+j} f}{\partial^i x \partial^j y}(P^\star) \right|}{i! j!} < \sqrt{\epsilon} \cdot \|f\|.$$

In this conditions, since

$$q(t) = \sum_{i+j=0}^{s-1} a_{i,j} t^{jd+i(1+d)},$$

we may write

$$q(t) = \sqrt{\epsilon} \cdot p(t),$$

where

$$p(t) = \sum_{i+j=0}^{s-1} b_{i,j} t^{jd+i(1+d)},$$

and

$$|b_{i,j}| = \frac{\left| \frac{\partial^{i+j} f}{\partial^i x \partial^j y}(P^\star) \right|}{i! j! \cdot \sqrt{\epsilon}} < \|f\|, \quad \text{for} \quad i = 0, \ldots, s - 1.$$

In particular, observe that

$$|p(0)| = |b_{0,0}| = \frac{|f(P^\star)|}{\sqrt{\epsilon}} < \frac{\epsilon \cdot \|f\|}{\sqrt{\epsilon}} = \sqrt{\epsilon} \cdot \|f\|.$$

In addition, we also have that for $i_2, j_2 \in \mathbb{N}$ with $s \leq i_2 + j_2 \leq d$ and i_2, j_2 not both simultaneously equal to i_1, j_1, it holds that

$$j_2 d + i_2(1 + d) > j_1 d + i_1(1 + d).$$

Indeed; if $i_2 + j_2 = s$ then $j_2 d + i_2(1 + d) > j_1 d + i_1(1 + d)$ if and only if $ds + i_2 > ds + i_1$. This inequality holds because $i_1 < i_2$ (by definition of i_1, and taking into account that $i_1 \neq i_2$ since i_2, j_2 not both simultaneously equal to i_1, j_1). Now let $i_2 + j_2 = s + \ell$, $\ell \geq 1$. Then $j_2 d + i_2(1 + d) > j_1 d + i_1(1 + d)$ if and only if $d\ell + i_2 > i_1$. This last inequality holds since $1 \leq i_1 \leq s - 1 \leq d - 1$, $0 \leq i_2 \leq d$ and then $i_1 - i_2 \leq s - 1 < d \leq d\ell$.

Thus, we have that for $k = j_1 d + i_1(1 + d) = i_1 + ds$, it holds that 0 is an exact root of multiplicity k of the polynomial $h^\star(t)$, since

$$\frac{\partial^i h^\star}{\partial^i t}(0) = 0, \quad i = 0, \ldots, k - 1,$$

and,

$$\frac{\left|\frac{\partial^k h^\star}{\partial^k t}(0)\right|}{k!} = |a_{i_1,j_1}| = \frac{\left|\frac{\partial^s f}{\partial^{i_1} x \partial^{j_1} y}(P^\star)\right|}{i_1! j_1!} \geq \sqrt{\epsilon} \cdot \|f\| > 0.$$

Now, by Lemma 11 one deduces that there exists a root $t_1 \in \mathbb{C}$ of $h(t)$ such that

$$|t_1| \leq \left(\frac{k! \cdot \sqrt{\epsilon} \cdot |p(0)|}{\left|\frac{\partial^k h^\star}{\partial^k t}(0)\right|}\right)^{1/k} =$$

$$\left(\frac{k! \cdot \epsilon \cdot \|f\|}{\left|\frac{\partial^k h^\star}{\partial^k t}(0)\right|}\right)^{1/k} \leq \left(\frac{k! \cdot \epsilon \cdot \|f\|}{\sqrt{\epsilon} \cdot k! \cdot \|f\|}\right)^{1/k} =$$

$$\left(\frac{\epsilon}{\sqrt{\epsilon}}\right)^{1/k} = \epsilon^{\frac{1}{2k}} = \epsilon^{\frac{1}{2(i_1 + ds)}}.$$

Therefore,

$$P = (t_1^{1+d} + a^\star, t_1^d + b^\star) \in \mathbb{C}^2$$

is an exact point of the curve $\mathcal{C}$, and

$$\|P^\star - P\|_2 = \|(t_1^{1+d}, t_1^d)\|_2 = |t_1|^d \sqrt{|t_1|^2 + 1} \leq \epsilon^{\frac{d}{2k}} \sqrt{\epsilon^{\frac{1}{k}} + 1} \leq$$

$$\epsilon^{\frac{d}{2(i_1 + ds)}} \sqrt{\epsilon^{\frac{1}{i_1 + ds}} + 1} \leq \sqrt{2} \sqrt{\epsilon^{\frac{d}{ds+s-1}}} \leq \sqrt{2} \sqrt{\epsilon^{\frac{1}{2s-1}}} \leq \sqrt{2} \sqrt{\epsilon^{\frac{1}{2s}}} \leq \sqrt{2} \sqrt{\epsilon^{\frac{1}{2d}}}.$$

$\square$

From Theorem 18, one deduces the following corollary.

Corollary 19. *Let C be a plane algebraic curve over $\mathbb{C}$ of proper degree $d > 0$, and let $P^\star \in \mathbb{C}^2$ be an ε–point of C. Then, there exists an exact point $P \in \mathbb{C}^2$ of C such that*

$$\|P^\star - P\|_2 \leq \sqrt{2}\sqrt{\epsilon^{\frac{1}{2d}}}.$$

$\square$

In Corollary 17, we have seen the local offset behavior of two curves when there exist ε–strong simple points. The next corollary analyzes the phenomenon when ε–points appear. The proof of Corollary 20 is similar and we omit it.

Corollary 20. *Let C, and $C^\star$ be two plane algebraic curves over $\mathbb{C}$ of proper degree $d > 0$, defined by the polynomials $f(x,y)$ and $f^\star(x,y)$ respectively. If every exact point of C is an ε–point of $C^\star$ then, the curve C is contained in the offset region of the curve $C^\star$ at distance at most $2\sqrt{2}\sqrt{\epsilon^{\frac{1}{2d}}}$.* $\square$

The previous corollary can be specialized to the case of perturbed curves by adding an small constant.

Corollary 21. *Let C, and $C^\star$ be two plane algebraic curves over $\mathbb{C}$ of proper degree $d > 0$, defined by the polynomials $f(x,y)$ and $f^\star(x,y)$ respectively, such that $f(x,y) = f^\star(x,y) - \epsilon \cdot \|f^\star(x,y)\|$. Then, the curve C is contained in the offset region of the curve $C^\star$ at distance at most $2\sqrt{2}\sqrt{\epsilon^{\frac{1}{2d}}}$.* $\square$

References

1. Arrondo E., Sendra J., Sendra J.R., (1997). *Parametric Generalized Offsets to Hypersurfaces.* J. of Symbolic Computation vol. 23, pp 267-285.
2. Bulirsch, R., Stoer, J., (1993). *Introduction to Numerical Analysis.* Springer Verlag, New York.
3. Emiris, I.Z., Galligo, A., Lombardi, H., (1997). *Certified Approximate Univariate GCDs* J. Pure and Applied Algebra, Vol.117 and 118, pp. 229–251.
4. Farouki, R.T., Rajan V.T., (1988). *On the Numerical Condition of Algebraic Curves and Surfaces. 1. Implicit Equations.* Computer Aided Geometric Design. Vol. **5** pp. 215–252.
5. Marotta, V., (2003). *Resultants and Neighborhoods of a Polynomial.* Symbolic and Numerical Scientific Computation, SNSC'01, Hagemberg, Austria. Lectures Notes in Computer Science 2630. Springer Verlag.
6. Mignotte. M., (1992). *Mathematics for Computer Algebra.* Springer-Verlag, New York, Inc.
7. Pan V.Y. (2001). *Univariate polynomials: nearly optimal algorithms for factorization and rootfinding* . ISSAC 2001, London, Ontario, Canada. ACM Press New York, NY, USA. pp. 253 - 267.
8. Pérez–Díaz, S., Sendra , J., Sendra , J.R., *Parametrization of Approximate Algebraic Curves by Lines* . Theoretical Computer Science Vol.315/2-3. pp.627-650.
9. Pérez–Díaz, S., Sendra , J., Sendra , J.R., *Parametrization of Approximate Algebraic Surfaces by Lines.* Preprint.
10. Sasaki, T., Terui, A., (2002). *A Formula for Separating Small Roots of a Polynomial.* ACM, SIGSAM Bulletin, Vol.36, N.3.
11. Winkler, J.R., (2001). *Condition Numbers of a Nearly Singular Simple Root of a Polynomial.* Applied Numerical Mathematics, 38, 3; pp.275-285.

Distance Separation Measures
Between
Parametric Curves and Surfaces
Toward
Intersection and Collision Detection Applications

Gershon Elber

Technion, Israel Institute of Technology, Haifa 32000, Israel,
gershon@cs.technion.ac.il,
WWW home page: http://www.cs.technion.ac.il/~gershon

Abstract. This paper investigates the use of separation measures for parametric curves and surfaces toward the resolution of interference and intersections between curves and surfaces as well as collision detection. Two types of distance separation measures are discussed.

While the trivial distance function can be derived quite efficiently, it is shown in this work that this trivial distance function is not the optimal approach, in general. A better and more efficient scheme that projects the distance onto the normal field of either manifold is demonstrated to be superior in correctly detecting highly coupled non-intersecting arrangements as such.

Finally, a few extensions that further ease the detection of intersection-free arrangements, for both planar arrangements and arrangements in R^3, are also discussed.

1 Introduction and Background

Collision detection tests and intersection computations between curves and surfaces are fundamental necessities in numerous fields, from robotics and navigation, through animation and simulation, all the way to modelling and Boolean operations. In curve-curve and surface-surface intersection applications [4], the subdivision, divide-and-conquer approach is a common solution due to its robustness and reasonable efficiency. A key component in these divide-and-conquer schemes is efficient detection of interference-free arrangements (IFA), when two curves or surfaces do not intersect. Early detection of an IFA as such could yield great savings in computation times by directly pruning large potential branches from the subdivision tree.

Many schemes exist to devise IFA guarantees. If the geometry is represented in the Bézier or B-spline form, the axes-aligned bounding box of the control points is also guaranteed to hold the geometry itself, taking into consideration positional information only, from the input geometry. Then, during the subdivision process, if the bounding boxes of the two manifolds undergoing the interference testing do not intersect, clearly no intersection could occur between the two curves or surfaces. Similarly, two curves or surfaces, if represented in the Bézier or B-spline form, are also known to be contained in the convex hull of their control points. Hence, the detection of no intersection between their two convex hulls will either ensure an IFA scenario or guarantee that this branch of the subdivision tree is completely purged.

The computation of bounding boxes as well as intersection tests between bounding boxes is simple and efficient. In contrast, the computation of the convex hull of a set of (control) points, and more so, the interference test between convex hulls, is far more computationally intensive. While a convex hull bound is tighter compared to that of a bounding box, in practice, the use of convex hulls as bounding interference tests is far less common compared to axes-aligned bounding boxes' use.

Other bounding schemes have also been developed. Slabs or oriented bounding boxes (OBB) are bounding regions that are aligned along the major axis of the geome-try in hand [2, 8], taking into consideration both positional and directional information from the input geometry. Consider a horizontal or vertical line compared to a diagonal one. Clearly, an axes-aligned bounding box will do miserably in bounding the diagonal line. The OBB orients itself along the geometry and will be equally tight to any line in an arbitrary orientation. The tighter the bounding geometry is, the lower the chance we will mistakenly declare a potential intersection. Computationally, the OBBs are moder-ately more expensive compared to axis-aligned regular bounding boxes, but due to the fact that they are far tighter, they are used in many intersection and collision detection tools.

In [7], pairs of parallel lines are used as bounding regions for curves. These parallel lines, yet another variant of the OBBs, are denoted as "fat-lines". In [6], "fat-arcs" of certain width are proposed as an even tighter bound on planar curves. This fat arc scheme is a generalization of the fat lines idea. For a fat arc with a radius that approaches infinity, the fat arc converges to the fat line scheme. Nonetheless, fat arcs also take into consideration second order curvature information and can tightly bound any arc or any curve of almost constant curvature. The difficulty in using fat arcs stems from the need for efficient computation of fat arcs and effective interference tests between fat arcs.

In [3], a bounding scheme to a given curve or surface in the form of a piecewise linear sleeve (sleve in [3] terms) was devised. This scheme is unique in its conversion of the freeform shape into a piecewise linear tight sleeve that closely holds the original geometry. One clear application to sleeves is collision detection.

The problem of devising efficient bounds for surfaces is more difficult, and in many cases, a simple bounding box is all that is needed. Even the extension of OBBs to three-dimensions introduces many special cases that must be handled, such as the intersection of two six-faced polyhedra, A and B, in a general position. One can reduce this interfer-ence problem to the problem of a single shrunken face, $\overline{A}$, versus an expanded six-faced polyhedra, $\overline{B}$, using the following process. Compute the minimal dimension of A and B, d_0, and assume without loss of generality that this minimal dimension is in A. Then, offset A inside by $d_0/2$ and offset B outside by $d_0/2$. A will collapse into a single face (or into a single edge or even into a point if two or more dimensions of A are the same), $\overline{A}$, and B will be expanded into a rounded box $\tilde{B}$. Alternatively, we can handle this case by simply offsetting the six faces of B by $d_0/2$ into a larger box. The revised offset of B, $\overline{B}$ is somewhat larger than the rounded box, $\tilde{B} \subset \overline{B}$, yet $\overline{B}$ is also much simpler to further process. Nevertheless, this entire process is obviously more complex than a comparison between two axis-aligned bounding boxes.

This paper presents an examination of a different separation scheme. Given two manifolds we need to test for potential interference, we compute a distance related function, $\mathcal{D}$, between them, so that $\mathcal{D}$ vanishes if and only if the two objects intersect. By examining $\mathcal{D}$, we will be able to detect cases where the input manifolds are free of intersection. Once we have shown this, we also consider a few ideas to further extend these intersection-free detection tests for both planar arrangements and arrangements in $\mathbf{R}^3$.

The rest of this paper is organized as follows. In sect. 2, we develop the theory behind the distance functions' approach and show that the simple distance square function one can derive is not necessarily the approach to use. In sect. 3, we present few simple examples using this distance approach. Then, in sect. 4, we consider several extensions to further improve upon the intersection-free detection tests, and finally, in sect. 5, we conclude.

2   Distance Maps and Projected Distance Maps

We start by considering parametric curves in the plane, though all we discuss here can easily be extended into surfaces in $\mathbf{R}^3$. Consider two C^1 continuous, regular parametric curves, $C_1(t)$ and $C_2(r)$. Then, the first distance function (square) we consider is:

Lemma 1. *Two regular planar parametric curves $C_1(t)$ and $C_2(r)$ do not intersect if the following holds:*

$$\mathcal{D}_1(r,t) : \langle C_1(t) - C_2(r), C_1(t) - C_2(r) \rangle \neq 0, \quad \forall t, r.$$

Proof: Trivial by construction. Because condition $\mathcal{D}_1$ measures the distance (square) between curve $C_1(t)$ and curve $C_2(r)$, for all t and r, and this distance is never allowed to vanish, clearly there exist no t and r for which $C_1(t)$ and $C_2(r)$ coalesce.   □

While trivial, $\mathcal{D}_1(r,t)$ deserves some discussion. If the two input C_i curves are polynomials of degrees m and n, $\mathcal{D}_1(r,t)$ is also a polynomial with degrees $(2m \times 2n)$. Further, one can symbolically express the coefficients of $\mathcal{D}_1(r,t)$ as functions of the control points of the two C_i curves (see Appendix A). As an example, given two cubic Bézier curves ($m = n = 3$), one is required to plug two sets of four control points into the coefficients functions of $\mathcal{D}_1$ and extract the $7^2 = 49$ coefficients. If all the coefficients of $\mathcal{D}_1$ are positive, $C_1(t)$ and $C_2(r)$ never intersect.

Unfortunately, in many cases where $C_1(t)$ and $C_2(r)$ do not intersect, $\mathcal{D}_1$ will continue to possess negative coefficients. This is because, while $\mathcal{D}_1(r,t)$ is a non-negative function, it indeed may contain negative coefficients, especially when $\mathcal{D}_1$ needs to represent vanishing distances. This case is, in fact, quite common as will be shown in sect. 3. We will further show in sect. 3 that the insertion of a few knots could also significantly improve this convergence and offer a superior interference test. The knot insertion function could also be a-priori computed as an Alpha-refinement matrix that maps the coefficients of $\mathcal{D}_1$ into their refined equivalent [1].

While $\mathcal{D}_1(r, t)$ is the straightforward approach to take toward collision detection, we also consider a second approach that not only reduces the degree of the computed distance function but, in fact, also provides superior performance in most cases:

Lemma 2. *Two C^1 planar regular parametric curves $C_1(t)$ and $C_2(r)$ do not intersect if either of the following holds:*

$$\mathcal{D}_2(r, t) : \langle C_1(t) - C_2(r), N_1(t)\rangle \neq 0, \quad \forall t, r,$$

or

$$\mathcal{D}_3(r, t) : \langle C_1(t) - C_2(r), N_2(r)\rangle \neq 0, \quad \forall t, r,$$

where N_i denotes the unnormalized normal field of C_i.

Proof: Consider condition $\mathcal{D}_2(r, t) : \langle C_1(t) - C_2(r), N_1(t)\rangle \neq 0$. The locus of points $\{P \mid \exists t \text{ such that } \langle C_1(t) - P, N_1(t)\rangle = 0\}$ in the plane is the set of points in the tangent space of $C_1(t)$ or the set of points on one or more of tangent lines of $C_1(t)$. Because we assume $\mathcal{D}_2$ never vanishes, no point of $C_2(r)$ can be on a tangent line of $C_1(t)$ for all t and r. Because $C_1(t)$ is contained in its tangent space, $C_2(r)$ coalesces with no point at $C_1(t)$. The proof for condition $\mathcal{D}_3$ is similar. □

As we have done for $\mathcal{D}_1$, one can symbolically express the coefficients of $\mathcal{D}_2(r, t)$ or $\mathcal{D}_3(r, t)$ as functions of the control points of the two C_i curves (see Appendix A).

Let $C_1(t)$ and $C_2(r)$ be two (piecewise) polynomials of degrees m and n. Recognizing that no unit normal field is required here, for (piecewise) polynomial curves, the degrees of $\mathcal{D}_2(r, t)$ and $\mathcal{D}_3(r, t)$ are going to be $((2m-1) \times n)$ and $(m \times (2n-1))$, respectively. This is due to the fact that normal field $N_i(t)$, of curve $C_i(t) = (x_i(t), y_i(t))$ can be computed as $(-y'(t), x'(t))$.

Assume C_i are linear segments. Lemma 2 reduces to the following conditions for two line segments to be intersection-free. Consider the segments $\overline{P_0 P_1}$ and $\overline{P_2 P_3}$, $P_i = (x_i, y_i)$. Then,

$$\mathcal{D}_2(r, t) : \langle (P_1 - P_0)t + P_0 - (P_3 - P_2)r - P_2, (y_0 - y_1, x_1 - x_0)\rangle \neq 0$$
$$\mathcal{D}_3(r, t) : \langle (P_1 - P_0)t + P_0 - (P_3 - P_2)r - P_2, (y_2 - y_3, x_3 - x_2)\rangle \neq 0. \quad (1)$$

Having the inner product with the normal equate with the cross product in the tangent field (dealing with planar curves, we consider only the Z coefficient of the cross product), one can rewrite $\mathcal{D}_2(r, t)$ in eq. (1) as

$$0 = ((P_1 - P_0)t + P_0 - (P_3 - P_2)r - P_2) \times (P_1 - P_0),$$
$$= (P_0 \times P_1) - ((P_3 - P_2)r - P_2) \times (P_1 - P_0). \quad (2)$$

Because condition (2) must hold for all r, condition $\mathcal{D}_2$ in essence examines whether one segment is completely contained in one-half space of the infinite line through the other segment. This linear case provides some intuition of the general case. Using

$\mathcal{D}_2(r, t)$ or $\mathcal{D}_3(r, t)$, we can check if one curve interferes with the tangent space of the other curves.

The extension of these distance functions to surfaces in $\mathbf{R}^3$ is straightforward:

$$\mathcal{D}_1(u, v, r, t) : \langle S_1(u, v) - S_2(r, t), S_1(u, v) - S_2(r, t) \rangle \neq 0,$$
$$\mathcal{D}_2(u, v, r, t) : \langle S_1(u, v) - S_2(r, t), N_1(u, v) \rangle \neq 0,$$
$$\mathcal{D}_3(u, v, r, t) : \langle S_1(u, v) - S_2(r, t), N_2(r, t) \rangle \neq 0. \tag{3}$$

Lemmas 1 and 2 hold for surfaces as well. For Lemma 2, $\mathcal{D}_2$ and $\mathcal{D}_3$ examine if points in one surface are contained in the two parameters family of tangent planes, or the tangent space, of the other surface.

While the use of $\mathcal{D}_2(r, t)$ and $\mathcal{D}_3(r, t)$ is valid regardless of the shape of the curves, the distance test offered by $\mathcal{D}_2(r, t)$ and $\mathcal{D}_3(r, t)$ is advantageous for inflection-free curves. In many interference and/or intersection applications, the input curves are already subdivided at inflection points as a preprocessing stage, possibly by detecting the locations on the curves where the curvature vanishes. Moreover, as will be shown in sect. 3, even with inflection points, $\mathcal{D}_2(r, t)$ and $\mathcal{D}_3(r, t)$ can do quite well. The same holds for two interacting surfaces where $\mathcal{D}_2(r, t)$ or $\mathcal{D}_3(r, t)$ are expected to shine compared to $\mathcal{D}_1(r, t)$, in non-hyperbolic regions. In our tests, presented in sect. 3, the distance functions of $\mathcal{D}_2(r, t)$ and $\mathcal{D}_3(r, t)$ were found, in most cases, to be much tighter than $\mathcal{D}_1(r, t)$, for both curves and surfaces. In Appendix A, the coefficients of the $\mathcal{D}_i$, $i = 1, 2, 3$, functions are computed and presented as functions of the control points of the given two polynomial curves. These are the functions one needs to evaluate in order to determine the strict positivity or negativity of the $\mathcal{D}_i$ functions, and using these functions, detect IFA scenarios.

3   Examples

We start our demonstration with a figure of quadratic Bézier arrangements, in fig. 1. In all presented figures, we first show the arrangement $((a)$ and (e) in fig. 1) and then draw the three $\mathcal{D}_i$ functions in order, using a $(r, t, \mathcal{D}_i)$ coordinate system $((b)$-(d) and (f)-(h) in fig. 1). By inserting a single knot in both curves at $1/2$, $\mathcal{D}_3$ is capable of detecting this case as an IFA.

Figure 2 presents a cubic and a quartic that are tightly coupled, while neither is completely convex. All three methods fail to detect this case as an IFA. Yet, after a single knot insertion in both curves at $1/2$, $\mathcal{D}_3$ successfully detects this case as an IFA.

Figure 3 presents our final example of a cubic and quadratic curve arrangement. While tightly coupled in (a), $\mathcal{D}_2$ is still capable of detecting this case as intersection-free. Further, in (e), an extremely tight arrangement is shown, so tight that in the drawing's resolution, the curves appear connected. With the refinement of both curves using a single knot, at $1/2$, $\mathcal{D}_2$ declares this case to be an IFA.

We continue and examine several examples of interacting surfaces. Here, equations (3) are computed as four-variate Bézier functions and their coefficients are examined for their signs. In fig. 4, two cases of quadratic surfaces, which are closely coupled, are presented. The distance function that is projected onto the normal field of

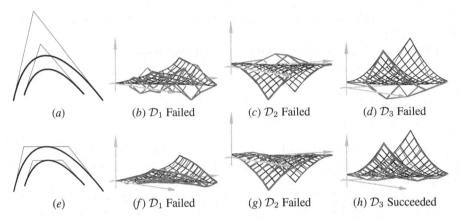

Fig. 1. Two quadratic polynomial curve arrangements. In (a) to (d), the original curves are considered. In (e) to (h), both curves are refined once at $t = 1/2$.

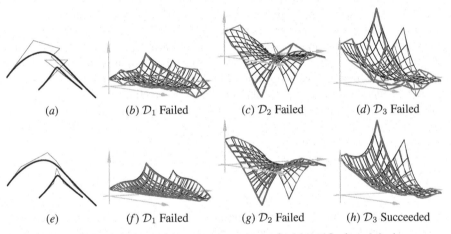

Fig. 2. Cubic and quartic polynomial curve arrangements. In (a) to (d), the original curves are considered. In (e) to (h), both curves are refined once at $t = 1/2$.

either surface does poorly for hyperbolic cases as can be seen in (a) and (b). (b) is a refinement of (a) at $1/4$, $1/2$, $3/4$ in both u and v. In hyperbolic regions, the tangent space spans both sides of the surface, vastly decreasing the probability of an IFA detection, when $\mathcal{D}_2$ or $\mathcal{D}_3$ are used. Only $\mathcal{D}_1$ is capable of detecting this case as an IFA, after the insertion of three knots in both surface directions, on both surfaces.

In contrast, when processing convex surfaces, the advantage of using either $\mathcal{D}_2$ or $\mathcal{D}_3$ is clearly revealed. In fig. 4 (c) and (d), two highly coupled convex shapes are presented. Yet $\mathcal{D}_2$ is capable of detecting this case as an IFA after a single knot insertion (see (d)) in both directions of both surfaces. The necessity for the geometry to be elliptic should, in fact, be imposed only on one of the surfaces, the surface that the distance vector is projected onto in its normal field. Hence, refinement will probably be more

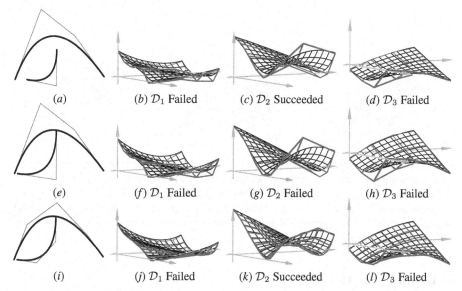

Fig. 3. Cubic and quadratic polynomial curve arrangements. In (a) to (d) and in (e) to (h), the original curves are considered. In (i) to (l), both curves are refined once at $t = 1/2$.

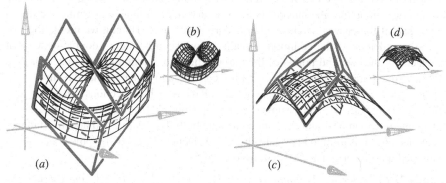

Fig. 4. Two biquadratic polynomial surface arrangements without refinement ((a) and (c)) and with refinement (b) at $1/4, 1/2, 3/4$ and (d) at $1/2$, in both u and v. In ((a) and (c)), $\mathcal{D}_1$, $\mathcal{D}_2$, and $\mathcal{D}_3$ all failed to find this case to be intersection-free. In (b) and (d), only $\mathcal{D}_1$ and $\mathcal{D}_2$ respectively, were successful in detecting this case as an IFA.

effective when applied to that surface. In fig. 4 (d), (only) distance function $\mathcal{D}_3$ was capable of declaring this an IFA, even when one of the surfaces is hyperbolic.

Distance functions $\mathcal{D}_2$ and $\mathcal{D}_3$ are also expected to do well in parabolic or almost parabolic surfaces. In fig. 5, (only) distance function $\mathcal{D}_2$ was capable of declaring this arrangement to be an IFA, after a single knot insertion in both surface directions, on both surfaces. Finally, fig. 5 (c) presents one case of having both elliptic and hyperbolic surfaces interact.

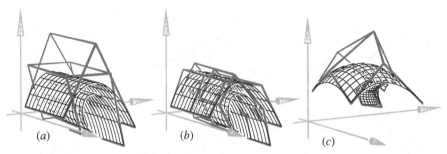

Fig. 5. A quadratic by linear (outer surfaces) and a quartic by linear (inner surface), almost developable polynomial surfaces' arrangement without refinement (*a*) and with refinement (*b*)at $1/2$ in both u and v. In (*a*), $\mathcal{D}_1$, $\mathcal{D}_2$, and $\mathcal{D}_3$ all failed to detect this case as intersection-free. In (*b*), $\mathcal{D}_2$ was successful in detecting this case as an IFA. (*c*) presents a large elliptic biquadratic and a small hyperbolic bicubic polynomial surface arrangement without refinement. As is, $\mathcal{D}_3$ was successful in detecting this case as an IFA.

4   Refining the IFA Tests

In [7], parallel lines are employed as bounding regions of Bézier curves. This bounding region was used in comparison with another curve, in an attempt to trivially eliminate intersection-free arrangements. Further, in [5], double-cones were used to bound the possible directions that the tangent field of a curve can assume, examining the directional span of the curves' Hodograph, the derivative curve. Herein, we merge the two processes and look at the common domain of two such bounding double-cones that are placed at the two end points of the curve. These two double-cones intersects in a double-wedge parallelogram shape that fully contain, and hence bound, the curve.

Let $\mathcal{C}_V^\alpha(C)$ be the double-cone that bounds all the possible directions that the tangent field of $C(t)$, $t \in [t_0, t_1]$ can assume (See fig. 6). Cone $\mathcal{C}_V^\alpha(C)$ has an axial direction V and an angular span of α and is assumed to span both the positive and negative tangent directions; else, it is infinite in both its directions. Denote by $\mathcal{C}_V^\alpha(C)[P]$ the cone $\mathcal{C}_V^\alpha(C)$ positioned with its origin at P. Then, following [5],

Lemma 3. *Consider a C^1 planar regular curve $C(t)$, $t \in [t_0, t_1]$, and $\mathcal{C}_V^\alpha(C)$, the bounding cone of the tangent field of $C(t)$. Then, $C(t) \subset \mathcal{C}_V^\alpha(C)$, $\forall t \in [t_0, t_1]$.*

Proof: See [5]. □

Let $\mathcal{C}_V^\alpha(C)^+$ be the positive half of $\mathcal{C}_V^\alpha(C)$ in the direction of the Hodograph and let $\mathcal{C}_V^\alpha(C)^-$ be the negative half, in the opposite direction.
Then,

Lemma 4. $C(t) \subset \mathcal{C}_V^\alpha(C)^+[C(t_0)] \bigcap \mathcal{C}_V^\alpha(C)^-[C(t_1)]$

Proof: At t_0, $C(t) \subset \mathcal{C}_V^\alpha(C)^+[C(t_0)]$ and similarly at t_1, $C(t) \subset \mathcal{C}_V^\alpha(C)^-[C(t_1)]$. Hence, $C(t)$ is also contained in their intersection. □

The intersection of the two cones, as presented in Lemma 4, creates a double-wedge or a parallelogram bounding region. Figure 7 shows a few examples.

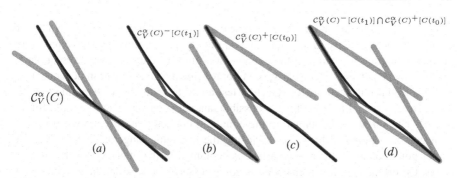

Fig. 6. The bounding cone, $\mathcal{C}_V^\alpha(C)$, in (a), is derived using the Hodograph of $C(t)$ and has the property that at the end of $C(t)$, $C(t)$ is completely contained in the negative infinite half of the cone, $\mathcal{C}_V^\alpha(C)^-[C(t_1)]$ (b), while at the beginning of $C(t)$, $C(t)$ is completely contained in the positive infinite half of the cone, $\mathcal{C}_V^\alpha(C)^+[C(t_0)]$ (c). (d) shows that the two half cones of (b) and (c) could serve as a finite double-wedge bounding region to $C(t)$.

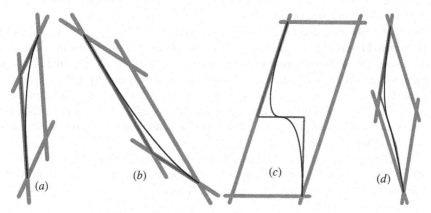

Fig. 7. Several examples of the double-wedge bounding regions, for planar quadratic $((a)$ and $(b))$ and cubic $((c)$ and $(d))$ curves

The parallelogram's bounding region is quite tight, especially in cases where the curve is almost straight. In a manner similar to oriented bounding boxes or the fat lines, and unlike the axes-aligned bounding box, the parallelogram is orientation independent. The difficulty in constructing and comparing two double-wedges is similar to that of the oriented bounding box. In fact, the idea of offsetting one bounding region inside and one bounding region outside, degenerating the first bounding region into a line, as presented in the introduction to this work, could be employed as well. Further, one can use this parallelogram bounding region to trivially reject intersection-free regions. If the second curve is on the opposite side of either one of the four boundary lines of the parallelogram region, that line serves as a separating line between the two curves. Testing on which side of a line a curve is, or whether they potentially intersect, is simple. It entails the evaluation of the signed distance of all the control points of the curves from the line. Hence, we exploit the four infinite lines defining the bounding parallelogram and try them as separation lines between the two curves. Each curve has its own bounding parallelogram, hence in all, eight lines should be tested. In practice, we found that this parallelogram test reduced the computation times of curve-curve intersections by several dozens percents. Yet clearly, such improvements are application dependent.

This trivial intersection-free rejection test could be extended to handling two surfaces. Here, a bounding cone for the possible tangent planes needs to be built, following [5]. The surface is contained in the bounding tangent planes' cone when its apex is placed on a point in the surface. Placing the bounding tangents' cone at the four corner points of the surface, the surface is contained in the intersection of these four cones. Unfortunately, this intersection is not necessarily convex and hence complex to compute. However, each of the faces of the four bounding tangents' cones, at the four corners of the surface, could be tested as a potential separation plane between the two surfaces. Each such face must hold its original surface on one side and if the second surface is completely on the other side of the face, that face serves as a seperation face. Once again, as the surface gets closer to being planar, the intersection of the four cones, at the four corners of the surface will tightly bound the surface and the above seperation scheme is more likely to succeed.

In [6], it is observed that for a cubic Bézier curve, $C(t) = \sum_{i=0}^{3} P_i \theta_i^3(t)$, the width of the oriented bounding box along edge $\overline{P_0 P_3}$ need not span all the width as asserted by the distance to the furthest control point, either P_1 or P_2. Instead, if P_1 and P_2 are on different sides of edge $\overline{P_0 P_3}$, the maximum width could be bound by $4/9$ of this maximal distance and if P_1 and P_2 are on the same side of edge $\overline{P_0 P_3}$, the maximum width could be bound by $3/4$ of this maximal distance. This is easily proven by computing the maximal contribution of $\theta_i^3(t)$, $i = 1, 2$, which is $4/9$, and the maximal contribution of $\theta_1^3(t) + \theta_2^3(t)$ together, which is $3/4$.

Generalizing this result, one could effortlessly provide slightly tighter bounds on the three functions of $\mathcal{D}_i(r, t)$, as follows. Compute the extreme values over all the coefficients of $\mathcal{D}_i(r, t)$, d_{jk}^i, as d_{min}^i and d_{max}^i. Now let,

$$d_{crnr}^i = \min \left(d_{00}^i, d_{0m}^i, d_{n0}^i, d_{nm}^i \right),$$

where n and m are the degrees of $\mathcal{D}_i(r, t)$, and let,

$$\theta_{crnr}^i = \min_{r, s} \left(\theta_0^m(r)\theta_0^n(s) + \theta_m^m(r)\theta_0^n(s) + \theta_0^m(r)\theta_n^n(s) + \theta_m^m(r)\theta_n^n(s) \right)$$

$$= \min_{r, s} \left((\theta_0^m(r) + \theta_m^m(r))(\theta_0^n(s) + \theta_n^n(s)) \right),$$

minimum that occurs at $r = s = 1/2$. Then,

$$\overline{d_{min}^i} = d_{min}^i * (1 - \theta_{crnr}^i) + d_{crnr}^i * \theta_{crnr}^i,$$

is one possible tighter bound on the minimum $\mathcal{D}_i$. This last result stems from having all four corner coefficients above or at level d_{crnr}^i and all other coefficients above or at level d_{min}^i. Therefore, assume, in the worst case, that the four corners are all at level d_{crnr}^i and the rest of the coefficients are all at level d_{min}^i, which yields this tighter bound.

5   Conclusion and Future Work

In this paper, we established a bounding scheme that is capable of detecting interference-free and/or intersection-free arrangements that are highly coupled, with the aid of distance functions that are projected onto the normal field of one of the manifolds. This variant of distance functions was shown to be superior to regular distance functions for (almost) convex planar curves and elliptic and parabolic surface regions. We further extended this approach to allow somewhat tighter bounds by carefully analyzing the extreme support locations of the corner coefficients of the scalar distance functions and offering a tight double-wedge bounding region.

It is plausible that other bounding schemes, such as [3], could also be merged in and help in establishing even tigher bounds compared to the presented appoarch in sect. 4. Such synergies should be further examined.

While not presented in this work, the interference between curves and surfaces could also be examined in a similar fashion, creating a trivariate distance function $\mathcal{D}_i(u, v, t)$ between curve $C(t)$ and surface $S(u, v)$. Refinement forces the control polygonal/mesh to quadratically converge to the shape of the curve or surface. Hence, one can insert several knots at once, achieving a tighter test for collision- or interference-free cases, at the cost of more computation and having more coefficients for which to evaluate their signs, in the $\mathcal{D}_i$ functions.

All presented schemes are suitable for both curves and surfaces, including the extension of the parallelogram double-wedges that offer separating lines as an interference-free detection scheme and extend it to separating planes in $\mathbf{R}^3$ between freeform surfaces. Similarly, and since the collision/intersection problem is reduced here to the existence of a non empty zero set of some constraint over the domain, any efficient scheme that can efficiently detect or deny the existence of roots in the domain could be equally employed. FInally, all the presented schemes could be applied to rational functions as well, sometimes at the cost of doubling the degree of the distance functions due to the quotient rule of differentiation and the addition of rationals.

References

1. E. Cohen, T. Lyche, and R. Riesenfeld. Discrete B-splines and Subdivision Techniques in Computer-Aided Geometric Design and Computer Graphics. Computer Graphics and Image Processing, 14, 87-111, October 1980.
2. S. Gottschalk, M. Lin and D. Manocha. "OBB-Tree: A Hierarchical Structure for Rapid Interference Detection." Siggraph 96 Conference Proceedings, pp 171-180, Aug. 1996.
3. J. Peters. "Efficient One-Sided Linearization of Spline Geometry." Mathematics of Surfaces, the 10th IMA International Conference, Michael J. Wilson and Ralph R. Martin (Eds.), pp 297-319, Leeds, UK, September 2003.
4. T. Sederberg and S. R. Parry. "A comparison of three curve intersection algorithms." Computer Aided Design, Vol 18, No. 1, pp 58-63, 1986.
5. T. W. Sederberg, R. J. Meyers. "Loop Detection in Surface Patch Intersections." Computer Aided Geometric Design, Vol. 5, No. 2, pp 161-171, 1988.
6. T. W. Sederberg, S. C. White and A. K. Zundel. "Fat Arcs: A Bounding Region with Cubic Convergence." Computer Aided Geometric Design, Vol. 6, No. 3, pp 205-218, 1989.

7. T. W. Sederberg, T. Nishita. "Curve Intersection using Bezier Clipping." Computer Aided Design, Vol. 22, No. 9, pp 337-345, 1990.
8. J. Yen, S. Spach, M. Smith and R. Pulleyblank. "Parallel Boxing in B-Spline Intersection." IEEE Computer Graphics and Applications, pp 72-79, Vol. 11, No. 1, Jan. 1991.

Appendix A: The Distance Functions for Two Bézier Curves

Given two Bézier curves $C_1(t) = \sum_{i=0}^{m} P_i \theta_i^m(t)$ and $C_2(r) = \sum_{j=0}^{n} Q_j \theta_j^n(r)$, consider $\mathcal{D}_1 = \langle C_1(t) - C_2(r), C_1(t) - C_2(r) \rangle$,

$$\mathcal{D}_1(t,r) = \left\langle \sum_{i=0}^{m} P_i \theta_i^m(t) - \sum_{j=0}^{n} Q_j \theta_j^n(r), \sum_{i=0}^{m} P_i \theta_i^m(t) - \sum_{j=0}^{n} Q_j \theta_j^n(r) \right\rangle.$$

Because,

$$\sum_{i=0}^{m} P_i \theta_i^m(t) - \sum_{j=0}^{n} Q_j \theta_j^n(r) = \sum_{i=0}^{m} P_i \theta_i^m(t) \sum_{j=0}^{n} \theta_j^n(r) - \sum_{i=0}^{m} \theta_i^m(t) \sum_{j=0}^{n} Q_j \theta_j^n(r)$$

$$= \sum_{i=0}^{m} \sum_{j=0}^{n} (P_i - Q_j) \theta_i^m(t) \theta_j^n(r),$$

we reduced the problem to a product of two vector surfaces. Consider two scalar Bézier surfaces $s_1(u,v)$ and $s_2(u,v)$. Their product equals,

$$s_1(u,v) s_2(u,v) = \sum_{i=0}^{m} \sum_{j=0}^{n} a_{ij} \theta_i^m(u) \theta_j^n(v) \sum_{k=0}^{p} \sum_{l=0}^{q} b_{kl} \theta_k^p(u) \theta_l^q(v)$$

$$= \sum_{i=0}^{m} \sum_{j=0}^{n} \sum_{k=0}^{p} \sum_{l=0}^{q} a_{ij} b_{kl} \theta_i^m(u) \theta_k^p(u) \theta_j^n(v) \theta_l^q(v)$$

$$= \sum_{i=0}^{m} \sum_{j=0}^{n} \sum_{k=0}^{p} \sum_{l=0}^{q} a_{ij} b_{kl} \frac{\binom{m}{i}\binom{p}{k}}{\binom{m+p}{i+k}} \theta_{i+k}^{m+p}(u) \frac{\binom{n}{j}\binom{q}{l}}{\binom{n+q}{j+l}} \theta_{j+l}^{n+q}(v)$$

$$= \sum_{r=0}^{m+p} \sum_{s=0}^{n+q} c_{rs} \theta_r^{m+p}(u) \theta_s^{n+q}(v)$$

where:

$$c_{rs} = \sum_{i=\max(0,r-p)}^{\min(r,m)} \sum_{j=\max(0,s-q)}^{\min(s,n)} a_{i,j} b_{r-i,s-j} \frac{\binom{m}{i}\binom{p}{r-i}}{\binom{m+p}{r}} \frac{\binom{n}{j}\binom{q}{s-j}}{\binom{n+q}{s}}. \quad (4)$$

Back to $\mathcal{D}_1(t,r)$, having $m = p$ and $n = q$ and interested only in the (simultaneous positivity/negativity of all the) coefficients, we get using eq. (4),

$$d_{rs}^1 = \sum_{i=\max(0,r-m)}^{\min(r,m)} \sum_{j=\max(0,s-n)}^{\min(s,n)} (P_i - Q_j)(P_{r-i} - Q_{s-j}) \frac{\binom{m}{i}\binom{m}{r-i}}{\binom{2m}{r}} \frac{\binom{n}{j}\binom{n}{s-j}}{\binom{2n}{s}},$$

$$r \in [0, 2m], \ s \in [0, 2n].$$

Now consider $\mathcal{D}_2 = \langle C_1(t) - C_2(r), N_1(t) \rangle$. It is easy to see that this constraint for planar curves is similar to the constraint of $\mathcal{D}_2 = (C_1(t) - C_2(r)) \times C_1'(t)$ or

$$\mathcal{D}_2(t,r) = \left(\sum_{i=0}^{m} P_i \theta_i^m(t) - \sum_{j=0}^{n} Q_j \theta_j^n(r) \right) \times \left(m \sum_{k=0}^{m-1} (P_{k+1} - P_k)\theta_k^{m-1}(t) \right)$$

$$= \left(\sum_{i=0}^{m} \sum_{j=0}^{n} (P_i - Q_j)\theta_i^m(t)\theta_j^n(r) \right) \times \left(m \sum_{k=0}^{m-1} (P_{k+1} - P_k)\theta_k^{m-1}(t) \right).$$

Note that although we use the cross-product notation, result $\mathcal{D}_2(t,r)$ is considered a scalar function as the curves are planar, considering only the Z coefficient of the cross product. Having $p = m-1$ and $q = 0$, $\mathcal{D}_2(t,r)$ is a function of degrees $((2m-1) \times n)$. Because we are interested again in the (simultaneous positivity/negativity of all the) coefficients, using eq. (4), these coefficients are,

$$d_{rs}^2 = m \sum_{i=\max(0,r-(m-1))}^{\min(r,m)} \sum_{j=\max(0,s-0)}^{\min(s,n)} (P_i - Q_j) \times (P_{r-i+1} - P_{r-i})$$

$$\frac{\binom{m}{i}\binom{m-1}{r-i}}{\binom{m+m-1}{r}} \frac{\binom{n}{j}\binom{0}{s-j}}{\binom{n+0}{s}},$$

$$= m \sum_{i=\max(0,r-m+1)}^{\min(r,m)} \sum_{j=s}^{s} (P_i - Q_j) \times (P_{r-i+1} - P_{r-i}) \frac{\binom{m}{i}\binom{m-1}{r-i}}{\binom{2m-1}{r}} \frac{\binom{n}{j}}{\binom{n}{s}},$$

$$= m \sum_{i=\max(0,r-m+1)}^{\min(r,m)} (P_i - Q_s) \times (P_{r-i+1} - P_{r-i}) \frac{\binom{m}{i}\binom{m-1}{r-i}}{\binom{2m-1}{r}},$$

$$r \in [0, 2m-1], \; s \in [0, n].$$

Elementary Theory of Del Pezzo Surfaces

Josef Schicho

Radon Institute for Computational and Applied Mathematics,
Austrian Academy of Sciences
Josef.Schicho@oeaw.ac.at

Abstract. Del Pezzo surfaces are certain algebraic surfaces in projective n-space of degree n. They contain an interesting configuration of lines and have a rational parametrization. We give an overview of the classification with an emphasis on algorithmic constructions (e.g. of the parametrization), on explicit computations, and on real algebraic geometry.

1 Introduction

This paper is elementary in the sense that it does not use the concepts and terminology of modern algebraic geometry, such as sheaves, schemes, divisors, or vector bundles. My personal opinion is that these concepts belong more to the "algebraic" than to the "geometric" part of "algebraic geometry", and the goal was to write an introduction to Del Pezzo surfaces for geometers and not for algebraists. This is also the reason why the paper is of survey type, but it cannot be used as an introduction to the modern theory of Del Pezzo surfaces. From that point of view, the main interest in Del Pezzo surfaces is related to birational classification of algebraic varieties of higher dimension (e.g. Calabi-Yau threefolds) or to arithmetic questions, and these relations are not even touched upon here. Our main intention was to collect material about this classical topic which could be of some interest to applied geometers. The main emphasis has been put on algorithmic techniques and on examples. For this reason, it would have been more justified to give the title "a very biased look at Del Pezzo surfaces".

The paper does assume a good familiarity with projective geometry, and the described algorithmic techniques can only be carried out if one can solve systems of algebraic equations in several unknowns.

The definition of Del Pezzo surfaces given in sect. 4 is not the usual one (which uses canonical divisors), but it follows Del Pezzo [5], who encounters this class of surfaces in his investigation of surfaces of degree n in $\mathbb{P}^n$. In the course of arriving at this definition, we give some theorems (Theorem 2, Theorem 5, Theorem 8, and Theorem 11) and occasionally proofs. Of course, these theorems are classical facts whose origins date back by centuries. A proof of Theorem 5 can be found in [7].

The unprojection algorithm in sect. 5 is original. Its advantage is that it makes a uniform treatment of parametrization algorithms (see sect. 6) possible.

The classification of Del Pezzo surfaces in sect. 6 is due to [5]; a modern proof can be found in [12]. No proof is contained in this paper because it would be too long and too technical. A complete elementary proof of Theorem 17 would also be surprisingly complicated because the finiteness of resolution of singularities is not a priori clear. Of course, the theorem also follows from the classification given in [5].

In the chosen approach, Del Pezzo surfaces of degree 2 and 1 are certainly unnatural (they also do not arise in [5]). But as early as in [10], these cases are discussed together with the other Del Pezzo surfaces, in the context of the classification of linear systems of elliptic curves in the plane. Theorem 24 and Theorem 28 can be found in [4].

The real classification of Del Pezzo surfaces, especially Theorem 30, is due to [3]. Modern treatments can be found in [20, 15, 21]. The technique used in example 35 to compute an improper parametrization (see also remark 36) is also mentioned in [3, 13, 17].

The author was supported by the Austrian science fund (FWF) in the frame of the special research area SFB 013 and of project P15551.

2 Projective Varieties, Degree, and Projection

Let $\mathbb{P}^n$ denote complex projective space of dimension n. Let $X \subset \mathbb{P}^n$ be a projective algebraic variety, i.e. the zero set of a homogeneous prime ideal. The *dimension* of X can be defined as the smallest integer m such that there exists an $n-m-1$-dimensional linear subspace disjoint from X. A generic linear subspace of dimension $n-m$ intersects X in a finite number of points. If we count with multiplicities, then this number depends only on X, and this is a way to define the *degree* of X (following [8]).

Let $p \in \mathbb{P}^n$ be a point, e.g. $p = (x_0 : \ldots : x_n) = (1:0:\ldots:0)$ (the affine origin). Let H be a linear hyperplane not containing p, e.g. the plane $x_0 = 0$. The projection $\pi_{p,H}$ with center p onto H is defined for all points except p. In the example, this is just the omission of the first projective coordinate x_0. - Let Y be the closure of the image of X under this projection. It is again a projective variety. Its dimension is either m or $m-1$. The second is the case if and only if X is a cone and p is its vertex, or X is a linear space and p is a point on X.

Remark 1. The choice of H is not essential. A different choice leads to another projective image Y' which is projectively equivalent to Y. In the following, we will often omit any explicit references to H.

If $\dim(Y) = \dim(X) = m$, then there is a positive integer f such that the preimage of a generic point of the projection map $\pi_p : X \to Y$ has f points. In case $f = 1$, then $\pi_p : X \to Y$ is birational. The number f is called the *tracing index* of the projection.

A generic linear $n-m-1$-subspace L of H intersects Y in $\deg(Y)$ points. The linear span of L and p intersects X in $f \cdot \deg(Y)$ points plus an intersection at p, that has to be counted with multiplicity

$$r := \deg(X) - f \cdot \deg(Y). \tag{1}$$

The number r is also called the *multiplicity* of X at p, and p is also called an r-fold point of X. Nonsingular points have multiplicity 1, and points outside X have multiplicity 0.

Theorem 2. *Let $X \subset \mathbb{P}^n$ be a projective variety of dimension m and degree d. Assume that X is not contained in a proper linear subspace. Then $d \geq n - m + 1$.*

Proof. We proceed by induction on n, fixing m. If $n = m$ (obviously the smallest possible value for n), then $X = \mathbb{P}^m$ and $d = 1$. The inequality is fulfilled.

Assume $n > m$. Let p be a nonsingular point of X. Let $Y \subset \mathbb{P}^{n-1}$ be the image of X under the projection from p. If Y were contained in a proper linear subspace L, then X would be contained in the linear span of L and p, contradicting the assumption. Therefore Y is not contained in a proper linear subspace.

Let f be the tracing index of the projection. Then

$$d = f \cdot \deg(Y) + 1 \geq f \cdot (n - m) + 1 \geq n - m + 1,$$

where the first inequality is a consequence of the induction hypothesis.

Remark 3. A closer look at the proof reveals that if equality holds, then the variety is rational (i.e. birationally equivalent to a projective space). Indeed, in this case the tracing index is always 1 in each projection step, so that we get a birational map from X to $\mathbb{P}^m$.

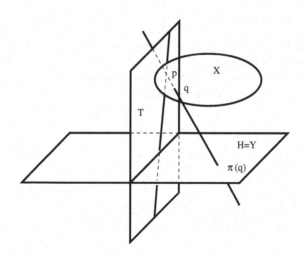

Fig. 1. Projection from a nonsingular point

Projections from nonsingular points are of special interest. Let $X \subset \mathbb{P}^n$ be a variety of dimension m, and let p be a nonsingular point of X. The projection $\pi_p : X \to Y$ is not defined at p. But for any differentiable curve $C : [0, 1] \to X$ with $C(0) = p$, the limit $\lim_{t \to 0} \pi_p(C(t))$ exists and lies on Y. The set of all these limits is equal to the intersection of the tangent space T of X at p with the projection hyperplane H (see fig. 1).

Conversely, assume that we have a variety $Y \subset \mathbb{P}^n$ of dimension m and an $m - 1$-dimensional linear subspace L on it. Can we construct a variety $X \subset \mathbb{P}^{n+1}$ and a nonsingular point $p \in X$ such that Y is the image of X under the projection by p? We will give partial answers to this question below.

3   Varieties of Minimal Degree

Let $C \subset \mathbb{P}^n$ be a curve of degree n. (From now on, a statement such as $C \subset \mathbb{P}^n$ implicitly implies the assumption that C is not contained in a proper linear subspace.) By remark 3, C is rational. Therefore, C has a parametrization $(P_0(t):\ldots:P_n(t))$ with polynomials $P_0,\ldots,P_n$ of degree at most n (and the maximum is reached by at least one of the P_i).

Since C is not contained in a proper linear subspace, the P_i are linearly independant. But the vector space of all polynomials of degree at most n has dimension $n+1$, and so the P_i form a basis. We can apply a projective transformation in order to transform this basis into the standard basis $P_i = t^i$, $i = 0,\ldots,n$. This implies that, up to projective transformations, there is precisely one curve $C \subset \mathbb{P}^n$ of degree n, which is also called the *rational normal curve* of degree n.

Remark 4. The Steiner construction (see [7], p. 528–533), shows that for any $n + 3$ points in general position, there is a unique rational normal curve passing through them.

For surfaces, we have a similar classification (see [7], p. 525).

Theorem 5. *Let $S \subset \mathbb{P}^n$ be a surface of degree $n-1$. Then S is either a rational scroll $R_{n,r}$ with parametrization $(1{:}t{:}\ldots{:}t^{n-r-1}{:}s{:}st{:}\ldots{:}st^r)$ for some $r \leq \frac{n-1}{2}$ (up to projective transformation), or $n = 5$ and S is the Veronese surface V with parametrization $(1{:}t{:}t^2{:}s{:}st{:}s^2)$ (up to projective transformation).*

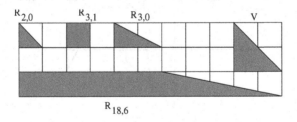

Fig. 2. Surfaces of minimal degree in lattice representation

Remark 6. Note that in both cases, the surface S is *toric*, i.e. parametrized by monomials. Toric surfaces have recently been used in [11] in order to generate multi-sided surface patches; they can be represented by lattice polygons in an obvious way. Actually, the surfaces of minimal degree are precisely the toric surfaces associated to a polygon without interior lattice points (see fig. 2).

Clearly, $R_{2,0}$ is the projective plane, and $R_{n,0}$ is the cone over the rational normal curve of degree $n - 1$. The quadric surface $R_{3,1} \subset \mathbb{P}^3$ has two rulings (families of lines), namely the ruling of lines given by $s = $ constant and the ruling of lines given by $t = $ constant.

Remark 7. It is easy to see that the rational scroll $R_{n,r}$ is a projection of $R_{n+1,r}$ – just omit the coordinate corresponding to t^{n-r}. The Veronese surface is not the projection of a surface in $\mathbb{P}^6$ of minimal degree, because it does not contain any lines.

4 Curves of Almost Minimal Degree

We say that a variety $X \subset \mathbb{P}^n$ has *almost minimal degree* if $\deg(X) = n - \dim(X) + 2$.

For any $n \geq 2$, we can produce almost minimal curves $C \subset \mathbb{P}^n$ by "spoiling rational normal curves". Take a rational normal curve $C' \subset \mathbb{P}^{n+1}$ of degree $n + 1$, and a point p outside C. Let C be the image of C' under the projection from p. Since p is a point of multiplicity 0, the degree of C is a divisor of $n + 1$. By Theorem 2, the degree is greater than or equal to $n + 1$, therefore it is equal to $n + 1$.

Conversely, any rational curve of almost minimal degree $C \subset \mathbb{P}^n$ is a spoiled rational normal curve. To show this, we write down a parametrization $(P_0(t) : \cdots : P_n(t))$ be a sequence of polynomials with maximal degree $n + 1$. Let $Q(t)$ be a polynomial of degree at most $n + 1$ that is linearly independent of $P_0, \ldots, P_n$. Then the curve defined by $(P_0(t) : \ldots : P_n(t) : Q(t))$ is a rational normal curve, and C is the image of the projection from $(0 : \ldots : 0 : 1)$. (This point must be a point outside the rational normal curve because its multiplicity is zero by the degree formula 1.)

There are also irrational curves of almost minimal degree; the first examples are the nonsingular cubic plane curves. It is well-known that the nonsingular plane cubic curves are *elliptic*, i.e. of genus one. Here is a general theorem on irrational curves of almost minimal degree.

Theorem 8. *Let $C \subset \mathbb{P}^n$ be an irrational curve of almost minimal degree. Then C is elliptic and nonsingular.*

Proof. We proceed by induction. If $n = 2$, it suffices to state that cubic plane curves are either rational or elliptic, and the elliptic ones are nonsingular.

Let $n \geq 3$. Let p be a nonsingular point on C. Let C' be the image of C under projection from p. Let $d := \deg(C')$. Then $d \mid n$ by the degree formula 1, and $d \geq n - 1$ by Theorem 2. This implies that $d = n$, i.e. C' has almost minimal degree, and the projection gives a birational map $C \to C'$. By induction hypothesis, C' is elliptic. Since the genus is a birational invariant, C is also elliptic.

In order to show that C is nonsingular, let q be an arbitrary point of C, and let r be its multiplicity. Let D be the image of C under projection from q. Then $\deg(D)$ is a divisor of $n + 1 - r$, which is greater than or equal to $n - 1$. This leaves only the cases $r = 1$ and $\deg(D) = n$, or $r = 2$ and $\deg(D) = n - 1$. In both cases, the tracing index of the projection must be one, so that C and D are birationally equivalent. But this rules out the second case, because D would then have minimal degree and therefore be rational. Hence $r = 1$, and we showed that C has only points with multiplicity one.

Example 9. For any $n \geq 2$, we have an elliptic curve $C \subset \mathbb{P}^{n+1}$ of almost minimal degree. Here is an example for $n = 3m - 1$.

Let C be the plane cubic with equation $x^3 + y^3 + z^3 = 0$. Let $f : \mathbb{P}^2 \to \mathbb{P}^n$ be the embedding given by

$$(x{:}y{:}z) \mapsto (x^m{:}\ldots{:}y^m{:}x^{m-1}z{:}\ldots{:}y^{m-1}z{:}x^{m-2}z^2{:}\ldots{:}y^{m-2}z^2).$$

The image of C is of degree $3m = n + 1$.

For all other n, examples can be constructed by one or two steps of point projection of the above example.

Remark 10. In general, it is not true that the absence of singularities of a curve of almost minimal degree implies that the curve is elliptic. An example of a spoiled rational normal curve without singularities is the "twisted quartic" in $\mathbb{P}^3$ with parametrization $(1:t:t^3:t^4)$.

5   Surfaces of Almost Minimal Degree

We can produce surfaces of almost minimal degree by spoiling surfaces of minimal degree, as we did in the previous section for curves. These surfaces are rational, and the projections of rational scrolls are ruled surfaces.

As a base for some proofs on induction, we need to have a rough classification of the cubic surfaces in $\mathbb{P}^3$. We distinguish the following types.

1. Cubic surfaces with a double line. These are the projections of cubic rational scrolls in $\mathbb{P}^4$.
2. Cones over nonsingular cubic plane curves. These are irrational. They have a triple point and no other singularities.
   (Note that the cones over singular cubic plane curves are already falling into type 1 above.)
3. Cubic surfaces with isolated double points. These are rational. Indeed, projection from a double point gives a birational map onto $\mathbb{P}^2$.
4. Nonsingular cubic surfaces. These are also rational.

A much finer classification can be found in [2, 1].

Type 2 can easily be generalized to arbitrary dimension: the cone over an elliptic curve of almost minimal degree is an irrational surface of almost minimal degree. It is well-known [5, 6] that every irrational surface of almost minimal degree is a ruled surface with elliptic base.

We define a *Del Pezzo surface* as a rational surface of almost minimal degree that is not a spoiled surface of minimal degree. The cubic Del Pezzo surfaces are the surfaces of type (3) and (4) above.

Theorem 11. *Let S be a Del Pezzo surface.*
a) S has at most isolated double points.
b) If S has degree at least 4, then the image of S under projection from a nonsingular point $p \in S$ is a Del Pezzo surface.
c) A generic hyperplane section of S is an elliptic curve of almost minimal degree.
d) The number of lines on S is finite.

Proof. (c): It is obvious that the generic hyperplane section has almost minimal degree. They are not rational, because then the surface would be a spoiled minimal surface (this is a consequence of the discussion of surfaces with rational hyperplane sections in [4]). Hence they are elliptic.

(b): Let S' be the image of the projection. By the degree formula 1, S' has almost minimal degree and is birationally equivalent to S. The hyperplane sections are projections from intersections of S with hyperplanes through p. Because of (c), these are elliptic curves. So, S' is rational and has generic hyperplane sections of genus one. On the other hand, S' cannot be a spoiled surface of minimal degree, because these have generic hyperplane sections of genus zero. Hence S' is a Del Pezzo surface.

(a): By the degree formula 1, S cannot have points of multiplicity 3 or more. We prove that the number of double points is finite, by induction on the degree. For degree 3, this follows from the classification of cubic surfaces above. For $n > 3$, choose a nonsingular point and project; the image is again a Del Pezzo surface S' of degree $n-1$, by (b). Therefore S' has only finitely many double points, by the induction hypothesis. It follows that the number of double points on S is also finite, since the image of a double point is a double point.

(d): We proceed by induction. For degree 3, it is well-known that any cubic surface of type (3) or (4) has only finitely many lines. For $n \geq 4$, assume indirectly that S has infinitely many lines. Let p be a nonsingular point on S. Since S is not a cone with vertex p, there are only finitely many lines through p. Hence there remain infinitely many lines on the image S' of the projection from p. But S' is a Del Pezzo surface, contradicting the induction hypothesis.

The lines on a Del Pezzo surface are interesting for several reasons. One of them is that they can be used to construct a Del Pezzo surface of degree one higher which projects to the given Del Pezzo surface.

Here is an explicit *unprojection algorithm*. It assumes that we have given a Del Pezzo surface $S \subset \mathbb{P}^n$ and a line l lying on S.

1. Choose a generic linear form $L(x_0, \ldots, x_n)$ vanishing on l.
2. Compute the intersection of the hyperplane defined by L with S. As we will show in Theorem 14, it consists of two components: the line l and a rational normal curve C of degree $n - 1$.
3. Choose a generic quadratic form $Q(x_0, \ldots, x_n)$ vanishing on C. (We will show in Theorem 14 that there exist such quadratic forms.)
4. Compute the image of S under the map given by

$$(x_0 : \ldots : x_n) \mapsto \left(x_0 : \ldots : x_n : \frac{Q(x_0, \ldots, x_n)}{L(x_0, \ldots, x_n)} \right).$$

Example 12. Let $S \subset \mathbb{P}^3$ be the surface given by

$$3x_0 x_1^2 + 3x_0 x_2^2 + 3x_0 x_3^2 - 3x_0^3 - 10 x_1 x_2 x_3 = 0.$$

This cubic has 27 lines on it (see fig. 3). Let l be the line $x_0 = x_3 = 0$.

We choose the linear form $L := x_3$. It intersects S in l and in the plane conic C defined by $x_3 = x_1^2 + x_2^2 - x_0^2 = 0$.

Now we choose the quadric $Q := x_1^2 + x_2^2 - x_0^2$.

To compute the image of the map defined in the unprojection algorithm, we introduce a new variable x_4. The equation $L x_4 - Q = 0$ holds on the image. A second

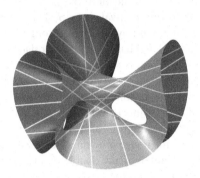

Fig. 3. A cubic Del Pezzo surface with 27 real lines (picture courtesy of O. Labs)

equation can be found by writing the equation of S as linear combination of L and Q and dividing by L, replacing Q/L by the new variable x_4:

$$\frac{3x_0(x_1^2 + x_2^2 - x_0^2) + (3x_0x_3 - 10x_1x_2)x_3}{x_3} = 3x_0x_4 + 3x_0x_3 - 10x_1x_2 = 0.$$

The image is the intersection of these two quadratic forms in $\mathbb{P}^4$, which is indeed a surface of degree 4.

Remark 13. How do we know whether our choice of the linear or quadratic form in steps 1 and 3 were general enough? In practice, the best strategy is just to try an arbitrary one. There is a chance that the choice does not work, but the bad choices are of measure zero in the set of all choices.

Theorem 14. *The unprojection algorithm is correct.*

Proof. We begin by proving the statement claimed in step 2: the hyperplane defined by L intersects S in l and a rational normal curve of degree $n - 1$. In fact, it is clear that l is a component of the intersection, and that the degrees of the remaining irreducible components add up to $n - 1$, but we have to show that there is only one remaining component.

Let p be a nonsingular point on l. The projection from p is a Del Pezzo surface S' by theorem 11. The line l projects to a point $q \in S'$, which is either a single or a double point (in fact, it is always a double point, as we will see in remark 15 below). Projection from q is birational by the degree formula 1: let S'' be the image. The remaining components project to a generic hyperplane section of S''. By Bertini's theorem (see [9], Thm. 8.18, p. 179; Rem. 8.18.1, p. 180), generic hyperplane sections are irreducible, and the statement is proven.

The ideal of the rational normal curve C is generated by quadratic forms. Therefore the generic quadratic form Q through C does not vanish identically on the line l. Consequently, Q is not contained in the vector space generated by the quadratic multiples of L and the quadratic forms vanishing on S. This implies that the image S_0 of the unprojection constructed in step 4 is not contained in a linear subspace.

The degree of S_0 is the number of intersections of S' with two generic hyperplanes H_1, H_2. We can assume that the form defining H_1 does not contain the new variable x_{n+1} (by linear algebra). It defines a hyperplane $H_3 \subset \mathbb{P}^n$. The intersection points of H_1 and H_2 and S_0 correspond to the intersection points of S and H_3 and some quadric surface Q_0, which we get when we multiply the equation of H_2 by L and replace L times the new variable by Q, minus the intersection points of S, L, and H_3. This number is $2n - (n-1) = n+1$. Therefore S_0 is a surface of almost minimal degree.

There is an obvious projection from S_0 to S (omitting the last coordinate). The center is a nonsingular point, by the degree formula 1. Because S_0 is rational and is not a spoiled surface of minimal degree, S_0 is a Del Pezzo surface.

Remark 15. Revisiting the above proof again, we can now show that if p is a nonsingular point lying on a line l contained in S, then the image q of l under the projection is a double point on the image S' of S. Let S'' be the image of the projection from q. The generic hyperplane section of S'' is a birational image of the rational normal curve which forms together with L the intersection of S with a general hyperplane through L. Hence S'' is not a Del Pezzo surface, and q cannot be a nonsingular point.

6 Classification of Del Pezzo Surfaces

For the theory of Del Pezzo surfaces, the techniques of projection and unprojection are very useful because they allow induction proofs (upward and downward). We can draw an (infinite) directed graph of all Del Pezzo surfaces up to projective transformations, with an edge from S_1 to its images under projections from nonsingular points. The natural question arises: is this graph connected?

It is clear that it would suffice to show that there is a path connecting any two cubic Del Pezzo surfaces, because we can always do projection steps down to degree 3, and these are the minimal vertices of the graph.

Another possible approach is to locate the maximal vertices of the graph.

Theorem 16. *Let S be a Del Pezzo surface without a line. Then S is one of the following three surfaces:*

1. *the nonsingular surface $F_9 \subset \mathbb{P}^9$ with parametrization $(1{:}s{:}t{:}s^2{:}st{:}t^2{:}s^3{:}s^2t : st^2{:}t^3)$;*
2. *the nonsingular surface $F_8 \subset \mathbb{P}^8$ with parametrization $(1{:}s{:}s^2{:}t{:}st{:}s^2t{:}t^2{:}st^2 : s^2t^2)$;*
3. *the surface $G_8 \subset \mathbb{P}^8$ with parametrization $(1{:}s{:}s^2{:}s^3{:}s^4{:}st{:}s^2t{:}s^3t{:}s^2t^2)$, which has a double point at $(0{:}\ldots{:}0{:}1)$.*

For the proof, which is beyond the scope of this paper, we refer to [12] or [18].

Theorem 17. *Every sequence of successive unprojections terminates.*

If the sequence contains a nonsingular surface, then all subsequent unprojections are also nonsingular, because we cannot get rid of double points by projection. Then it is also clear that the sequence terminates, because by unprojecting nonsingular Del Pezzo surfaces we cannot create new lines (as lines always project to lines), but we will erase at least one line. This follows from the fact that all lines not passing through the center of projection are also there on the image of projection. If the image is nonsingular, then there is no line passing through the center, because such a line would project to a double point (see remark 15).

Unprojection can create new lines if the exceptional line contains double points. We do not give a termination proof for this case, because this would require a deeper analysis of the type of double points of Del Pezzo surfaces. For a full proof of termination (using a different approach), we refer to [18].

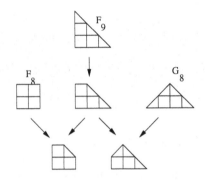

Fig. 4. A piece of the graph of Del Pezzo surfaces

Corollary 18. *The graph of Del Pezzo surfaces is connected. For every Del Pezzo surface except F_8 and G_8, there is a sequence of unprojections terminating with F_9.*

Proof. To show that the graph is connected, it suffices to show that there are paths of projections and unprojections connecting F_9, F_8, and G_8. These paths are shown in fig. 4. The lattice polygons represent monomial parametrizations of Del Pezzo surfaces (see also remark 6).

The surfaces F_8 and G_8 have a group of projective automorphisms that acts transitively on the nonsingular points. Therefore, there is, up to projective isomorphism, just one projection image of F_8 and one from G_8, namely those depicted in fig. 4.

Let S be a Del Pezzo surface different from F_8 and G_8. Then there is a sequence of unprojections terminating with F_9, F_8, or G_8. If it terminates with F_8 or G_8, then the last but one surface can also be unprojected to F_9, as can be seen in fig. 4.

Corollary 19. *Every Del Pezzo surface $S \subset \mathbb{P}^n$ except F_8 and G_8 has a parametrization by cubic polynomials through $9 - n$ base points (infinitely near base points are allowed).*

Proof. Every such Del Pezzo surface is a projection from S_9, which is parametrized by cubic polynomials. If S has degree n, then we need n projection steps, each introducing one base point.

Here is an algorithm to construct a parametrization of an implicitly given Del Pezzo surface. It assumes that we have given a Del Pezzo surface $S \subset \mathbb{P}^n$ in implicit form.

1. Reduce to the case $n = 4$ by some projection or unprojection steps.
2. Compute a line on S.
3. Project from the line. This is a birational map onto $\mathbb{P}^2$.
4. Compute the inverse of the map.
5. Reverse the projection/unprojection steps from step 1.

Example 20. Let S be the cubic surface from example 12. We already did the unprojection to a surface $S_0 \subset \mathbb{P}^4$ with equations

$$x_3x_4 - x_1^2 - x_2^2 + x_0^2 = 3x_0x_4 + 3x_0x_3 - 10x_1x_2 = 0.$$

The surface S_0 contains the line $(3{:}3{:}3p{:}9p{:}p)$. The projection from this line is given by $(x_0{:}\ldots{:}x_4) \mapsto (x_0 - x_1 : 3x_2 - x_3 : x_2 - 3x_4)$.
   We do the linear coordinate change

$$(x_0, \ldots, x_4) = (y_3, y_0 + y_3, y_2 + 3y_4, y_1 + 9y_4, y_4)$$

in order to move l to a coordinate subspace. The transformed system is

$$y_1y_4 - y_0^2 - 2y_0y_3 - y_2^2 - 6y_2y_4 = 3y_1y_3 - 10y_0y_2 - 30y_0y_4 - 10y_2y_3 = 0.$$

This is a linear system for y_3, y_4. The inverse of the projection is given by the solution to this system. The parametrization of S_0 can then easily be computed by plugging into the above change of coordinates. The parametrization of S is then computed even more easy, we just have to truncate the last coordinate function.

Remark 21. In steps 1 and 2, we need some line on the surface. This can be done by plugging the parametrization of a general line into the equations and solving for the coefficients of the general line. It pays off to first project the surface into 3-space before, because this reduces the number of unknowns.
   For the inversion of a birational map, we refer to [16].

Remark 22. For nonsingular Del Pezzo surfaces of degree less than or equal to 7, the number of lines depends only on the degree (e.g. nonsingular cubic surfaces have 27 lines). The incidence graph of the lines is also determined by the degree. See [12] for details.

7   Del Pezzo Surfaces of Degree 2 and 1

In order to describe Del Pezzo surfaces of degree 2 and 1, we need to introduce a generalization of projective spaces, namely weighted projective spaces (see also the short note [14]).

Let $w := (w_0, \ldots, w_n)$ be a vector of positive integers. *Weighted projective space* $\mathbb{P}_w$ is defined as the quotient of $\mathbb{C}^{n+1} - \{0\}$ by the equivalence relation identifying $(x_0, \ldots, x_n)$ with $(\lambda^{w_0} x_0, \ldots, \lambda^{w_n} x_n)$, for any $\lambda \in \mathbb{C}^*$. This is an algebraic variety of dimension n. If $w = (1, \ldots, 1)$, then $\mathbb{P}_w$ is just $\mathbb{P}^n$.

Weighted projective varieties are algebraic subvarieties of $\mathbb{P}_w$. They are defined by *weighted homogeneous* polynomials. The weigthed degree of a monomial summand $x_0^{e_0} \ldots x_n^{e_n}$ is defined as $\sum_i w_i e_i$, and a polynomial is weighted homogeneous iff all its monomials have the same weighted degree.

For any projective variety $X \subset \mathbb{P}^n$, the Hilbert function $H : \mathbb{N} \to \mathbb{N}$ is defined by setting $H(m)$ as the dimension of the quotient vector space of all forms of degree m in $x_0, \ldots, x_n$ modulo the vanishing ideal of X. For large m, the function H is a polynomial. Its degree is the dimension of X. If the dimension is r, then the leading coefficient of the Hilbert polynomial times $r!$ is equal to the degree of X (see [9]).

Using the Hilbert function, we can define the degree also for varieties in weighted projective spaces. It is natural to define that a surface S has almost minimal degree d if the leading coefficient of the Hilbert polynomial is $\frac{d}{2}$, and the value of the Hilbert function at $m = 1$ is $d + 1$. (In the case of ordinary projective space, this is equivalent to saying that S has degree d and is contained in $\mathbb{P}^d$ but not in a linear subspace.)

When we add the restrictions that S is rational and not spoiled, we have defined weighted Del Pezzo surfaces. It turns out that we get two new types of Del Pezzo surfaces, namely those of degree 2 and those of degree 1.

By definition, a Del Pezzo surface of degree 2 is a surface $S \in \mathbb{P}_{1,1,1,2}$ defined by a polynomial F of weighted degree 4, subject to the following conditions. We can write $F(x_0, x_1, x_2, y)$ as $cy^2 + F_2(x_0, x_1, x_2)y + F_4(x_0, x_1.x_2)$ for a suitable constant c and polynomials F_2, F_4 of degree 2 and 4, and we define the quartic polynomial $D(x_0, x_1, x_2) := \mathrm{disc}_y(F) = F_2^2 - 4cF_4$.

1. The discriminant D is squarefree. (One can show that otherwise S is a spoiled surface of minimal degree.)
2. The discriminant D has no four-fold point. This just excludes 4 lines meeting in a point. (One can show that otherwise S is not rational.)

Let S be a Del Pezzo surface of degree 2. The projection onto the first three projective coordinates projects S onto $\mathbb{P}^2$. This map is actually defined everywhere (because the point $(0{:}0{:}0{:}1)$ does not lie on S), and has tracing index 2. The inverse image of a line $l \subset \mathbb{P}^2$ is in general an elliptic curve on S. If l is a tangent to the discriminant curve $D = 0$, then the inverse image is rational. If l is a bitangent, i.e. l is tangent at two points, then the inverse image has two components, both of which are rational. In such a case, the two components are called *pseudo-lines*. They play a similar role as the lines of Del Pezzo surfaces in ordinary projective space.

Any pseudo-line l can be defined by a linear equation $L(x_0, x_1, x_2) = 0$ and a weighted quadratic equation of type $y - Q(x_0, x_1, x_2) = 0$. The unprojection is given as the image of S under the rational map defined by $\left(x_0, x_1, x_2, \frac{y-Q}{L}\right)$.

Example 23. Let S be the Del Pezzo surface given by the equation

$$y^2 - 10x_1 x_2 y + 9x_0^2 x_1^2 + 9x_0^2 x_2^2 - 9x_0^4 = 0$$

in $\mathbb{P}_{1,1,1,2}$. The discriminant is $100x_1^2x_2^2 - 36x_0^2x_1^2 - 36x_0^2x_2^2 + 36x_0^4$. The line $x_0 = 0$ is a bitangent. For computing the inverse image, we set x_0 to 0 and get the equation $y^2 - 10x_1x_2y$, which factors into $y(y - 10x_1x_2)$. Each of the two factors give one pseudo-line.

We use the pseudo-line $x_0 = y = 0$ for unprojection. The unprojection map is $\left(x_0 : x_1 : x_2 : \frac{y}{3x_0} \right)$, and the image is the cubic surface

$$3x_0x_1^2 + 3x_0x_2^2 + 3x_0x_3^2 - 3x_0^3 - 10x_1x_2x_3 = 0.$$

Projection from a nonsingular point of a cubic Del Pezzo surface $S_0 \subset \mathbb{P}^3$ (say the point $(0{:}0{:}0{:}1)$) is more than just omitting the last coordinate: we also need to give a value for the additional coordinate y of weight 2. This value is not uniquely determined. It is the product of x_3 with the leading coefficient of the cubic equation with respect to x_3 (which is a linear polynomial because $(0{:}0{:}0{:}1)$ is a nonsingular point), plus an arbitrary quadratic form in x_0, x_1, x_2.

Theorem 24. *Let $S \subset \mathbb{P}_{1,1,1,2}$ be a Del Pezzo surface of degree 2. Then S has a parametrization with the first three coordinate functions being cubics through 7 base points, and the fourth coordinate function being a sextic vanishing doubly at the 7 base points.*

Proof. Every quartic has a bitangent. So, take one, and use one of the two pseudo-lines in the preimage for unprojection. Let S_0 be the resulting cubic Del Pezzo surface. By Corollary 19, S_0 has a parametrization $(C_0{:}C_1{:}C_2{:}C_3)$ by cubic through 6 base points $p_1, \ldots, p_6$. By projection, we introduce an additional base point p_7. The first three coordinate functions C_0, C_1, C_2 (which are part of the parametrization of S) pass also through p_7. The fourth component of the parametrization of S can be computed as $F := L(C_0, C_1, C_2)C_3 + Q(C_0, C_1, C_2)$, where L is the equation of the tangent plane to the projection center $(0{:}0{:}0{:}1)$, and Q is an arbitrary quadratic form. Hence F has degree 6, and vanishes doubly at $p_1, \ldots, p_6$. But $L(C_0, C_1, C_2)$ vanished doubly at p_7, therefore F also has a double point at p_7.

Remark 25. A nonsingular plane quartic has exactly 28 bitangents. Because a Del Pezzo surface of degree 2 is nonsingular iff its discriminant is nonsingular, we see that the number of pseudo-lines on a nonsingular Del Pezzo surface of degree 2 is 56.

Let us now turn to Del Pezzo surfaces of degree 1. By definition, this is a surface $S \in \mathbb{P}_{1,1,2,3}$ defined by a polynomial F of weighted degree 6, subject to the following conditions. We can write $F(x_0, x_1, y, z)$ as $c_1z^2 + c_2y^3 + F_1yz + F_2y^2 + F_3z + F_4y + F_6$ for a suitable constants c_1, c_2 and polynomials F_1, F_2, F_3, F_4, F_6 in x_0, x_1 of degree 1, 2, 3, 4, 6, and we define the polynomial $D(x_0, x_1, y) := \mathrm{disc}_z(F)$ (a weighted polynomial of degree 6).

1. The discriminant D is squarefree, and $c_2 \neq 0$. (One can show that otherwise S is a spoiled surface of minimal degree.)
2. The discriminant D has at most one triple point. (One can show: if D is squarefree, and $c_2 \neq 0$, then it has at most two triple points, and if it has two triple points, then S is not rational.)

Similar as for Del Pezzo surfaces of degree 2, chopping of the coordinate z gives a rational map of tracing index 2. The image is the weighted projective plane $\mathbb{P}_{1,1,2}$. There are two kinds of pseudo-lines. When the inverse image of a curve of weighted degree 2, not passing through the point $(0{:}0{:}1)$, splits into two components, both of them are pseudo-lines of the first kind. The second type arises as the inverse image of a curve of weighted degree 1, if this inverse image contains a singular point.

Example 26. Let $S \subset \mathbb{P}_{1,1,2,3}$ be given by the equation

$$z^2 - y^3 - x_0^4 x_1^2 - 2x_0^3 x_1^3 - x_0^2 x_1^4 = 0.$$

The inverse image of $y = 0$ splits into two pseudo-lines $y = z \pm (x_0^2 x_1 + x_0 x_1^2) = 0$ of the first kind.

The unprojection map with respect to one of them is

$$(x_0{:}x_1{:}y{:}z) \mapsto \left(x_0{:}x_1{:}\frac{z + x_0^2 x_1 + x_0 x_1^2}{y}{:}y \right).$$

Its image is the surface in $\mathbb{P}_{1,1,1,2}$ with equation

$$(x_3 y - 2x_0^2 x_1 - 2x_0 x_1^2)x_3 - y^2 = 0.$$

Example 27. Let $S \subset \mathbb{P}_{1,1,2,3}$ be the surface in example 26. The point $p := (1{:}0{:}0{:}0)$ is a double point of S. There is a unique form of weighted degree 1 vanishing at p, namely x_1. This gives the pseudo-line $z^2 - y^3 = x_1 = 0$. Its unprojection map is

$$(x_0{:}x_1{:}y{:}z) \mapsto \left(x_0{:}x_1{:}\frac{y}{x_1}{:}\frac{z}{x_1} \right),$$

and the equation of the image is

$$y^2 - x_2^3 x_1 - x_0^4 - 2x_0^3 x_1 - x_0^2 x_1^2 = 0.$$

Theorem 28. *Let $S \subset \mathbb{P}_{1,1,2,3}$ be a Del Pezzo surface of degree 1. Then S has a parametrization with the first two coordinate functions being cubics through 8 base points, the third coordinate function being a sextic vanishing doubly at the 7 base points, and the fourth coordinate function being a ninetic vanishing triply at the 7 base points.*

The proof is similar to the proof of Theorem 24.

Remark 29. The number of pseudo-lines on a nonsingular Del Pezzo surface of degree 1 is 240. See [12] for a proof.

The parametrization algorithm in sect. 6 can easily be generalized to Del Pezzo surfaces of degree 2 and 1. The so constructed parametrizations are of the type described in the theorems 24 and 28.

8   Real Del Pezzo Surfaces

If the system of equations defining a complex Del Pezzo surface are real numbers, then set of real solutions – if not empty – form a real algebraic surface, which we call a *real Del Pezzo surface*.

Projection from real nonsingular points and unprojection using real lines (or pseudo-lines in degree 2 or 1) works exactly as in the complex case. A new construction is the projection from a pair (p, p') of complex conjugate points. Both points must be nonsingular, and not lying on a common line on S. The result is over the complex numbers isomorphic to the result of two subsequent projections. The result can be realized as a real algebraic surface, because it is the projection from the line pp', and this is a real line.

Similarily, we have a new construction of unprojection using a pair of complex conjugate lines (or pseudo-lines). The two lines must not meet in a nonsingular point, because otherwise unprojection from one line would delete the other line.

Projection and unprojection are real birational maps. The number of connected components is invariant under real birational maps. But this number is not always the same for all real Del Pezzo surfaces. For instance, there are cubics with one component and cubics with two component. Other examples are given below. Therefore, the real graph of Del Pezzo surfaces is not connected.

Here is the classification of maximal vertices of this graph. The proof is again beyond the scope of this paper; we refer to [20].

Theorem 30. *Let S be a real Del Pezzo surface without a real (pseudo-)line and without a pair of complex conjugate (pseudo-)lines that do not intersect each other. Then S is one of the following.*

1. *one of the surfaces F_9, F_8, or G_8, appearing in the complex classification Theorem 16. All these surfaces have one component;*
2. *a surface in $\mathbb{P}^8$ with parametrization $(1{:}s{:}s^2{:}t{:}st{:}(s^2 + t^2)s{:}t^2{:}(s^2 + t^2)t{:}(s^2 + t^2)^2)$, which has one component;*
3. *a Del Pezzo surface of degree 4 with two components;*
4. *a Del Pezzo surface of degree 2 with three or four components;*
5. *a Del Pezzo surface of degree 1 with five components.*

Remark 31. It is easy to see that surface 2 in the above classification is isomorphic to F_8 over the complex numbers. Over the reals, they are not isomorphic. In order to see this, note that F_8 has two one-parameter-families of conics, setting either s or t to a constant parameter. But surface 2 has no real conic at all.

Example 32. Let $S \subset \mathbb{P}^4$ be the Del Pezzo surface

$$x_1^2 + x_2^2 - x_0^2 = x_3^2 + x_4^2 - x_1 x_2 = 0.$$

There are no real lines on S, and 8 complex lines. These are the lines $p_i q_j$, $i = 1, \ldots, 4$, $j = 1, 2$, where p_1, p_2, p_3, p_4 are the four real intersection points of the conic C : $x_1^2 + x_2^2 - x_0^2 = x_3 = x_4 = 0$ with the hyperplanes $x_1 = 0$ and $x_2 = 0$, and q_1, q_2

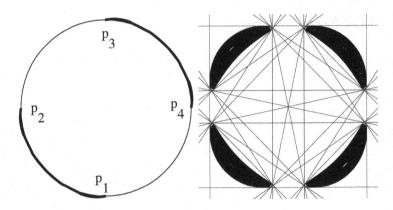

Fig. 5. *Left*: Planar picture of a Del Pezzo surface with 2 components. *Right*: A quartic curve with 28 real bitangents.

are the conjugate complex points $(0 : 0 : 0 : 1 : \pm i)$. For each $i = 1, \dots, 4$, the line $p_i q_1$ is conjugate to the line $p_i q_2$, and this pair of conjugates meets in p_i. Since p_i is a nonsingular point of S (the only singularities on S are q_1 and q_2), unprojection is not possible.

In order to see the two connected components, we project the surface onto the first three projective coordinates. The complex image is the conic C. The real image is the subset of points on the conic for which the form $x_1 x_2$ is positive or zero. This subset of conics has two components, namely the arc connecting $p_1 p_2$ and the arc $p_3 p_4$ in the notation as in fig. 5 (*left*).

Example 33. Let $F(x_0, x_1, x_2)$ be the quartic equation

$$F = 17(x_1^4 + x_2^4) + 30x_1^2 x_2^2 - 160(x_1^2 + x_2^2)x_0^2 + 380x_0^4,$$

and let S be the Del Pezzo surface with equation $y^2 + F$ in $\mathbb{P}_{1,1,1,2}$. The quartic F has 28 real bitangents (see fig. 5, right). Each preimage splits into pair of complex conjugate pseudo-lines, intersecting each other in two real points on the surface, namely the preimages of the tangential points. Therefore unprojection is not possible.

The surface has 4 components, each projecting onto one of the four components of the subset of the plane defined by $F \leq 0$ (the black regions in fig. 5, right).

Example 34. An example of a Del Pezzo surface with 5 components is given by the equation

$$x_0^6 + x_1^6 + 2(x_0^4 + x_1^4)y - 0.9x_0^2 x_1^2 y - y^3 + z^2 = 0$$

in $\mathbb{P}_{1,1,2,3}$.

Since a real parameterization of tracing index 1 is a real birational map, only the surfaces with one component may have such a parametrization. For all maximal vertices in the real graph of Del Pezzo surfaces that have only one component, we have given a parametrization in Theorem 30. It follows that a real Del Pezzo surface has a birational parametrization if and only if it has one component.

We can say a bit more: projection does not increase the degree of the parametrization; and for the maximal vertices, we have given parametrizations of degree 3 and 4. It follows that every real Del Pezzo surface with one component has a birational parametrization of degree 3 or 4.

For real Del Pezzo surfaces with two, three, or four components, one can construct parametrizations which are not birational. No such construction is known for Del Pezzo surfaces with five components. In particular, we do not know if the surface in example 34 has a real parametrization or not.

Example 35. In order to construct a parametrization of the surface S in example 32, we first give a parametrization of the arc p_1p_2 of the conic C:

$$(x_0:x_1:x_2) = \left(1:\frac{2(t^2+1)}{(t^2+1)^2+1}:\frac{(t^2+1)^2-1}{(t^2+1)^2+1}\right),$$

by composing a well-known parametrization of the conic with the function $t \mapsto t^2+1$. This is an algebraic way of restricting the parameter space to the interval $[1,\infty)$.

This parametrization is plugged into the equation $x_3^2 + x_4^2 - x_1x_2$, leaving the problem of parametrizing a circle with radius $\frac{2t^2(t^2+1)(t^2+2)}{((t^2+1)^2+1)^2}$. Such a parametrization can be computed by a projection from the point $\left(\frac{t(\sqrt{2}t^2-2)}{t^2+2t+2}, \frac{(\sqrt{2}+2)t^2}{t^2+2t+2}\right)$ to a line followed by an unprojection:

$$(x_0:x_3:x_4) = \left(1:\frac{t(-\sqrt{2}t^2+2-4ts-2ts\sqrt{2}+s^2\sqrt{2}\sqrt{2}t^2-2s^2)}{(t^4+2t^2+2)(1+s^2)}:\right.$$
$$\left.\frac{-t(-2t-\sqrt{2}t+2s\sqrt{2}t^2-4s+2ts^2+ts^2\sqrt{2})}{(t^4+2t^2+2)(1+s^2)}\right)$$

The computation was done with the help of the computer algebra system Maple. Concatenation of these two parametrizations gives a parametrization of S with tracing index 2.

Remark 36. The technique used in example 35 can be used to parametrize arbitrary Del Pezzo surfaces of degree 4 with two components (and therefore all Del Pezzo surfaces with two components, because we can reduce to degree 4 by unprojection): compute a projection with conic fibers, restrict the parameter space algebraically, parametrize the parametric family of conics. A similar technique can also be used for Del Pezzo surfaces with 3 components (see also [13, 17]).

In order to parametrize Del Pezzo surfaces with 4 components, it is theoretically possible to use the construction in [19] which works over arbitrary fields. Unfortunately, the so constructed parametrization has tracing index 24 and is very complicated. The author computed a parametrization of example 33 with this method, but the output fills several pages.

References

1. Bruce, J. W., and Wall, C. T. C. On the classification of cubic surfaces. *J. London Math. Soc. (2) 19* (1979), 245–256.

2. Cayley, A. A memoir on cubic surfaces. *Phil. Trans. Roy. Soc. London 159* (1869), 231–326.
3. Comesatti, A. Fondamenti per la geometria sopra le superficie rationali del punto di vista reale. *Math. Ann. 73* (1912), 1–72.
4. Conforto, F. *Le superficie razionali*. Zanichelli, 1939.
5. del Pezzo, P. On the surfaces of order n embedded in n-dimensional space. *Rend. mat. Palermo 1* (1887), 241–271.
6. Fujita, T. On the structure of polarized manifolds with total deficiency one. *J. Math. Soc. Japan 33* (1981), 415–434.
7. Griffiths, P., and Harris, J. *Principles of algebraic geometry*. John Wiley, 1978.
8. Harris, J. *Algebraic geometry, a First Course*. Springer, 1992.
9. Hartshorne, R. *Algebraic Geometry*. Springer–Verlag, 1977.
10. Jung, G. Ricerche sui sistemi lineari di curve algebriche di genere qualunque. *Annali di Mat. 2* (1888), 277–312.
11. Krasauskas, R. Toric surface patches. In *Advances in geometrical algorithms and representations* (2002), vol. 17, pp. 89–113.
12. Manin, Y. *Cubic Forms*. North-Holland, 1974.
13. Peternell, M. *Rational parametrizations for envelopes of quadric families*. PhD thesis, Techn. Univ. Vienna, 1997.
14. Reid, M. Graded rings and varieties in weighted projective space. Tech. rep., Math. Institute, University of Warwick, 2002. downloadable via www.maths.warwick.ac.uk/˜miles/surf/more/grad.pdf.
15. Russo, F. The antibirational involutions of the plane and the classification of real del Pezzo surfaces. In *Algebraic Geometry*, M. B. et al., Ed. de Gruyter, 2002, pp. 289–312.
16. Schicho, J. Inversion of rational maps with Gröbner bases. In *Gröbner bases and applications* (1998), B. Buchberger and F. Winkler, Eds., Cambridge Univ. Press, pp. 495–503.
17. Schicho, J. Rational parameterization of real algebraic surfaces. In *Proc. ISSAC'98* (1998), ACM Press, pp. 302–308.
18. Schicho, J. Rational parametrization of surfaces. *J. Symb. Comp. 26*, 1 (1998), 1–30.
19. Sermenev, A. M. On some unirational surfaces. *Mat. Zametki 5* (1969), 155–159.
20. Silhol, R. *Real Algebraic Surfaces*. Springer, 1980.
21. Wall, C. T. C. Real forms of smooth del Pezzo surfaces. *J. Reine Angew. Math. 375/376* (1987), 47–66.

The Geometry of the Tangent Developable

Pål Hermunn Johansen

Centre of Mathematics for Applications & Department of Mathematics,
University of Oslo, P.O. Box 1053 Blindern, NO-0316 Oslo, Norway
hermunn@math.uio.no

Abstract. The tangent developable of a curve $C \subset \mathbb{P}^3$ is a singular surface with a cuspidal edge along C and the flex tangents of C. It also contains a multiple curve, typically double, and we express the degree of this curve in terms of the invariants of C. In many cases we can calculate the intersections of C with the multiple curve, and pictures of these cases are provided.

1 Introduction

If we have a curve on which tangents can be defined, then the associated tangent developable is the surface swiped out by the tangents. Tangent developables have a cuspidal edge, and are easy to generate. Since most developable surfaces are tangent developables, the Computer Aided Geometric Design community should be interested in their properties. This article describes the local and global geometry of tangent developables.

For the local study of tangent developables we consider analytic real curves. Cleave showed in [1] that for most curves the tangent developable has a cuspidal edge along most of the curve. This was extended by Mond in [4] and [5] where he analyzed the tangent developable of more special curves. This work was further extended by Ishikawa in [3], and results from that article are used in sect. 3.

The following section contains figures illustrating the local behavior of tangent developables, and one may want to have a brief look at these before reading the rest of the text.

In sect. 5 the tangent developables of complex projective algebraic curves are described. Algebraic geometrical invariants are introduced and relations between these invariants are taken from [6]. We also show that tangent developables of rational curves of degree ≥ 4 have a double curve.

Many thanks goes to Ragni Piene for lots of good advice and considerable help with this article.

2 Tangent Developables

Given a curve in some space, its tangent developable is the union of the tangent lines to the curve. The tangent line at a singular point is defined as the limit of tangent lines at non-singular points. If the curve is algebraic, then its tangent developable will be an algebraic surface.

Assume we have a parameterization of a curve with a non-vanishing derivative. Then we can make a map that parameterizes the corresponding tangent developable. Let $U \subset \mathbb{R}$, $\gamma : U \rightarrow \mathbb{R}^3$ be a map with a non-vanishing derivative. Define the map $\Gamma : U \times \mathbb{R} \rightarrow \mathbb{R}^3$ by

$$\Gamma(t, u) = \gamma(t) + u\gamma'(t) \tag{1}$$

In this case the tangent developable of $\gamma(U)$ is the image of Γ. The following example uses this technique to calculate the implicit equation of a tangent developable.

Example 1 (The tangent developable of the twisted cubic). Consider the twisted cubic curve parameterized by $\gamma : \mathbb{R} \rightarrow \mathbb{R}^3$ where $\gamma(t) = (t, t^2, t^3)$. The tangent developable is then the image of $\Gamma : \mathbb{R}^2 \rightarrow \mathbb{R}^3$ where $\Gamma(t, u) = (t + u, t^2 + 2ut, t^3 + 3ut^2)$. The algebra program Singular [2] can calculate the implicit equation of the surface:

$$z^2 - 6xyz + 4x^3z + 4y^3 - 3x^2y^2 = 0.$$

In this case the implicit equation describe the same set of points as the the image of Γ. However, when dealing with real parameterizations this is not always true.

Calculating the Jacobian ideal shows us that the tangent developable is singular exactly at $\gamma(\mathbb{R})$. Moreover, if the surface is intersected with a general plane, the resulting curve will have a cusp singularity at each intersection point with $\gamma(\mathbb{R})$.

Definition 2 (The type of a germ). *Let γ be a smooth (C^∞) curve germ, $\gamma : (\mathbb{R}, p) \rightarrow (\mathbb{R}^3, q)$. We say that the germ is of finite type if the vectors*

$$\gamma'(p), \gamma''(p), \gamma'''(p), \gamma^{(4)}(p), \dots$$

span $\mathbb{R}^3$. In this case, let $a_i = \min\{k \mid \dim\langle\gamma'(p), \gamma''(p), \dots, \gamma^{(k)}(p)\rangle = i\}$ and define the type of the germ to be the triple (a_1, a_2, a_3).

In this article we will only look at parameterizations where all the germs are of finite type.

What does a tangent developable look like? Along most of the curve, the tangent developable has a cuspidal edge singularity, so it is never smooth.

3   Local Properties of a Real Tangent Developable

We now want to study the local properties of the tangent developable close to the curve. Now we are no longer forced to use complex numbers, so we choose to study only real tangent developables. Since this is a local study, we now look at germs of curves $\gamma : (\mathbb{R}, 0) \rightarrow (\mathbb{R}^3, 0)$, as in definition 2.

Cleave shows in [1] that the tangent developable of most smooth curves γ have a cuspidal edge along most of the curve. That is, the cuspidal edge exists at intervals of points of type $(1, 2, 3)$. We have already decided only to look at curves where all the points are of finite type, and for all of these curves we will have a cuspidal edge along most of the curve.

In the language of Cleave: Given a curve with nonzero curvature and torsion at a point $\gamma(t_0)$. If the tangent developable is intersected with a general plane through $\gamma(t_0)$, the resulting curve will have a cusp at that point. In [5] Mond provides drawings of the tangent developable at points of type $(1, 2, k)$ for $3 \leq k \leq 7$. This is (in the language of differential geometry) when the torsion vanishes to order ≤ 4.

This was extended by Goo Ishikawa in [3], where he proves the following: The local diffeomorphism class of the tangent developable is determined by the type of the point if and only if the type is one of the following: $(1, 2, 2 + r)$ where r is a positive integer, $(1, 3, 4)$, $(1, 3, 5)$, $(2, 3, 4)$ or $(3, 4, 5)$.

In other words, for these types we can restrict our study to curves on the form

$$
\begin{aligned}
x &= t^{l_1+1} =: t^a \\
y &= t^{l_2+2} =: t^b \\
z &= t^{l_3+3} =: t^c
\end{aligned}
$$

at the origin. For other types we have to include more terms (of the power series) in the local parameterizations to study the point. In these cases we can get several different real pictures, but since points of other types are quite exotic, they will not be analyzed here.

Knowing this we can calculate local self intersection curves at points of type $(1, 2, k)$ quite easily:

Example 3 $((a, b, c) = (1, 2, k)$ *for* $k \geq 3)$. To find local self intersection curves we need to solve the equation $\Gamma(t, u) = \Gamma(s, v)$ where Γ is defined as in eq. (1), $\Gamma(t, u) = (t + u, t^2 + 2tu, t^k + kt^{k-1}u)$. Some straightforward calculations leads us to solving

$$
\begin{aligned}
-(t^2 - s^2) + 2w(t - s) &= 0 \\
(1 - k)(t^k - s^k) + kw(t^{k-1} - s^{k-1}) &= 0,
\end{aligned}
$$

where $w = t + u = s + v$.

Assuming $s \neq t$ we (eventually) get

$$
\begin{aligned}
0 &= 2(1 - k)(t^k - s^k) + k(t + s)(t^{k-1} - s^{k-1}) \\
&= (2 - k)(t^{k-1} + s^{k-1}) + 2(t^{k-2}s + t^{k-3}s^2 + \ldots + ts^{k-2}).
\end{aligned}
$$

It is not hard to prove that $s = -t$ is the only possible *real* self intersection by analyzing the polynomial $f(t) = (2-k)(t^{k-1}+1)+2(t^{k-2}+t^{k-3}+\ldots+t)$ and its derivative. The real self intersection occurs exactly when k is even. This is compatible with what Mond found in [4], but since Mond looked at C^∞ curves he could only draw the conclusion for $k \leq 7$. Note that we have complex self intersections for all $k \geq 5$.

Example 4 (Types $(1, 3, 4)$, $(1, 3, 5)$, $(2, 3, 4)$ *and* $(3, 4, 5))$. Points of types $(1, 3, 4)$, $(1, 3, 5)$ and $(2, 3, 4)$ each have one local real self intersection curve, while points of type $(3, 4, 5)$ have no real self intersection curves. This was calculated using Singular [2].

The following section contains pictures of all of these types.

4   Illustrations

This section contains figures of tangent developables of different curves, each parameterized by a map $t \to (t^a, t^b, t^c)$ for some triple (a, b, c). For each of the curves, the origin is of type (a, b, c) and all other points (close to the origin) are of type $(1, 2, 3)$. For all the figures, we have drawn the points that are at a distance of ≤ 2 from the origin, so the figures illustrate the local properties of the tangent developable.

The first five figures show points of type $(1, 2, k)$. We can see that we have self intersection curves exactly when k is even, as calculated in example 3.

In the first figure, all points are of type $(1, 2, 3)$:

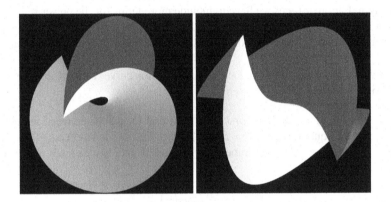

For most curves, the only types are $(1, 2, 3)$ and $(1, 2, 4)$. The following figure shows a point of type $(1, 2, 4)$:

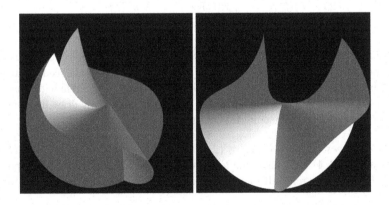

The following figures show points of type $(1, 2, k)$.

The tangent developable of the curve (t, t^2, t^5)

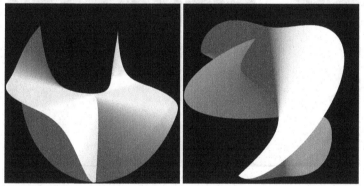

The tangent developable of the curves (t, t^2, t^6) and (t, t^2, t^7)

The rest of the figures come from example 4. Note that for the points where $k_1(0) = 1$ (types $(1, 3, 4)$ and $(1, 3, 5)$) the line which is a cuspidal edge, but not part of the curve, is an inflectional tangent line. This corresponds to the Plücker formula mentioned in sect. 5, $c = r_0 + k_1$, where c is the degree of the cuspidal edge.

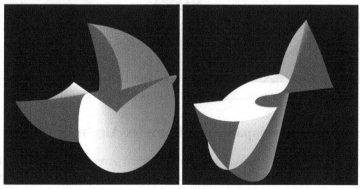

The tangent developable of the curve (t, t^3, t^4)

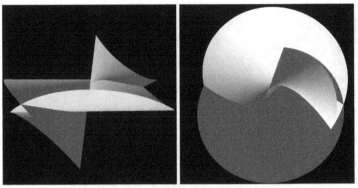

The tangent developable of the curve (t, t^3, t^5)

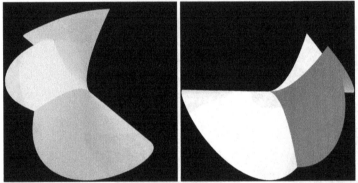

The tangent developable of the curve (t^2, t^3, t^4)

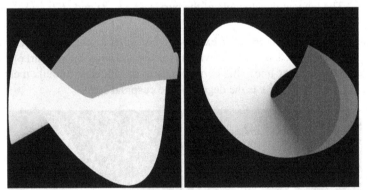

The tangent developable of the curve (t^3, t^4, t^5)

5   The Tangent Developable of a Complex Algebraic Curve

To any projective algebraic curve, there are associated several invariants, most importantly the degree and genus of the curve. Classical algebraic geometry gives many relations between these values and the geometry of the curve. In [6] Piene obtained results for the tangent developable, and the formulas have been taken from that article.

In this section a curve will be a reduced algebraic curve C_0 in the projective complex 3-space, $\mathbb{P}^3_{\mathbb{C}}$. We also assume that the curve spans the space. Let $X \subset \mathbb{P}^3_{\mathbb{C}}$ denote the tangent developable of C_0.

Let $h : C \to C_0$ be the normalization map, so that C is the desingularization of C_0. Let g denote the (geometric) genus of the curve and r_0 the degree. The rank r_1 is defined as the number of tangents that intersect a general line. Clearly this is the same as the degree of the tangent developable. The class r_2 is defined as the number of osculating planes to C_0 that contain a general point. The osculating plane at a point on the curve is the plane with the highest order of contact with the curve at that point. Another point of view is that the osculating plane at a point x_0 is the limit of the planes containing x_0, x_1 and x_2 as $x_1, x_2 \to x_0$.

For each point $p \in C$, we can choose affine coordinates around $h(p)$ such that the branch of C_0 determined by p has a (formal) parameterization at $h(p)$ equal to

$$x = t^{l_1+1} + \ldots$$
$$y = t^{l_2+2} + \ldots$$
$$z = t^{l_3+3} + \ldots$$

with $l_0 := 0 \le l_1 \le l_2 \le l_3$. This (formal) parameterization is also a curve germ $\gamma : (\mathbb{C}, 0) \to (\mathbb{C}^3, 0)$. Because of this we extend the notion of the type to the complex domain, and say that the type of the germ determined by p is equal to (l_1+1, l_2+2, l_3+3).

The coordinates are chosen such that p is the origin, the tangent is the line $y = z = 0$, and the osculating plane is $z = 0$. We call $k_i(p) = l_{i+1} - l_i$ the ith stationary index of p. Since $k_i(p) \ne 0$ only for a finite number of points p we can define $k_i = \sum_{p \in C} k_i(p)$.

If $l_1 = 0$, then the germ is nonsingular. If $l_1 \ge 1$ we say that the germ has a cusp, and if $l_1 = 1$ the cusp is said to be ordinary. If $l_1 = 0$ and $l_2 \ge 1$ we call the point $h(p)$ an inflection point or flex, and if $l_2 = 1$ the flex is ordinary. If $l_1 = l_2 = 0$ and $l_3 \ge 1$ we say that the curve has a stall or a point of hyperosculation. For most curves we will have no cusps and no flexes.

Now it is time to state the relations between these values, all taken from [6]:

$$r_1 = 2r_0 + 2g - 2 - k_0 \tag{2}$$
$$r_2 = 3(r_0 + 2g - 2) - 2k_0 - k_1 \tag{3}$$
$$k_2 = 4(r_0 + 3g - 3) - 3k_0 - 2k_1 \tag{4}$$

Note that $r_1 \ge 3$ since since r_1 is the degree of the tangent developable, and no quadric surface with a cuspidal edge exists. Furthermore, $r_2 \ge 3$ since r_2 is the degree of the dual curve, and the dual curve must span the space. From the definition we get $k_2 \ge 0$.

The tangent developable X of C_0 has degree $\mu_0 = r_1$, rank $\mu_1 = r_2$ (defined as the class of the intersection of the tangent developable with a general plane, a plane curve) and class $\mu_2 = 0$ (defined as the number of tangent planes containing a general line). Its cuspidal edge consists of C_0 and the flex tangents of C_0. The cuspidal edge has degree $c = r_0 + k_1$.

Formulas involving algebraic invariants, as those above, are often called Plücker formulas, and such formulas is central in enumerative algebraic geometry. There are lots of Plücker formulas, relating many different algebraic invariants.

In addition to the cuspidal edge, X has a double (or higher order multiple) curve, some times called the nodal curve of C_0. It consists of points that are on more than one tangent of C_0. Eventual bitangents are part of the nodal curve.

Let b denote the degree of the nodal curve. If the nodal curve is double and the flexes of C_0 are ordinary, then [6] gives the following expressions for b:

$$2b = \mu_0(\mu_0 - 1) - \mu_1 - 3c = r_1(r_1 - 1) - r_2 - 3(r_0 + k_1)$$
$$= r_1(r_1 - 4) - k_0 - 2k_1$$
$$= (2r_0 + 2g - 2 - k_0)(2r_0 + 2g - 6 - k_0) - k_0 - 2k_1.$$

For rational curves, $g = 0$, so then

$$2b = (2r_0 - 2 - k_0)(2r_0 - 6 - k_0) - k_0 - 2k_1.$$

In this case we see that

$$k_2 = 4(r_0 - 3) - 3k_0 - 2k_1 \geq 0$$

implies

$$k_0 = \tfrac{4}{3}r_0 - 4 - \tfrac{2}{3}k_1 - \tfrac{1}{3}k_2 \leq \tfrac{4}{3}r_0 - 4$$

We can find a lower bound for b for rational curves of degree $r_0 \geq 4$ by first eliminating k_1 (using eq. (4)) in the expression for b:

$$2b = (2r_0 - 2 - k_0)(2r_0 - 6 - k_0) - k_0 - 2k_1$$
$$= (2r_0 - 2 - k_0)(2r_0 - 6 - k_0) + 2k_0 + k_2 - 4r_0 + 12$$
$$\geq (2r_0 - 2 - k_0)(2r_0 - 6 - k_0) + 2k_0 - 4r_0 + 12 \text{ (using } k_2 \geq 0).$$

As a function in k_0 the expression above is strictly decreasing (for $k_0 \leq \tfrac{4}{3}r_0 - 4$). In other words, we can set $k_0 = \tfrac{4}{3}r_0 - 4$ and not break the inequality:

$$2b \geq (2r_0 - 2 - (\tfrac{4}{3}r_0 - 4))(2r_0 - 6 - (\tfrac{4}{3}r_0 - 4)) + 2(\tfrac{4}{3}r_0 - 4) - 4r_0 + 12$$
$$= \tfrac{4}{9}r_0(r_0 - 3).$$

We conclude that rational curves with $b = 0$ must have degree ≤ 3, and the twisted cubic is the only one of these that is not planar. It follows that every rational curve C_0 of degree greater than 3 gives a tangent developable with a nodal curve of positive degree.

We want to check if $b = 1$ is possible. If $g = 0$ and $b = 1$, then $2b \geq \tfrac{4}{9}r_0(r_0 - 3)$ implies $r_0 = 4$. Also, $k_0 \leq \tfrac{4}{3}r_0 - 4 = \tfrac{4}{3}$. This leads us to consider two cases, $k_0 = 0$ and $k_0 = 1$. If $k_0 = 0$ the formula for b implies $k_1 = 5$ and eq. (4) gives $k_2 = -6$. If $k_0 = 1$ the formula for b implies $k_1 = 1$ and eq. (4) give $k_2 = -1$. The second stationary index k_2 cannot be negative, so $b = 1$ is impossible.

The following example shows that $b = 2$ actually can occur for $g = 0$ and $r_0 \geq 4$:

Example 5 (A singular curve of degree 4). Let the curve $\gamma_0 : \mathbb{C} \to \mathbb{C}^3$ be given by $\gamma_0(t) = (t, t^2, t^3 + t^4)$. This is an imbedding that is one to one on points, so the degree is 4. Note that γ_0 is nonsingular, but if we take the projective completion $\gamma : \mathbb{P}^1 \to \mathbb{P}^3$ given by

$$\gamma(s;t) = (s^4; s^3t; s^2t^2; st^3 + t^4)$$

we get a singular curve. In fact, setting $t = 1$ yields the local parameterization at $(0;1)$, $s \to (s^4; s^3; s^2; s+1)$. Let $(w;x;y;z)$ be the projective coordinates for $\mathbb{P}^3_{\mathbb{C}}$. Since $1/(1+s) = 1 - s + s^2 - s^3 + \dots$ in a neighborhood of 0, setting $z = 1$ gives the local parameterization

$$w = s^2 - s^3 + s^4 - s^5 + \dots$$
$$x = s^3 - s^4 + s^5 - s^6 + \dots$$
$$y = s^4 - s^5 + s^6 - s^7 + \dots.$$

We see that the type of the local parameterization is $(2, 3, 4)$, and thus $k_0(\gamma(0;1)) = 1$ and $k_1(\gamma(0;1)) = k_2(\gamma(0;1)) = 0$. At any other point we see that the first and second derivative are linearly independent, so each of them are of type $(1, 2, n)$ for some value of n. This means that we have $k_0 = 1$ and $k_1 = 0$. The degree of the curve is $r_0 = 4$, and the genus of the curve is $g = 0$ since the curve is rational. Now we can calculate the rest of the invariants mentioned above.

From the formulas we get the rank of the curve, $r_1 = 5$, the class of the curve, $r_2 = 4$, the second stationary index, $k_2 = 1$, the degree of the surface $\mu_0 = r_1 = 5$, the rank of the surface $\mu_1 = r_2 = 4$, and finally the degree of the nodal curve, $b = 2$.

Using Singular [2], we can verify some of the results. A Gröbner bases computation gives us the implicit equation of the surface:

$$F = 3wx^2y^2 + 12x^3y^2 - 4w^2y^3 - 14wxy^3 + 8x^2y^3 - 9wy^4 - 4wx^3z$$
$$-16x^4z + 6w^2xyz + 24wx^2yz - 6w^2y^2z - w^3z^2$$

This equation is, predictably, of degree $\mu_0 = r_1 = 5$. We can find the singular locus by setting the four partial derivatives equal to zero. The last one,

$$\frac{1}{2} \cdot \frac{\partial F}{\partial z} = -2wx^3 - 8x^4 + 3w^2xy + 12wx^2y - 3w^2y^2 - w^3z,$$

leads us to consider two cases, $w = 0$ and $w \neq 0$.

The first case implies $x = 0$ from $\partial F/\partial z = 0$, and then $\partial F/\partial w = 0$ gives $y = 0$. This leaves us with one point, namely $(0;0;0;1) = \gamma(0;1)$, the singular point of the curve.

If $w \neq 0$ we can choose $w = 1$ and solve the system of equation quite easily. This is because $\partial F/\partial z = 0$ becomes

$$0 = -2x^3 - 8x^4 + 3xy + 12x^2y - 3y^2 - z, \tag{5}$$

so we can substitute z into the other equations. In other words, assuming $\partial F/\partial z = 0$, the equation $\partial F/\partial y = 0$ gives

$$0 = 16x^6 + 8x^5 - 32x^4y + x^4 - 16x^3y + 16x^2y^2 - 2x^2y + 8xy^2 + y^2$$
$$= (4x + 1)^2(x^2 - y)^2.$$

If $x^2 - y = 0$, then eq. (5) gives $z = x^3 + x^4$, as expected.

Setting $x = -1/4$ in the rest of the equations gives us a solution for every y, so z is a polynomial of degree 2 in y given by (5). This is the degree of the nodal curve that we calculated earlier.

Note that most curves will have $k_0 = k_1 = 0$, with a nodal curve of degree $b = 2(r_0 + g - 1)(r_0 + g - 3)$. Unless $r_0 = 3$ and $g = 0$, the nodal curve will not be empty.

The cuspidal edge and the nodal curve may both be singular, and they will usually intersect. If the nodal curve is double and the flexes are ordinary, X will have a finite number of points with multiplicity ≥ 3. These points can be of different types.

If the nodal curve has a node at q outside the cuspidal edge, then q must lie on at least 3 tangents, and therefore the nodal curve must have at least multiplicity 3 at q since any selection of two out of three tangents will give a branch in the nodal curve.

The total number T of triple points of the tangent developable X of C_0 is given in [6] and is

$$T = \tfrac{1}{6}(r_1 - 4)((r_1 - 3)(r_1 - 2) - 6g). \tag{6}$$

The formula (6) is valid when the nodal curve is double. When the nodal curve is more than double we have to use a generalized formula for the degree of the multiple curves (also found in [6]). If the nodal curve consists of curves D_j, where D_j is ordinary j-multiple, then the degrees b_j of D_j satisfy

$$\sum_j j(j-1)b_j = r_1(r_1 - 1) - r_2 - 3(r_0 + k_1), \tag{7}$$

still assuming the flexes to be ordinary. Note that this is a very special case, and that producing interesting examples with high j may be hard.

An example where the nodal curve is triple can be found in [7, p. 65], and we have calculated, using Singular [2], the details[1].

Example 6 (The equianharmonic rational quartic). Let $\alpha = \tfrac{1}{3}\sqrt{-3}$, let C_0 be the rational curve defined by the map $\gamma : \mathbb{P}^1_{\mathbb{C}} \to \mathbb{P}^3_{\mathbb{C}}$ where

$$\gamma(s;t) = (\alpha s^4 - s^2 t^2; \alpha s^3 t; \alpha s t^3; \alpha t^4 - s^2 t^2),$$

and let X be its tangent developable. A Gröbner bases computation gives us the implicit equation $F = 0$ of the surface X. Here F is a polynomial of degree 6 in the projective coordinates $(w; x; y; z)$:

$$\begin{aligned}
F = {} & 12w^2 x^3 y + 3w^4 y^2 - 72\alpha w^2 x^2 y^2 + 12w^2 xy^3 - 256\alpha x^3 y^3 \\
& + 18\alpha w^3 xyz + 24wx^3 yz + 6w^3 y^2 z + 48\alpha wx^2 y^2 z + 24wxy^3 z \\
& + 3w^2 x^2 z^2 - 12\alpha w^2 xyz^2 + 12x^3 yz^2 + 3w^2 y^2 z^2 - 72\alpha x^2 y^2 z^2 \\
& + 12xy^3 z^2 + 4\alpha w^3 z^3 + 6wx^2 z^3 + 18\alpha wxyz^3 + 3x^2 z^4
\end{aligned}$$

---

[1] There is an error in [7], m is not supposed to be $\sqrt{-3}$, but the same as α in the example, $m = \tfrac{1}{3}\sqrt{-3}$.

Taking a primary decomposition of the Jacobian ideal of F, we find that the singular locus of X consists of two components, the curve C_0 and the conic D defined by $z^2 + 4xy = 0$ in the plane $w + z = 0$.

We want to show that D is a triple curve of X. The conic D can be parameterized by $\theta : \mathbb{P}^1_{\mathbb{C}} \to \mathbb{P}^3_{\mathbb{C}}$ where

$$\theta(u; v) = (-2uv; -v^2; u^2; 2uv).$$

Using this parameterization we find the following: The point $\theta(u; v)$ lies on the tangent to C_0 at $\gamma(s; t)$ if and only if

$$G(s, t, u, v) := s^3 u - 3\alpha s t^2 u + 3\alpha s^2 t v - t^3 v = 0.$$

For a fixed $(u; v) \in \mathbb{P}^1_{\mathbb{C}}$, zeros of $G(s, t, u, v)$ corresponds to points on C_0 whose tangent contain $\theta(u; v)$. For most $(u; v) \in \mathbb{P}^1_{\mathbb{C}}$ we will get three distinct tangents. In fact, let $\Delta(u, v)$ denote the discriminant of G with respect to $(s; t)$. In this case

$$\Delta(u, v) = (u^2 + (3\alpha + 1)uv - v^2)(u^2 - (3\alpha + 1)uv - v^2).$$

If $\Delta(u, v) \neq 0$, then the point $\theta(u; v)$ lies on three distinct tangents to C_0.

Let A denote the four points on D corresponding to $\Delta(u, v) = 0$. We conclude that each point on D not in A lies on exactly three tangents of C_0. This means that D is a triple curve of X.

Moreover, A is exactly the intersection of D and C_0, and these four points are the only points on C_0 whose local parameterization is not of type $(1, 2, 3)$. In fact, the local parameterizations in each point of A is of type $(1, 2, 4)$. This means that $k_0 = k_1 = 0$ and $k_2 = 4$. Furthermore, the degree of C_0 is $r_0 = 4$ and the formulas give the rank $r_1 = 6$ and the class $r_2 = 6$ of the curve. The multiple curve only have one component, the triple curve D, and this corresponds to $b_3 = 2$ in eq. (7).

The set A form an equianharmonic set on D, and that is why C_0 is called the equianharmonic rational quartic. Note that this example is very special and arise from the thorough study [7] of the rational normal curve in $\mathbb{P}^4_{\mathbb{C}}$. The curve C_0 is constructed by projecting the rational normal curve in $\mathbb{P}^4_{\mathbb{C}}$ from a general point on a quadric called the nucleus of the polarity. The equation of the nucleus of the polarity is $x_0 x_4 - 4 x_1 x_3 + 3 x_2^2$ and the projection centre of this example is $(1, 0, \alpha, 0, 1)$.

All the formulas in this section holds for curves in $\mathbb{P}^3_{\mathbb{C}}$. We can not make similar equalities for real curves, but the projective invariants of the complex curve give results for the real part in the form of inequalities. However, these inequalities will not be made explicit in this article.

References

1. J. P. Cleave. The form of the tangent-developable at points of zero torsion on space curves. *Math. Proc. Cambridge Philos. Soc.*, 88(3):403–407, 1980.
2. G.-M. Greuel, G. Pfister, and H. Schönemann. SINGULAR 2.0. A Computer Algebra System for Polynomial Computations, Centre for Computer Algebra, University of Kaiserslautern, 2001. http://www.singular.uni-kl.de.

3. Goo Ishikawa. Topological classification of the tangent developables of space curves. *J. London Math. Soc. (2)*, 62(2):583–598, 2000.
4. David Mond. On the tangent developable of a space curve. *Math. Proc. Cambridge Philos. Soc.*, 91(3):351–355, 1982.
5. David Mond. Singularities of the tangent developable surface of a space curve. *Quart. J. Math. Oxford Ser. (2)*, 40(157):79–91, 1989.
6. Ragni Piene. Cuspidal projections of space curves. *Math. Ann.*, 256(1):95–119, 1981.
7. H. G. Telling. *The Rational Quartic Curve*. Cambridge Tracts in Mathematics and Mathematical Physics No. 34. Cambridge University Press, Cambridge, 1936.

Numerical and Algebraic Properties of Bernstein Basis Resultant Matrices

Joab R. Winkler

Sheffield University, Sheffield, United Kingdom
j.winkler@dcs.shef.ac.uk
WWW home page: http://www.dcs.shef.ac.uk/~joab

Abstract. Algebraic properties of the power and Bernstein forms of the companion, Sylvester and Bézout resultant matrices are compared and it is shown that some properties of the power basis form of these matrices are not shared by their Bernstein basis equivalents because of the combinatorial factors in the Bernstein basis functions. Several condition numbers of a resultant matrix are considered and it is shown that the most refined measure is NP–hard, and that a simpler, sub–optimal measure is easily computed. The transformation of the companion and Bézout resultant matrices between the power and Bernstein bases is considered numerically and algebraically. In particular, it is shown that these transformations are ill–conditioned, even for polynomials of low degree, and that the matrices that occur in these basis transformation equations share some properties.

1 Introduction

Many branches of science and engineering, including robotics [3], computer–aided geometric design [5, 10], computer graphics [9] and computer vision [11] require the computation of the resultant of two polynomials. The power basis is the natural representation of the polynomials in these applications, apart from computer graphics and computer–aided geometric design, for which the Bernstein basis is the preferred representation. This requirement has motivated the development of resultant matrices for Bernstein polynomials and an investigation into their algebraic and numerical properties. In particular, the Bézout, companion and Sylvester resultant matrices for Bernstein polynomials are developed in [2, 15, 17] respectively, and some of their properties are considered.

A review of the companion, Sylvester and Bézout resultant matrices for Bernstein basis polynomials is contained in sect. 2 and it is shown that the algebraic differences between the power and Bernstein forms of each of these matrices arise from the combinatorial factors in the Bernstein basis functions. A discussion of the condition numbers that may be used to assess the numerical stability of a resultant matrix is contained in sect. 3. It is shown that the computation of an accurate and refined condition number is NP–hard, but that it is easy to compute a sub–optimal condition number. The transformation of a resultant matrix between the power and Bernstein bases is considered in sect. 4 and it is shown that the condition number of the transformation increases rapidly with the degrees of the polynomials. A summary of the paper is contained in sect. 5.

2   Bernstein Basis Resultant Matrices

This section reviews the companion, Sylvester and Bézout resultant matrices for Bernstein polynomials and compares them to their power basis equivalents. The matrices are constructed for the polynomials

$$p(x) = \sum_{i=0}^{m} a_i \binom{m}{i} (1-x)^{m-i} x^i \text{ and } r(x) = \sum_{i=0}^{n} c_i \binom{n}{i} (1-x)^{n-i} x^i. \qquad (1)$$

The Bézout resultant matrix requires that the polynomials be of the same degree, from which it follows that degree elevation must be applied to the lower degree polynomial.

2.1   The Companion Resultant Matrix

The companion resultant matrix of the polynomials $p(x)$ and $r(x)$ in (1) is [15]

$$p(M) = \sum_{i=0}^{m} a_i \binom{m}{i} (I-M)^{m-i} M^i, \qquad (2)$$

where $M = E^{-1}A$ is the companion matrix of $r(x)$ and of order $n \times n$. The matrices A and E, both of which are also of order $n \times n$, are

$$A = \begin{bmatrix} 0 & 1 & 0 & 0 & \cdots & 0 & 0 \\ 0 & 0 & 1 & 0 & \cdots & 0 & 0 \\ \cdot & \cdot & \cdot & \cdot & \cdots & \cdot & \cdot \\ 0 & 0 & 0 & 0 & \cdots & 0 & 1 \\ -c_0 & -c_1 & -c_2 & -c_3 & \cdots & -c_{n-2} & -c_{n-1} \end{bmatrix}, \text{ and} \qquad (3)$$

$$E = \begin{bmatrix} \frac{\binom{n}{1}}{\binom{n}{0}} & 1 & 0 & 0 & \cdots & 0 & 0 \\ 0 & \frac{\binom{n}{2}}{\binom{n}{1}} & 1 & 0 & \cdots & 0 & 0 \\ \cdot & & \cdot & \cdot & \cdots & \cdot & \cdot \\ 0 & 0 & 0 & 0 & \cdots & \frac{\binom{n}{n-1}}{\binom{n}{n-2}} & 1 \\ -c_0 & -c_1 & -c_2 & -c_3 & \cdots & -c_{n-2} & -c_{n-1} + \frac{\binom{n}{n}}{\binom{n}{n-1}} \end{bmatrix}.$$

The normalisation $c_n = 1$ is required and it is noted that a companion matrix for a power basis polynomial requires a similar normalising constraint. It is immediately obvious that a companion matrix for a Bernstein polynomial has a more complex structure than its power basis equivalent.

The differences between the companion matrices in these bases are readily apparent by noting that if C is the power basis companion matrix of the monic polynomial

$$f(x) = \sum_{i=0}^{n} f_i x^i, \qquad f_n = 1, \qquad (4)$$

with the same structure as A in (3), then

$$TC = C^T T, \qquad (5)$$

where T is the upper triangular Hankel matrix with entries $\{f_i\}_{i=0}^n$,

$$
T = \begin{bmatrix}
f_1 & f_2 & f_3 & \cdots & f_{n-1} & 1 \\
f_2 & f_3 & \cdot & \cdots & 1 & 0 \\
f_3 & \cdot & \cdot & \cdots & 0 & 0 \\
\cdot & \cdot & \cdot & \cdots & \cdot & \cdot \\
f_{n-1} & 1 & 0 & \cdots & 0 & 0 \\
1 & 0 & 0 & \cdots & 0 & 0
\end{bmatrix}.
\tag{6}
$$

Furthermore, it is easily verified that if

$$
g(x) = \sum_{i=0}^{n} g_i x^i,
\tag{7}
$$

then $g(C)$ is a power basis companion resultant matrix for $f(x)$ and $g(x)$ that satisfies

$$
Tg(C) = (Tg(C))^T = (g(C))^T T,
\tag{8}
$$

since T is symmetric. This equation shows that T defines a similarity transformation between $g(C)$ and its transpose.

It is shown in [15] that (8) is not satisfied by M, that is, $Tp(M) \neq (p(M))^T T$ where the entries of T are the coefficients c_i of $r(x)$. This difference between the companion resultant matrix for the power and Bernstein bases is due to the combinatorial factors in the Bernstein basis functions because it is shown in [12] that (5) and (8) are satisfied by the scaled Bernstein basis, which is defined by the functions $\{(1-x)^{n-i}x^i\}_{i=0}^n$.

2.2   The Sylvester Resultant Matrix

The Sylvester resultant matrix $S(p,r)$ for the polynomials (1) is developed in [17], but a different method for its construction is now given. In order to simplify the derivation, the matrix will be developed for specified m and n, but it will be readily seen that its extension to general m and n follows easily.

A standard result in the theory of polynomials states that a necessary and sufficient condition for the polynomials (1) to have a non–constant common divisor is that there exist polynomials $F(x)$ and $G(x)$ such that

1. $\deg F(x) < \deg r(x)$
2. $\deg G(x) < \deg p(x)$
3. $F(x)p(x) + G(x)r(x) = 0$,

where deg denotes degree. As an example, assume that $m = 2$ and $n = 3$, in which case the polynomials $F(x)$ and $G(x)$ are given by

$$
F(x) = F_0 \binom{2}{0} (1-x)^2 + F_1 \binom{2}{1} (1-x)x + F_2 \binom{2}{2} x^2, \quad \text{and}
$$

$$
G(x) = G_0 \binom{1}{0} (1-x) + G_1 \binom{1}{1} x,
$$

respectively. The expression $F(x)p(x) + G(x)r(x)$ is therefore equal to

$$\binom{4}{0}(1-x)^4 \left(\frac{F_0\binom{2}{0}a_0\binom{2}{0} + G_0\binom{1}{0}c_0\binom{3}{0}}{\binom{4}{0}} \right)$$

$$+ \binom{4}{1}(1-x)^3 x \left(\frac{F_0\binom{2}{0}a_1\binom{2}{1} + F_1\binom{2}{1}a_0\binom{2}{0} + G_0\binom{1}{0}c_1\binom{3}{1} + G_1\binom{1}{1}c_0\binom{3}{0}}{\binom{4}{1}} \right)$$

$$+ \binom{4}{2}(1-x)^2 x^2 \left(\frac{F_0\binom{2}{0}a_2\binom{2}{2} + F_1\binom{2}{1}a_1\binom{2}{1} + F_2\binom{2}{2}a_0\binom{2}{0} + G_0\binom{1}{0}c_2\binom{3}{2} + G_1\binom{1}{1}c_1}{\binom{4}{2}} \right)$$

$$+ \binom{4}{3}(1-x)x^3 \left(\frac{F_1\binom{2}{1}a_2\binom{2}{2} + F_2\binom{2}{2}a_1\binom{2}{1} + G_0\binom{1}{0}c_3\binom{3}{3} + G_1\binom{1}{1}c_2\binom{3}{2}}{\binom{4}{3}} \right)$$

$$+ \binom{4}{4}x^4 \left(\frac{F_2\binom{2}{2}a_2\binom{2}{2} + G_1\binom{1}{1}c_3\binom{3}{3}}{\binom{4}{4}} \right),$$

and since it must be zero for all values of x, it follows that

$$\begin{bmatrix} \frac{1}{\binom{4}{0}} & 0 & 0 & 0 & 0 \\ 0 & \frac{1}{\binom{4}{1}} & 0 & 0 & 0 \\ 0 & 0 & \frac{1}{\binom{4}{2}} & 0 & 0 \\ 0 & 0 & 0 & \frac{1}{\binom{4}{3}} & 0 \\ 0 & 0 & 0 & 0 & \frac{1}{\binom{4}{4}} \end{bmatrix} \begin{bmatrix} a_0\binom{2}{0} & 0 & 0 & c_0\binom{3}{0} & 0 \\ a_1\binom{2}{1} & a_0\binom{2}{0} & 0 & c_1\binom{3}{1} & c_0\binom{3}{0} \\ a_2\binom{2}{2} & a_1\binom{2}{1} & a_0\binom{2}{0} & c_2\binom{3}{2} & c_1\binom{3}{1} \\ 0 & a_2\binom{2}{2} & a_1\binom{2}{1} & c_3\binom{3}{3} & c_2\binom{3}{2} \\ 0 & 0 & a_2\binom{2}{2} & 0 & c_3\binom{3}{3} \end{bmatrix} \begin{bmatrix} F_0\binom{2}{0} \\ F_1\binom{2}{1} \\ F_2\binom{2}{2} \\ G_0\binom{1}{0} \\ G_1\binom{1}{1} \end{bmatrix} = \begin{bmatrix} 0 \\ 0 \\ 0 \\ 0 \\ 0 \end{bmatrix}.$$

The Sylvester resultant matrix of the polynomials (1) with $m = 2$ and $n = 3$ is defined as the transpose of the product

$$\begin{bmatrix} \frac{1}{\binom{4}{0}} & 0 & 0 & 0 & 0 \\ 0 & \frac{1}{\binom{4}{1}} & 0 & 0 & 0 \\ 0 & 0 & \frac{1}{\binom{4}{2}} & 0 & 0 \\ 0 & 0 & 0 & \frac{1}{\binom{4}{3}} & 0 \\ 0 & 0 & 0 & 0 & \frac{1}{\binom{4}{4}} \end{bmatrix} \begin{bmatrix} a_0\binom{2}{0} & 0 & 0 & c_0\binom{3}{0} & 0 \\ a_1\binom{2}{1} & a_0\binom{2}{0} & 0 & c_1\binom{3}{1} & c_0\binom{3}{0} \\ a_2\binom{2}{2} & a_1\binom{2}{1} & a_0\binom{2}{0} & c_2\binom{3}{2} & c_1\binom{3}{1} \\ 0 & a_2\binom{2}{2} & a_1\binom{2}{1} & c_3\binom{3}{3} & c_2\binom{3}{2} \\ 0 & 0 & a_2\binom{2}{2} & 0 & c_3\binom{3}{3} \end{bmatrix},$$

and its extension to arbitrary m and n yields the Sylvester resultant matrix for the polynomials (1),

$$S(p,r) = \begin{bmatrix} a_0\binom{m}{0} & a_1\binom{m}{1} & a_2\binom{m}{2} & \cdot & & \\ & a_0\binom{m}{0} & a_1\binom{m}{1} & a_2\binom{m}{2} & \cdot & \\ & & a_0\binom{m}{0} & a_1\binom{m}{1} & \cdot\cdot & \\ & & & & \cdot & \\ & & & & & \cdot\cdot \\ c_0\binom{n}{0} & c_1\binom{n}{1} & c_2\binom{n}{2} & \cdot & & \\ & c_0\binom{n}{0} & c_1\binom{n}{1} & c_2\binom{n}{2} & \cdot & \\ & & c_0\binom{n}{0} & c_1\binom{n}{1} & \cdot\cdot & \end{bmatrix} D, \qquad (9)$$

where

$$D = \text{diag} \left[\frac{1}{\binom{m+n-1}{0}} \ \frac{1}{\binom{m+n-1}{1}} \ \cdots \ \frac{1}{\binom{m+n-1}{m+n-2}} \ \frac{1}{\binom{m+n-1}{m+n-1}} \right].$$

It follows immediately from (9) that the striped pattern of the Sylvester resultant matrix for power basis polynomials is not shared by its Bernstein basis equivalent $S(p, r)$, and that this is due to the combinatorial factors in the diagonal matrix D. Furthermore, it is shown in [17] that this striped pattern is present in the Sylvester resultant matrix for polynomials expressed in the scaled Bernstein basis because D reduces to the identity matrix for this basis. This anomalous feature is analogous to the effect of the combinatorial factors on the companion resultant matrix for Bernstein polynomials, as discussed in sect. 2.1.

2.3 The Bézout Resultant Matrix

The Bézout resultant matrix $Z(p, r)$ requires that the polynomials (1) be of the same degree, and degree elevation must therefore be applied to the polynomial of lower degree. If it is assumed that $m < n$, then degree elevation must be applied to $p(x)$ a total of $(n - m)$ times.

The matrix $Z(p, r)$ is defined by

$$\frac{p(x)r(y) - p(y)r(x)}{x - y} = \sum_{i,j=1}^{n} z_{ij} \binom{n-1}{i-1} (1-x)^{(n-1)-(i-1)} x^{i-1}$$

$$\times \binom{n-1}{j-1} (1-y)^{(n-1)-(j-1)} y^{j-1},$$

where the entries $z_{ij} = z_{ji}, i, j = 1, \ldots, n$, of $Z(p, r)$ are [2]

$$z_{i,1} = \frac{n}{i} \left(a_i c_0 - a_0 c_i \right), \qquad 1 \le i \le n,$$

$$z_{i,j+1} = \frac{n^2}{i(n-j)} \left(a_i c_j - a_j c_i \right) + \frac{j(n-i)}{i(n-j)} z_{i+1,j}, \qquad 1 \le i, j \le n-1,$$

$$z_{n,j+1} = \frac{n}{(n-j)} \left(a_n c_j - a_j c_n \right), \qquad 1 \le j \le n-1.$$

It was shown in sect.s 2.1 and 2.2 that some properties of the companion and Sylvester resultant matrices of two power basis polynomials are not shared by their Bernstein basis equivalents because of the combinatorial factors in the Bernstein basis functions. This divergent characteristic extends to the Bézout resultant matrix of two Bernstein basis polynomials. In particular, it is shown in [1] that the Bézout resultant matrix $Z(f, g)$ of the power basis polynomials $f(x)$ and $g(x)$ in (4) and (7), respectively, is related to the resultant matrix $g(C)$, where C is the companion matrix of $f(x)$ with the same pattern as A in (3),

$$Z(f, g) = T g(C), \qquad (10)$$

where T is defined in (6). It follows immediately from (8) that $Z(f, g)$ is symmetric, as required by the definition of the Bézout resultant matrix.

It was shown in sect. 2.1 that the symmetry property (8) does not extend to the Bernstein basis, and it therefore follows that (10) does not extend to the Bernstein basis, that is, the equation that links the Bézout and companion resultant matrices for this basis is not of the form (10).

Finally, it is interesting to note that since

$$\begin{vmatrix} af(x) + bg(x) & af(y) + bg(y) \\ cf(x) + dg(x) & cf(y) + dg(y) \end{vmatrix} = (ad - bc) \begin{vmatrix} f(x) & f(y) \\ g(x) & g(y) \end{vmatrix},$$

it follows that the resultant of $af(x) + bg(x)$ and $cf(x) + dg(x)$ is proportional to the resultant of $f(x)$ and $g(x)$ [4].

3   The Numerical Condition of a Resultant Matrix

The determination of the computational reliability of a resultant requires that a condition number of a resultant matrix be developed. These condition numbers for a resultant matrix for polynomials in an arbitrary basis are considered in this section and it is shown that the most refined condition number yields a difficult structure–preserving matrix approximation problem.

Resultants were originally developed for the computation of the greatest common divisor (GCD) of two polynomials, and this application enables several condition numbers of a resultant matrix to be developed and compared. In particular, consider the companion and Bézout resultant matrices $R = R(h, l)$ of the polynomials $h = h(x)$ and $l = l(x)$, both of which are expressed in an arbitrary polynomial basis.[1] This general representation is appropriate because the discussion in this section is valid for all bases of this class.

It is known that the degree of the GCD of two polynomials is equal to the rank loss of their resultant matrix R, and that the coefficients of the GCD are obtained by reducing R to row echelon form. It is appropriate, therefore, to define the condition number of R as the reciprocal of the distance to unit loss of rank. This is intuitively appealing because if R is near singularity, then a small perturbation in R is required to reduce its rank by unity, and R is ill–conditioned. Conversely, if R is well–removed from singularity, then a relatively large perturbation in R is required to reduce its rank by unity, and R is well–conditioned. This argument is simple but there are several more issues that must be considered for a complete discussion of the condition numbers of R.

Let

$$R^{(r)}(h, l) = \left\{ r_{ij}^{(r)}(h, l) \right\}_{i,j=1}^{n} \quad \text{and} \quad R^{(r-1)}(h^*, l^*) = \left\{ r_{ij}^{(r-1)}(h^*, l^*) \right\}_{i,j=1}^{n},$$

be, respectively, a resultant matrix of rank r for the polynomials $h = h(x)$ and $l = l(x)$, and a resultant matrix of the same type (companion or Bézout) of rank $(r - 1)$ for the polynomials $h^* = h^*(x)$ and $l^* = l^*(x)$. The perturbations in the elements of $R^{(r)}(h, l)$ are denoted by

$$\delta R^{(r)}(h, l) = \left\{ \delta r_{ij}^{(r)}(h, l) \right\}_{i,j=1}^{n}.$$

---

[1] The reason for the exclusion of the Sylvester resultant matrix is given at the end of the section.

Four different condition numbers of $R^{(r)}(h, l)$ can be defined:

1. **CN 1:** Let $\varepsilon^{(c)}$ be the minimum constant such that

$$\left| \delta r_{ij}^{(r)}(h, l) \right| \leq \varepsilon^{(c)} \left| r_{ij}^{(r)}(h, l) \right|, \qquad i, j = 1, \ldots, n,$$

and the perturbed matrix $R^{(r)}(h, l) + \delta R^{(r)}(h, l)$ and $R^{(r-1)}(h^*, l^*)$ have the same structure, that is, the perturbed matrix is a resultant matrix for the polynomials $h^*(x)$ and $l^*(x)$, it is of the same type as $R^{(r)}(h, l)$ and it has rank $(r - 1)$. It follows from the discussion above of the inverse relationship between the numerical condition of a resultant matrix and its distance to singularity that this definition of the perturbation of a matrix yields a condition number $1/\varepsilon^{(c)}$.

2. **CN 2:** Let $\varepsilon^{(n)}$ be the minimum constant such that

$$\left\| \delta R^{(r)}(h, l) \right\|_2 \leq \varepsilon^{(n)} \left\| R^{(r)}(h, l) \right\|_2,$$

and the perturbed matrix $R^{(r)}(h, l) + \delta R^{(r)}(h, l)$ is a resultant matrix for the poly-nomials $h^*(x)$ and $l^*(x)$, it is of the same type as $R^{(r)}(h, l)$ and it has rank $(r - 1)$. Clearly, this definition of the perturbation of a resultant matrix yields a condition number $1/\varepsilon^{(n)}$.

3. **CN 3:** Let $\varepsilon_{approx}^{(c)}$ be the minimum constant such that

$$\left| \delta r_{ij}^{(r)}(h, l) \right| \leq \varepsilon_{approx}^{(c)} \left| r_{ij}^{(r)}(h, l) \right|, \qquad i, j = 1, \ldots, n,$$

and the perturbed matrix $R^{(r)}(h, l) + \delta R^{(r)}(h, l)$ has rank $(r - 1)$. The condition number of a resultant matrix that is associated with this perturbation is $1/\varepsilon_{approx}^{(c)}$.

4. **CN 4:** Let $\varepsilon_{approx}^{(n)}$ be the minimum constant such that

$$\left\| \delta R^{(r)}(h, l) \right\|_2 \leq \varepsilon_{approx}^{(n)} \left\| R^{(r)}(h, l) \right\|_2,$$

and the perturbed matrix $R^{(r)}(h, l) + \delta R^{(r)}(h, l)$ has rank $(r - 1)$. This definition of the perturbation of a resultant matrix yields a condition number $1/\varepsilon_{approx}^{(n)}$.

Condition number CN1 is the most refined measure of the condition of a resultant ma-trix because it is computed from the minimum componentwise distance to singular-ity, such that the perturbed matrix is a resultant matrix of the same type that has unit loss of rank. It follows, therefore, that this condition number requires that a structure–preserving perturbation be computed. Condition number CN2 is slightly less refined because the perturbation is measured in the normwise sense rather than the componen-twise sense. It is, however, an accurate measure of the condition of a resultant matrix because the perturbed matrix is a resultant matrix of the same type. Condition numbers CN3 and CN4 are only approximate measures of the condition of a resultant matrix because the structure of the perturbed matrix is not defined in these measures. In par-ticular, the perturbed matrices are of rank $(r - 1)$ but otherwise unspecified. Since a normwise measure of a perturbation is less refined than a componentwise measure, it follows that the condition number CN4 is the least refined of the four condition numbers specified above.

The computation of the componentwise condition numbers CN1 and CN3 is NP–hard [7] and thus they are not practical measures of the numerical condition of a resultant matrix. The condition number CN2 is difficult to compute because it requires a structure–preserving perturbation, but condition number CN4 is easy to compute in the 2–norm. It follows, therefore, that this measure is the most practical of the four condition numbers specified above, but it is the least accurate because it is measured in the normwise sense rather than the componentwise sense, and the structure of the perturbed matrix is not specified.

The condition number CN4 is easily computed from the singular value decomposition of $R^{(r)}(h, l)$. In particular, if $\sigma_i(R^r), i = 1, \ldots, n$, are the singular values of $R^{(r)}(h, l)$, arranged in non–increasing order, then $\sigma_r(R^r)$ is the minimum normwise distance between $R^{(r)}(h, l)$ and the set of all matrices of rank $(r - 1)$ [6]. It is shown in [13–15] that it is necessary to normalise this distance because the singular values of $\alpha p(M)$, where $p(M)$ is defined in (2), are $\alpha \sigma_i(p(M))$, and the singular values of the Bézout resultant matrix $Z(\alpha p, \beta r)$ are $\alpha \beta \sigma_i(Z(p, r))$. Specifically, since scaling the coefficients of one or both polynomials does not change the roots of the polynomials, it follows that this arbitrary scale factor must be removed, and this leads to the normalised distance to singularity of a resultant matrix. This is defined as σ_r / σ_1, and thus the condition number CN4 of a resultant matrix of rank r that is associated with this normalised distance is equal to σ_1 / σ_r.

This property of scale–invariance follows immediately for the companion resultant matrix because

$$\frac{\sigma_1(\alpha p(M))}{\sigma_r(\alpha p(M))} = \frac{\sigma_1(p(M))}{\sigma_r(p(M))},$$

for all $\alpha \neq 0$, and for the Bézout resultant matrix because

$$\frac{\sigma_1(Z(\alpha p, \beta r))}{\sigma_r(Z(\alpha p, \beta r))} = \frac{\sigma_1(Z(p, r))}{\sigma_r(Z(p, r))},$$

for all $\alpha, \beta \neq 0$. This measure is used in [13] and [14, 15] to compare the numerical condition of the companion resultant matrix for power basis and, respectively, scaled Bernstein and Bernstein basis polynomials. Although this measure is not optimal, the improvement in the numerical condition of the Bernstein form of the resultant matrix over its power basis equivalent is large, which is in accord with the improved numerical stability of the Bernstein basis with respect to the power basis.

It is important to note that the ratio σ_1 / σ_r cannot be used for the Sylvester resultant matrix because

$$\frac{\sigma_1(S(\alpha p, \beta r))}{\sigma_r(S(\alpha p, \beta r))} = \frac{\sigma_1(S(p, r))}{\sigma_r(S(p, r))} \quad \text{only if } \alpha = \beta \neq 0, \tag{11}$$

but this equation is not satisfied if $\alpha \neq \beta$. This feature of the Sylvester resultant matrix arises because the coefficients of the polynomials are decoupled in this matrix. In particular, it follows from (9) that each element of this matrix is a function of the coefficients of only one, and not both, polynomials, and this must be compared with the companion and Bézout resultant matrices, each of whose elements is a function of the coefficients of both polynomials. This anomalous characteristic of the Sylvester resultant matrix implies that the normalised distance to singularity can be made arbitrarily small or large, merely by scaling one or both polynomials. It is instructive to note that the Bézout resultant matrix is preferred to the Sylvester resultant matrix in [5] because it yields results that are independent of the coordinate system that is imposed on curves and surfaces when this matrix is used in geometric modelling applications. By contrast, the Sylvester resultant matrix yields results that are dependent upon this coordinate system, which is clearly unsatisfactory.

It is shown in [2] that the computation of the GCD from the Bézout resultant matrix can be reduced to its block triangular factorisation. A fast algorithm that performs this factorisation is described and it is shown that the numerical performance of the method is strongly dependent upon the condition numbers of its trailing submatrices. Computational experiments show that the Bernstein form of the Bézout resultant matrix is numerically superior to its power basis equivalent because the growth of these condition numbers for the Bernstein form of the matrix is much slower than it is for its power basis equivalent. This result is therefore consistent with the results in [14, 15] for the power and Bernstein forms of the companion resultant matrix.

4   The Transformation of a Resultant Matrix Between the Power and Bernstein Bases

It was shown in sect. 3 that a resultant matrix of two Bernstein polynomials is numerically superior to its equivalent power basis form. Since the polynomials that generate resultant matrices are, in many applications, expressed in the power basis, it is natural to determine the numerical condition of the transformation of a resultant matrix between these bases. In particular, if this transformation is well–conditioned, then it is adequate to construct the power basis resultant matrix, and then perform a basis transformation in order to compute the resultant in the Bernstein basis.

It is shown [16] that the transformation of the companion resultant matrix $p(M)$, which is defined in (2), to its power basis form is

$$p(M) = B^{-1}q(N)B, \tag{12}$$

where M is the companion matrix of $r(x)$, which is defined in (1), N is the companion matrix of the power basis form of $r(x)$ with the same pattern as A in (3), and $q(x)$ is the power basis form of $p(x)$. The elements of B and B^{-1}, b_{jk} and b_{jk}^{-1}, respectively, are given by [16]

$$b_{jk} = \begin{cases} \dfrac{n-k}{n} \dfrac{\binom{k}{j}}{\binom{n-1}{j}} & j = 0, \ldots, n-1; \ k = j, \ldots, n-1, \\ 0 & k < j, \end{cases}$$

and

$$b_{jk}^{-1} = \begin{cases} (-1)^{k-j}\binom{n}{j}\binom{n-1-j}{k-j} & j = 0, \ldots, n-1; \ k = j, \ldots, n-1, \\ 0 & k < j. \end{cases}$$

It is shown in [16] that B is totally non–negative, and that BB^T and $B^T B$ are oscillatory. It follows, therefore, that the singular vectors of B have well–defined sign patterns.

Equation (12) can be cast into a linear algebraic equation by the Kronecker product and vec operator [7],

$$\left(B^{-T} \bigotimes B\right) \text{vec } p(M) = \text{vec } q(N). \tag{13}$$

The coefficient matrix of this equation, which is of order $n^2 \times n^2$, is non–singular since B is non–singular.

The transformation of the Bézout resultant matrix between the power and Bernstein bases [2] has a similar form to (12) and is given by

$$Z_b = W^T Z_p W, \tag{14}$$

where Z_p is the Bézout resultant matrix of two power basis polynomials, Z_b is the Bézout resultant matrix of the same polynomials but expressed in the Bernstein basis, and the elements $w_{ij}, i, j = 1, \ldots, n$, of W, the transformation matrix between the power and Bernstein bases of degree $(n - 1)$, are

$$w_{ij} = \begin{cases} \binom{j-1}{i-1}\binom{n-1}{i-1}^{-1} & i \leq j, \\ 0 & i > j. \end{cases}$$

It is shown in [18] that W has the same properties as B, that is, it is totally non–negative, and WW^T and W^TW are oscillatory.

The application of the Kronecker product and vec operator to (14) yields a linear algebraic equation whose coefficient matrix is non–singular and of order $n^2 \times n^2$,

$$\left(W^{-T} \bigotimes W^{-T}\right) \text{vec } Z_b = \text{vec } Z_p. \tag{15}$$

It is shown in [8] that if the singular values of X and Y are, respectively,

$$\sigma_i(X), \; i = 1, \ldots, n, \quad \text{and} \quad \sigma_j(Y), \; j = 1, \ldots, n,$$

then the singular values of the Kronecker product $X \bigotimes Y$ are $\sigma_i(X)\sigma_j(Y), i, j = 1, \ldots, n$. The application of this result to (13) and (15) shows that the condition numbers of the coefficient matrices of these equations are

$$\kappa_2\left(B^{-T}\bigotimes B\right) = \kappa_2(B)^2 = \left(\frac{\sigma_1(B)}{\sigma_n(B)}\right)^2, \quad \text{and}$$

$$\kappa_2\left(W^{-T}\bigotimes W^{-T}\right) = \kappa_2(W)^2 = \left(\frac{\sigma_1(W)}{\sigma_n(W)}\right)^2,$$

respectively. Figure 1 shows that the 2–norm condition numbers of the coefficient matrices in (13) and (15) increase rapidly with n, and that the difference between the condition numbers is small. The large value of the condition numbers, even for small values of n, implies that these basis transformations should not be performed because they may be severely ill–conditioned. The same transformation of the Sylvester resultant matrix cannot be considered because, as shown in (11), the ratios of the singular values of this matrix are not scale–invariant.

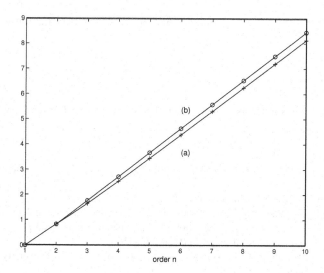

Fig. 1. The variation of (a) $\log_{10} \kappa_2 \left(B^{-T} \otimes B \right)$ and (b) $\log_{10} \kappa_2 \left(W^{-T} \otimes W^{-T} \right)$ with the order n

5   Summary

This paper has considered some algebraic and numerical properties of resultant matrices for Bernstein polynomials. It has been shown that Bernstein basis resultant matrices differ in some important aspects from their power basis and scaled Bernstein basis equivalents because of the combinatorial factors in the Bernstein basis functions. For example, it was shown that the Sylvester resultant matrix for Bernstein polynomials is not striped, an upper triangular Hankel matrix T does not define a similarity transform between a Bernstein basis companion resultant matrix and its transpose, and the Bézout and companion resultant matrices for Bernstein polynomials are not related by T.

It was shown that the Sylvester resultant matrix has some theoretical problems because a simple condition number that is appropriate for the other resultant matrices cannot be used for this matrix. This anomalous behaviour arises because the coefficients of the polynomials are decoupled in the matrix.

Four condition numbers of a resultant matrix were considered and it was shown that the componentwise measures are NP–hard, and that a structure–preserving normwise measure yields a difficult matrix approximation problem. The easiest condition number to compute is the least accurate because it is measured in the normwise manner rather than the componentwise manner, and the structure of the perturbed matrix that has unit loss of rank is not specified.

The condition numbers of the transformation of the companion and Bézout resultant matrices between the power and Bernstein bases were computed, and it was shown that they increase rapidly with the degrees of the polynomials. The combination of this result and the enhanced numerical stability of Bernstein basis resultant matrices with respect to their power basis equivalents implies that resultants should be evaluated from

Bernstein basis resultant matrices directly, and that a basis transformation should not be performed.

References

1. Barnett, S.: Polynomials and Linear Control Systems. Marcel Dekker. New York, USA, 1983.
2. Bini, D., Gemignani, L.: Bernstein-Bezoutian matrices. Theoretical Computer Science (to appear).
3. Canny, J. F.: The Complexity of Robot Motion Planning. The MIT Press. Cambridge, USA, 1988.
4. Goldman, R. N.: Private communication.
5. Goldman, R. N., Sederberg, T. W., Anderson, D. C.: Vector elimination: A technique for the implicitization, inversion and intersection of planar parametric rational polynomial curves. Computer Aided Geometric Design **1** (1984) 327–356.
6. Golub, G. H., Van Loan, C. F.: Matrix Computations. John Hopkins University Press. Baltimore, USA, 1989.
7. Higham, N. J.: Accuracy and Stability of Numerical Algorithms. SIAM. Philadelphia, USA, 2002.
8. Horn, R. A., Johnson, C. R.: Topics in Matrix Analysis. Cambridge University Press. Cambridge, England, 1991.
9. Kajiya, J. T.: Ray tracing parametric patches. Computer Graphics **16** (1982) 245–254.
10. Manocha, D., Canny, J. F.: Algorithms for implicitizing rational parametric surfaces. Computer Aided Geometric Design **9** (1992) 25–50.
11. Petitjean, S.: Algebraic geometry and computer vision: Polynomial systems, real and complex roots. Journal of Mathematical Imaging and Vision **10** (1999) 191–220.
12. Winkler, J. R.: A resultant matrix for scaled Bernstein polynomials. Linear Algebra and its Applications **319** (2000) 179–191.
13. Winkler, J. R.: Computational experiments with resultants for scaled Bernstein polynomials, in *Mathematical Methods for Curves and Surfaces: Oslo 2000*, Tom Lyche and Larry L. Schumaker (eds.). Vanderbilt University Press. Nashville, Tennessee, USA (2001) 535–544.
14. Winkler, J. R.: Properties of the companion matrix resultant for Bernstein polynomials, in *Uncertainty in Geometric Computations*, J. R. Winkler and M. Niranjan (eds.). Kluwer Academic Publishers. Massachusetts, USA (2002) 185–198.
15. Winkler, J. R.: A companion matrix resultant for Bernstein polynomials. Linear Algebra and its Applications **362** (2003) 153–175.
16. Winkler, J. R.: The transformation of the companion matrix resultant between the power and Bernstein polynomial bases. Applied Numerical Mathematics **48** (2004) 113–126.
17. Winkler, J. R., Goldman, R. N.: The Sylvester resultant matrix for Bernstein polynomials, in *Curve and Surface Design: Saint-Malo 2002*, Tom Lyche, Marie-Laurence Mazure and Larry L. Schumaker (eds.). Nashboro Press. Brentwood, Tennessee, USA (2003) 407–416.
18. Winkler, J. R., Ragozin, D. L.: A class of Bernstein polynomials that satisfy Descartes' rule of signs exactly, in *The Mathematics of Surfaces IX*, Roberto Cipolla and Ralph Martin (eds.). Springer-Verlag. London, England (2000) 424–437.

Polynomial C^2 Spline Surfaces Guided by Rational Multisided Patches

Kęstutis Karčiauskas[1] and Jörg Peters[2]

[1] Vilnius University, Vilnius 03225, Lithuania,
`kestutis.karciauskas@maf.vu.lt`,
WWW home page: `http://www.mif.vu.lt/cs2/cagl/`
[2] University of Florida, Gainesville, FL 32611, USA
`jorg@cise.ufl.edu`,
WWW home page: `http://www.cise.ufl.edu/~jorg/`

Abstract. An algorithm is presented for approximating a rational multi-sided M-patch by a C^2 spline surface. The motivation is that the multi-sided patch can be assumed to have good shape but is in nonstandard representation or of too high a degree. The algorithm generates a finite approximation of the M-patch, by a sequence of patches of bidegree $(5, 5)$ capped off by patches of bidegree $(11, 11)$ surrounding the extraordinary point.

The philosophy of the approach is (i) that intricate reparametrizations are permissible if they improve the surface parametrization since they can be precomputed and thereby do not reduce the time efficiency at runtime; and (ii) that high patch degree is acceptable if the shape is controlled by a guiding patch.

1 Introduction

When constructing C^2 spline surfaces using a finite number of tensor-product Bézier patches such as [4, 11, 13, 7], the shape is often unsatisfactory near extraordinary points where more or less than four patches meet since the curvature is not evenly distributed or shows local extrema not implied by the surrounding data [10]. For subdivision schemes, the cause of shape artifacts has recently been analyzed and made explicit (see [9], [6]).

By contrast, rational multisided M-patches [5], joined smoothly to a surrounding B-spline complex, appear to consistently yield good shape. This paper does not verify the empirical observation of good shape but explores the technical challenge of how to transfer a good M-patch into a standard spline framework.

M-patches are rational and can therefore be represented as a collection of rational tensor-product Bézier subpatches. If the number of sides is 5 or 6, a variant of the M-patch can be represented as a collection of rational tensor-product Bézier patches of bidegree $(8, 8)$. But for a general m-sided M-patch, the bidegree is $(4(m-2), 4(m-2))$.

The idea pursued in this paper is to capture the shape of the M-patch with a C^2 approximation of moderate bidegree. We describe a finite approximation of the M-patch, by a sequence of patches of bidegree $(5, 5)$ but with patches adjacent to the extraordinary point of bidegree $(11, 11)$. A key point is the definition of reparametrizations that improve the surface parametrization when approximating the M-patch. These reparametrization, maps $\mathbb{R}^2 \to \mathbb{R}^2$, decompose the domain m-gon into C^2-connected annuli or rings and a central piece. The surface approximation consists correspondingly of nested annuli and a final cap. Each annulus and the cap pick up second order Hermite data from the M-patch.

The bidegree of the central, extraordinary patches is still high, but all experiments with constructions of lower the bidegree have so far resulted in a considerably reduction of the surface quality. The paper focuses on the technical challenge of creating a C^2 surface that Hermite-interpolates a given M-patch.

One of the key ideas, construction of good reparametrizations and composition with a polynomial patch that determines the shape is a logical extension of similar ideas proposed in [11, 13, 7].

The paper is organized as follows.
Section 2 defines M-patches and the transition of the M-patch to an existing spline complex.
Section 3 defines the transitional reparametrizations and the patches of bidegree $(5, 5)$.
Section 4 describes the cap of patches of bidegree $(11,11)$ that approximates the M-patch near the extraordinary point.

The exposition expects familiarity with standard representations of geometric design. The tensor-product Bernstein-Bézier and (uniform) B-spline representations (see [1, 3, 12]) have the form $\sum_{i=0}^{d_1} \sum_{j=0}^{d_2} \mathbf{c}_{ij} b_i^{d_1}(u) b_j^{d_2}(v)$, where b_i^d is the ith basis function. In the case of the Bernstein-Bézier form, $b_i^d(u) := \binom{d}{i}(1-u)^{d-i}u^i$. In particular, one can choose the domain as a unit square and then associate layers of coefficients $\mathbf{c}_{ij}$, $j \leq k$ with kth derivatives perpendicular to an edge of the square. In the following, we will often refer to *the three boundary layers* of a polynomial to mean the layers of coefficients $\mathbf{c}_{ij}$, $j \leq 3$ that determine position, first and second derivative across a boundary. Catmull-Clark subdivision [2] generalizes the refinement of bicubic uniform splines to control nets with nodes of arbtrary valence as illustrated in Fig. 1

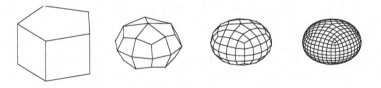

Fig. 1. Three steps of Catmull-Clark subdivision

2   M-patches and their tensor-border

In this section, we review the construction of rational multi-sided patches, called M-patches [5]. First, M-patches are defined. Then a control structure, called tensor-border, is defined that mimics the behavior of tensor-product patches along the boundary. Such a control structure can be derived from an M-patch by combining M-patch basis functions as in [5] or, as in our context, by reparametrization. A third subsection sketches such a reparametrization. That is, the M-patch composed with the reparametrization has tensor-border structure, suitable for smoothly joining the patch to a an existing spline complex and filling an m-sided hole.

2.1 Definition of M-Patches

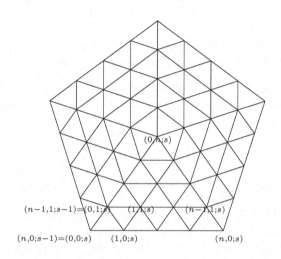

Fig. 2. A nested polygon net $\mathcal{NP}_5^n$

Let P_m be a regular m-gon, $m > 4$, with center $\mathbf{0}$ and vertices

$$\mathbf{v}_s := (\cos s\alpha, \sin s\alpha), \quad \alpha = 2\pi/m, s = 0, 1, \ldots, m - 1.$$

Subscripts are assigned in a cyclic fashion, e.g. $\mathbf{v}_{s+1} = \mathbf{v}_0$ if $s = m - 1$ and $\mathbf{v}_{s-1} = \mathbf{v}_{m-1}$ if $s = 0$. We abbreviate $c := \cos \alpha$.

An m-sided *nested polygon net* $\mathcal{NP}_m^n$ of *depth* n is a set of points in $\mathbb{R}^2$,

$$\frac{n - i - j}{n}\mathbf{v}_s + \frac{i}{n}\mathbf{v}_{s+1} + \frac{j}{n}\mathbf{0}, \quad 0 \le s \le m - 1, \ 0 \le i + j \le n$$

connected as shown in Fig.2. That is, each triangle $\triangle \mathbf{v}_s \mathbf{v}_{s+1} \mathbf{0}$ is triangulated in the manner of Greville abscissae of the Bézier form. There are $mn(n + 1)/2 + 1$ points in $\mathcal{NP}_m^n$. It is convenient to refer to the point

$$((n - i - j)\mathbf{v}_s + i\mathbf{v}_{s+1} + j\mathbf{0})/n \quad \text{by the triple } (i, j; s)$$

where $(n - j, j; s - 1) = (0, j; s)$.

An edge E_s is a line through the vertices $\mathbf{v}_s$, $\mathbf{v}_{s+1}$. We set

$$\ell_s(x, y) := -x \cos(s\alpha + \frac{\alpha}{2}) - y \sin(s\alpha + \frac{\alpha}{2}) + \cos \frac{\alpha}{2}.$$

The linear function ℓ_s is zero on the edge E_s and positive at the other points of P_m. The edges E_{s-1} and E_{s+1} intersect at a point $\mathbf{k}_s$.

Set $C(x, y) = \frac{\cos^2(\alpha/2)}{\cos^2 \alpha} - x^2 - y^2$. The equation $C(x, y) = 0$ defines a circle passing through the points $\mathbf{k}_s$. For $q = (i, j; s)$ we define

$$f_q := k_{ij}^n g_s^{n-i-j} g_{s+1}^i g^j,$$

where

$$g_s := C \prod_{\sigma \neq s-1, s} \ell_\sigma, \quad g := \prod_{\sigma=0}^{m-1} \ell_\sigma.$$

and k_{ij}^n are positive numbers that satisfy $k_{ij}^n = k_{n-i-j,j}^n$ and $k_{i0}^n = \binom{n}{i}$. Examples are given in [5]. The functions f_q are the *basis* functions of M-patch.

Definition 1. *The m-sided M-patch with the control points $\mathbf{q}_\sigma \in \mathbb{R}^3$ and their weights w_σ, $\sigma \in \mathcal{NP}_m^n$, is the image of the mapping $F : P_m \to \mathbb{R}^3$ defined by the formula*

$$F = \frac{\sum_{\sigma \in \mathcal{NP}_m^n} w_\sigma \mathbf{q}_\sigma f_\sigma}{\sum_{\sigma \in \mathcal{NP}_m^n} w_\sigma f_\sigma}. \tag{1}$$

2.2   Tensor-border Structure

To be able to tie an M-patch into a tensor-product spline complex and fill an m-sided hole, we define a *tensor-border net* with quadrilateral mesh cells. In fact, a tensor-border net is a collection of overlapping quadrilateral nets as shown in Fig. 3. The connectivity and enumeration is as follows.

For fixed integers $m, n, k \leq \lceil n/2 \rceil$ an m-sided *tensor-border net* $\mathcal{T}_m(n, k)$ of *degree* n and *order* k is indexed by the set of triples

$$[i, j; s], \quad s = 0, \ldots, m-1, \quad i = 0, \ldots, n, \quad j = 0, \ldots, k.$$

The triples are identified via

$$[i, j; s] = [j, n - i; s + 1], \text{ for } i \geq n - k.$$

Here, we use the notation $[i, j; s]$ to avoid confusion with the subscripts of the basis functions of M-patches.

Definition 2. *A patch F has a* tensor-border *of order k and of degree n with the vertices $\mathbf{q}_\sigma$, $\sigma = [i, j; s] \in \mathcal{T}_m(n, k)$, if, along each edge E_s, it can be locally reparametrized by maps $\rho^{k;s} : \mathbb{R}^2 \to \mathbb{R}^2$ with parameters (u, t) in such a way that the crossderivatives ∂_t^κ up to order k,*

$$\left. \frac{\partial^\kappa F \circ \rho^{k;s}}{\partial^\kappa t} \right|_{t=0}, \quad \kappa \leq k,$$

coincide with the crossderivatives of the tensor-product patch of bidegree (n, n) with the control points $\mathbf{q}_{ij} := \mathbf{q}_{[i,j;s]}$.

The reparametrization $\rho^{k;s}$ is defined in the next subsection.

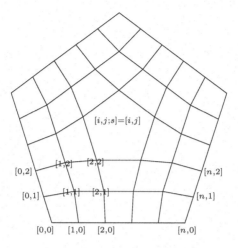

Fig. 3. Tensor-border net of order 2

2.3 Reparametrizations to Form a Tensor-Border

It is convenient to construct the reparametrizations $\rho^{k;s}$ in Bézier form. Since we are interested in crossderivatives up to order 2, the first three layers of Bézier control points are calculated. We construct the reparametrizations $\rho^{2;s}$ based on two simpler maps $\rho^{0;s}$ and $\rho^{1;s}$, as follows.

(1) $\rho^{0;s} := \mathbf{v}_s(1 - u) + \mathbf{v}_{s+1}u$ for u on the edge E_s;
(2) $\rho^{1;s}$ is of bidegree $(3,3)$ (see Fig. 4 *left*) and its Bézier coefficients $\rho_{i0}^{1;s}, i = 0, \ldots, 3$
     represent $\rho^{0;s}$ in degree-raised form;
     the coefficients $\rho_{01}^{1;s}$ and $\rho_{31}^{1;s}$ are defined by symmetry;
     $\rho_{11}^{1;s} := \mathrm{sc}_1\mathbf{v}_s, \rho_{21}^{1;s} := \mathrm{sc}_1\mathbf{v}_{s+1}$, where

$$\mathrm{sc}_1 := \frac{3 + 12c + 4c^2 + 8c^3}{9(1 + 2c)}.$$

(3) $\rho^{2;s}$ is of bidegree $(5,5)$ (see Fig. 4 *right*) and its Bézier coefficients $\rho_{ij}^{2;s}, i = 0, \ldots, 5, j = 0, 1$, represent $\rho^{1;s}$ in degree-raised form;
     the coefficients $\rho_{02}^{2;s}, \rho_{12}^{2;s}, \rho_{42}^{2;s}, \rho_{52}^{2;s}$ are defined by symmetry;
     $\rho_{22}^{2;s} := \mathrm{sc}_2\mathbf{v}_s, \rho_{32}^{2;s} := \mathrm{sc}_2\mathbf{v}_{s+1}$, where

$$\mathrm{sc}_2 := \frac{5 + 41c + 129c^2 + 168c^3 + 132c^4 + 120c^5 + 48c^6 + 32c^7}{25(1 + 2c)^3}.$$

Theorem 3. *After reparametrization with $\rho^{2;s}$ an M-patch of depth* 4 *has a tensor-border of order* 2 *and of degree* 7.

The proof is analogous to Proposition 11 of [5].

One further challenge of the construction is that the tensor-border is C^∞ along the domain edge E_s while the corresponding edge in the spline complex is a spline of finite smoothness. The two can be joined by a annulus of patches that match the spline data on the outside and form, along each inside edge, a single polynomial.

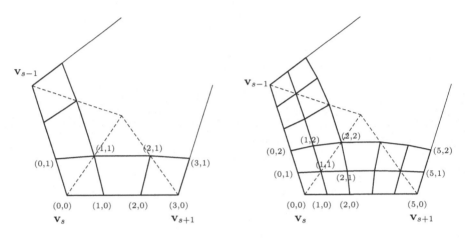

Fig. 4. Reparametrizations of M-patches: $\rho^{1;s}$ for the first cross-derivatives; $\rho^s := \rho^{2;s}$ for the first and second cross-derivatives

3   Approximating the M-Patch by Successive Annuli of Bidegree (5,5)

We now assume that the M-patch has been constructed by fixing the tensor-border to match the given spline complex in a C^2 fashion and fill its m-sided hole. The goal of the following construction is to approximate the M-patch by annuli of the tensor-product patches of bidegree $(5,5)$. Degree $(5,5)$ allows matching second order data from an outer annulus and, independently, using three inner layers of Bézier coefficients to pick up second order Hermite data from concentric curves surrounding the central point Q of the M-patch. An infinite sequence of these annuli can be constructed in the spirit of subdivision surfaces, but here we will generate only a few annuli before capping off with the construction detailed in sect. 4.

3.1   Outline of the Construction

The control points of Bézier patch $G(u, v)$ of bidegree (5,5) are fully determined by the values of its Taylor expansions

$$\partial_u^i \partial_v^j G, \quad 0 \le i, j \le 2,$$

at the corners $(0,0), (1,0), (0,1), (1,1)$. These (Hermite) data are derived by reparametrizing the domain polygon P_m and calculating the derivatives of the reparametrized M-patch. Calculating the derivatives of this composition is a routine job if the reparametrizations is given.

Free parameters are determined by minimizing the following functional.

Definition 4. *For a function h with parameters (u, v) in the unit square $I^2 = [0..1]^2$, we define the functional*

$$\mathcal{F}_5(h) := \int\!\!\int_{I^2} \sum_{i+j=5, i, j \geq 0} \binom{5}{i} (\partial_u^i \partial_v^j h)^2.$$

For a mapping $h = (h_1, h_2) : I^2 \to \mathbb{R}^2$,

$$\mathcal{F}_5(h) := \mathcal{F}_5(h_1) + \mathcal{F}_5(h_2).$$

3.2 The C^2 Transition Between the Reparametrizations $\rho^{2;s}$ and a Bicubic C^2 Reparametrization

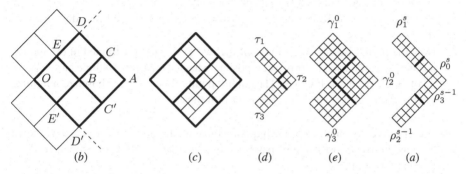

(b)   (c)   (d)   (e)   (a)

Fig. 5. Three inner layers, corresponding to second-order boundary data, of the bidegree $(3,3)$ reparametrization τ_i are constructed in $(b),(c),(d)$. Three outer layers of the $(5,5)$ partitioned tensor-border reparametrization $\rho^{2;s}$ are shown (a). The reparametrization γ^0 of bidegree $(5,5)$ in fig. (e) is the result of juxtaposing the three inner and outer layers.

The reparametrization $\rho^{2;s} : \mathbb{R}^2 \to \mathbb{R}^2$ is of bidegree $(5,5)$. This is an unnecessarily high bidegree for forming an m-sided C^2 annulus. Since the bidegree of the innermost patches increases with the degree of the reparametrization, it is convenient to transition to a C^2 annulus of bidegree $(3,3)$ whose properties are well-known and that can be refined by the well-known Catmull-Clark rules.

To transition between the high-degree reparametrization $\rho^{2;s}$ and a (piecewise) bicubic C^2 reparametrization $\tau : \mathbb{R}^2 \to \mathbb{R}^2$ of the first annulus, a reparametrization γ^0 of bidegree $(5,5)$ is stitched together from the boundary layers of Bézier coefficients of $\rho^{2;s}$ and τ: the outer three layers Hermite-interpolate $\rho^{2;s}$ up to second order and the inner three layers are taken from τ.

If, as is natural, we choose a bicubic τ with m-fold symmetry, we can define its relevant part by five spline control points A,B,C,D,E, each in $\mathbb{R}^2$ (see Fig. 5 b). Here, we say relevant part, since we only need the innermost three layers of Bézier coefficients of the bicubic annulus for γ^0. A third annulus would only be needed if we wanted to define the complete bicubic τ. The innermost three layers are defined by a central spline control point $0 = (0,0)$ and two surrounding annuli of spline control points. Under symmetry, with A fixed to establish scale, there are a total of 5 scalar degrees of freedom. Due to symmetry, it suffices to describe one sector of γ^0.

(a) The reparametrization $\rho^{2;s}(u, v)$ (for $s = 0$) of bidegree $(5, 5)$ is subdivided into four parts

$$\rho_k^s := \rho^{2;s}\left(\frac{k}{4}(1 - u) + \frac{1 + k}{4}u, \frac{1}{4}v\right), \quad k = 0, \ldots, 3.$$

Two pieces ρ_1^s, ρ_0^s of $\rho^{2;s}$ and two pieces of ρ_3^{s-1}, ρ_2^{s-1} of $\rho^{2;s-1}$ form an L-shape. Of this (outer) L-shape, we need only calculate three layers in Bézier form as shown in Fig. 5 a. By construction of $\rho^{2;s}$, the overlap of ρ_0^s and ρ_3^{s-1} is consistent.

(b,c) The (undetermined) control points $A, \ldots, E$ (Fig. 5 b) determine the three boundary layers of a second (inner) L-shape made up of bicubic patches (Fig. 5 c).

(d) The three layers of Bézier coefficients in (c) are degree-raised to bidegree $(5, 5)$ (Fig. 5 d).

(e) Now the partial Bézier control nets of the two L shapes constructed in (a) and (d) are simply juxtaposed to form a composite L shape, consisting of three parts γ_1^0, γ_2^0, γ_3^0 of bidegree $(5, 5)$ (Fig. 5 e). The control points $A, \ldots, E$ are chosen to minimize the functional $\mathcal{F}_5(\tau_1) + \mathcal{F}_5(\tau_2) + \mathcal{F}_5(\tau_3)$ of these parts.

One sector of the first annulus of the biquintic reparametrization γ^0 for $m = 8$ is shown in Fig. 6 as the rightmost triple of patches. By construction, γ^0 is non-singular along inner and outer annulus edges.

3.3 Subsequent Annuli of Bidegree (5,5)

The first annulus of reparametrizations was created with the help of B-spline control points $A, \ldots, E$. We apply Catmull-Clark subdivision rules to these points and convert the contracting control point nets to contracting annuli of bicubic C^2-reparametrization γ^k (Fig. 6).

The subdivided B-spline control point nets converge rapidly to the characteristic configuration. The characteristic configuration defines a bicubic annulus γ^c, the characteristic map. At the Kth iteration, we transition from γ^K to γ^c using a second single annulus of reparametrizations of bidegree (5,5) that Hermite-blends inner and outer parts of the two annuli exactly as in the construction of γ^0. The nth subsequent reparametrization can then be chosen as $\gamma^{K+n} = \lambda^n \gamma^c$ where

$$\lambda := \frac{1}{16}(c + 5 + \sqrt{(c + 9)(c + 1)})$$

is the subdominant eigenvalue of Catmull-Clark subdivision.

From basic facts of Catmull-Clark subdivision, it follows that the constructed reparametrizations γ^k are C^2 within each annulus and join C^2 with adjacent annuli.

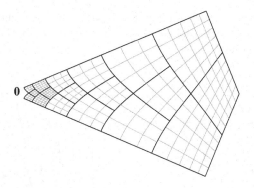

Fig. 6. A sector of the contracting sequence of reparametrizations γ^k. The grey parameter lines reveal the bidegree: $(3, 3)$ except for the outermost layer shown on the right which is of bidegree $(5, 5)$.

3.4 Annuli of Bézier Patches

We now construct annuli of Bézier patches of bidegree (5,5) that

- are internally C^2 and and join C^2 with the adjacent annuli and that
- approximate the M-patch.

Every reparametrization γ^k is composed with the map F that defines the M-patch. Denote the composition G with parameters (u, v). Each bidegree $(5, 5)$ patch is constructed as follows.

4	4	4	3	3	3
4	4	4	3	3	3
4	4	4	3	3	3
1	1	1	2	2	2
1	1	1	2	2	2
1	1	1	2	2	2

Fig. 7. Grouping of the Bernstein-Bézier coefficients of a patch of bidegree $(5, 5)$

(1) the values of $\partial_u^i \partial_v^j G$ for $0 \le i, j \le 2$. are calculated at the vertices $(0, 0)$, $(1, 0)$, $(0, 1)$, $(1, 1)$;
(2) the calculated data are converted to Bézier form as a part of a patch of bidegree $(5, 5)$;
(3) the four parts are merged into one patch of bidegree $(5, 5)$ (see Fig. 7);

(4) the first three layers – marked in Fig. 8 *left* – of a new, inner annulus are replaced by the subdivided C^2 extension of the last three layers – marked in Fig. 8 *right* – of the old, outer annulus.

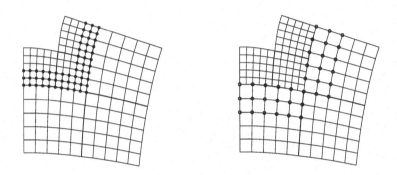

Fig. 8. C^2 extension of the data to inner annulus

Since the reparametrizations are (parametrically) C^2 and non-singular across patch boundaries,

- the patches are C^2 in each annulus and
- adjacent annuli join C^2.

Remark 5. If we make the sequence of patches infinite, we obtain a *guided subdivision* construction of degree (5,5). Such a construction can be shown to have a limit point Q, that is the image of the center of the domain polygon under the M-patch. Furthermore, the surface is G^1 and the limit of the tangent planes is the tangent plane of the M-patch at Q. Experiments show that any curvature oscillation is very small. A similar construction of bidegree (6,6) can be shown to be curvature continuous with the limit of Gaussian and mean curvatures equal the corresponding curvatures of the M-patch at Q. However, here, we are concerned only with the finite construction.

4   A Spline Cap of Bidegree (11,11)

In this section, we will develop a 'cap' of m patches of bidegree $(11, 11)$ that are G^2-joined around Q, the central point of the M-patch. While this bidegree is higher than recent C^2 constructions in the literature, it has well-controlled shape. Following the approach from [7], the three boundary layers of the patches are taken from the composition $PW_7 \circ \tilde{\gamma}^s$ where $\tilde{\gamma}^s : \mathbb{R}^2 \to \mathbb{R}^2$ is a reparametrization and $PW_7 : \mathbb{R}^2 \to \mathbb{R}^3$ defines the shape. The construction of the relevant layers of the maps is as follows (see Fig.s 9, 11):

(1) Set

$$\mathbf{v}'_s := \left(\cos \left(s\alpha - \frac{\alpha}{2} \right), \sin \left(s\alpha - \frac{\alpha}{2} \right) \right), \quad \overline{\mathbf{v}}_s := \left(\frac{\cos(s\alpha)}{\cos(\alpha/2)}, \frac{\sin(s\alpha)}{\cos(\alpha/2)} \right).$$

(2) the (three boundary layers of the) symmetric tensor-product reparametrizations $\tilde{\gamma}^s$ are defined in each sector $\mathbf{v}'_s\mathbf{0}\mathbf{v}'_{s+1}$;

(3) an auxiliary piecewise C^2 mapping PW_7 of degree 7 is constructed in each of the triangles $\triangle\mathbf{0}\mathbf{v}'_s\overline{\mathbf{v}}_s$ and $\triangle\mathbf{0}\overline{\mathbf{v}}_s\mathbf{v}'_{s+1}$;

(4) the composition $PW_7 \circ \tilde{\gamma}^s$ is a tensor-product of bidegree $(11, 11)$; its three boundary layers are well-defined from the earlier construction an can be extracted.

We now describe the relevant layers of maps PW_7 and $\tilde{\gamma}^s$ in detail.

4.1   The 135-Reparametrizations $\tilde{\gamma}^s$

In this section, we construct the reparametrization $\tilde{\gamma}^s$ shown in Fig. 9. The reparametrization is almost completely pinned down by the requirements that

- $\tilde{\gamma}^s$ is a piece of a control net with m-fold symmetry and
- the ith boundary layer of $\tilde{\gamma}^s$ in Bézier form represents a curve of degre $2i + 1$.

The remaining degrees of freedom are denoted by t_i and are set by minimizing a functional. Note that this calculation is done once and for all and we get specific Bernstein-Bézier coefficients $\tilde{\gamma}^s_{ij}$ that define the first three layers of a polynomial $\tilde{\gamma}^s$ of bidegree $(5,5)$.

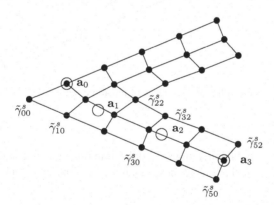

Fig. 9. Boundary layers of Bernstein-Bézier coefficients of the 135-reparametrization $\tilde{\gamma}^s$

- $\tilde{\gamma}^s_{i0} := (1 - i/5)\mathbf{0} + (i/5)\mathbf{v}'_s$, $i = 0,\ldots,5$; the points $\tilde{\gamma}^s_{0i}$, $i = 1,\ldots,5$, are symmetric to the points $\tilde{\gamma}^s_{i0}$;
- the points $\tilde{\gamma}^s_{i1}$, $i = 0,\ldots,5$, represent a cubic with the control points $\mathbf{a}_0 := \tilde{\gamma}^s_{01}$, $\mathbf{a}_1 := (2/3)\mathbf{0} + (1/3)\mathbf{v}'_s + (1/5)\tan(\alpha/2)\overrightarrow{\mathbf{N}}_s$, $\mathbf{a}_2 := (1/3)\mathbf{0} + (2/3)\mathbf{v}'_s + t_0\overrightarrow{\mathbf{N}}_s$, $\mathbf{a}_3 := \mathbf{v}'_s + t_1\overrightarrow{\mathbf{N}}_s$ in degree-raised form where $\overrightarrow{\mathbf{N}}_s := (-\sin(s\alpha - \alpha/2), \cos(s\alpha - \alpha/2))$ is the normal perpendicular to the edge $\overline{\mathbf{v}}_s, \overline{\mathbf{v}}_{s+1}$; the points $\tilde{\gamma}^s_{1i}$, $i = 2,\ldots,5$, are symmetric to the points $\tilde{\gamma}^s_{i1}$;

- $\tilde{\gamma}_{22}^s := (1 - t_2)\mathbf{0} + t_2\overline{\mathbf{v}}_s$, $\tilde{\gamma}_{i2}^s := (1 - t_i)\mathbf{0} + t_i\mathbf{v}_s' + 2(\tilde{\gamma}_{i1}^s - \tilde{\gamma}_{i0}^s)$, $i = 3, \ldots, 5$; the points $\tilde{\gamma}_{2i}^s$, $i = 3, \ldots, 5$, are symmetric to the points $\tilde{\gamma}_{i2}^s$.

The three boundary layers of the reparametrization just constructed define curves of degree 1, 3 and 5. Therefore $\tilde{\gamma}^s$ is denoted as 135-reparametrization. The reparametrizations $\rho^{2;s}$ from sect. 2.3 are also of type 135.

Suppose $R \geq 2$ annuli of Bézier surfaces of bidegree $(5, 5)$ are to be generated before the final cap of bidegree $(11, 11)$. We now fix the free parameters t_i and adjust the contracting reparametrizations of the last two annuli. This procedure is explained below and illustrated for the case $R = 3$ in Fig. 10. As usual, due to symmetry, only one sector need be considered.

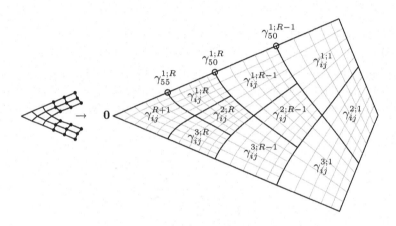

Fig. 10. Finite sequence of reparametrizations

(1) The contracting reparametrizations γ^r, defined in sect. 3, are degree-raised to bidegree $(5, 5)$; the control points of these reparametrizations are denoted by $\gamma_{ij}^{1;r}$, $\gamma_{ij}^{2;r}$, $\gamma_{ij}^{3;r}$.

(2) the points $\gamma_{ij}^{k;r}$, $r \leq R - 2$, remain unchanged;

(3) the part of $\gamma_{ij}^{k;R-1}$ that guarantees a C^2 join to the previous $(R - 2)$-th annulus remain unchanged;

(4) the points $\tilde{\gamma}_{ij}^s$ are denoted γ_{ij}^{R+1};

(5) the points $\gamma_{ij}^{1;R}$, $3 \leq i, j \leq 5$, are changed to Hermite-interpolate the three boundary layers of γ_{ij}^{R+1}; this guarantees a C^2 join between $\gamma^{1;R}$ and γ^{R+1} for those layers; the points $\gamma_{ij}^{3;R}$ are changed analoguously;

(6) the remaining points $\gamma_{ij}^{k;R-1}$, $\gamma_{ij}^{k;R}$ are changed to satisfy the conditions of symmetry and of a C^2 join;

(7) the reparametrizations defined by the new points $\gamma_{ij}^{k;R-1}$ and $\gamma_{ij}^{k;R}$ are $\gamma^{k;R-1}$ and $\gamma^{k;R}$, respectively; the remaining coefficients of $\gamma^{k;R-1}$ and $\gamma^{k;R}$ and the parame-

ters t_i, $i \in \{0, 1, 3, 4, 5\}$ are determined by minimizing the functional

$$\sum_{k=1}^{3} (\mathcal{F}_5(\gamma^{k;R-1}) + \mathcal{F}_5(\gamma^{k;R}));$$

(8) the points γ_{ij}^{R+1}, $3 \leq i, j \leq 5$, are determined via the C^2 conditions for joining γ^R to γ^{R+1}.

(9) the parameter t_2 is calculated by minimizing a functional $\mathcal{F}_5(\gamma^{R+1})$, where γ^{R+1} is the reparametrization controled by the points γ_{ij}^{R+1}, $0 \leq i, j \leq 5$.

4.2 A Total Degree 7 C^2 Approximation of the Geometry

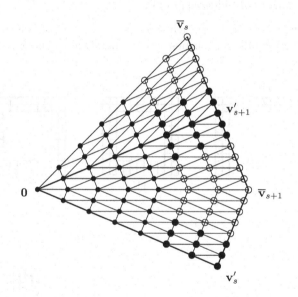

Fig. 11. Schematic view of the Bernstein-Bézier coefficients of three pieces of the auxiliary mapping PW_7 of total degree 7

The sector $\mathbf{v}'_s \mathbf{0} \mathbf{v}'_{s+1}$ is split into the triangles $\triangle \mathbf{0} \mathbf{v}'_s \bar{\mathbf{v}}_s$ and $\triangle \mathbf{0} \bar{\mathbf{v}}_s \mathbf{v}'_{s+1}$. The control points of a piecewise C^2 mapping PW_7 of degree 7 over $\triangle \mathbf{0} \mathbf{v}'_s \bar{\mathbf{v}}_s$, respectively over $\triangle \mathbf{0} \bar{\mathbf{v}}_s \mathbf{v}'_{s+1}$, are denoted by $\mathbf{q}_{ijk}^{2s}$ and $\mathbf{q}_{ijk}^{2s+1}$, $i + j + k = 7$. The common central point is $\mathbf{q}_{700}^{2s} = \mathbf{q}_{700}^{2s+1}$.

Let $rep := u\gamma_{50}^{R+1} + v\gamma_{05}^{R+1} + (1 - u - v)\mathbf{0}$ and $G = F \circ rep$, where F defines the M-patch. The values of

$$\partial_u^i \partial_v^j G, \quad 0 \leq i + j \leq 4 \quad \text{at } (0, 0)$$

are calculated and converted to Bézier form of total degree 4. These define a quartic mapping T_4. The points $\mathbf{q}_{ijk}^s$, $i \geq 3$, (in Fig. 11 marked as the small disks) are control points of T_4 after raising the degree to 7.

The remaining control points insure a C^2 join between adjacent sectors and are marked as big disks in Fig. 9. For these points the conditions of C^2 join are the same as for the control points of tensor-product surfaces. They are uniquely determined by the requirement that data derived from $PW_7 \circ \tilde{\gamma}^s$ be joined C^2 to the patches of bidegree $(5,5)$ from the R-th annulus.

The control points marked as the circles do not affect the construction.

Remark 6. Splitting the sector $\mathbf{v}_s' 0 \mathbf{v}_{s+1}'$ into the triangles $\triangle 0 \mathbf{v}_s' \overline{\mathbf{v}}_s$ and $\triangle 0 \overline{\mathbf{v}}_s \mathbf{v}_{s+1}'$ simplifies the intermediate expressions so that a computer algebra system can derive explicit formulas.

4.3   Capping Patches of Bidegree $(11, 11)$

We now describe how the Bézier patches G_Q^s, $s = 1, \ldots, m$ of bidegree $(11, 11)$ are constructed that cap off the construction at the extraordinary point Q.

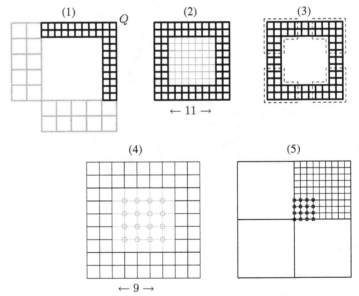

Fig. 12. Construction of the capping patch G_Q^s

The cross-derivatives of $PW_7 \circ \tilde{\gamma}^s(u, v)$,

$$PW_7 \circ \tilde{\gamma}^s|_{v=0}, \quad (PW_7 \circ \tilde{\gamma}^s)_v|_{v=0}, \quad (PW_7 \circ \tilde{\gamma}^s)_{vv}|_{v=0}$$

are of degrees 7, 9 and 11 respectively. For each edge of G_Q^s emmanating from Q, this data defines the first three layers of Bézier coefficients so that G_Q^{s-1} and G_Q^s are joined with curvature continuity, i.e. the cap is internally curvature continuous. Fig. 12 (1) shows these layers colored black.

The data of the ordinary patches from the R-th annulus, adjacent to the extraordinary patches, are colored grey. These data, degree-raised to bidegree $(11, 11)$ and C^2 extended, define three new layers for the remaining two edges of G_Q^s that are not connected to Q. The cap is thereby joined C^2 to the innermost ordinary annulus.

In Fig. 12 (2), the layers of both types are colored black while the remaining 36 inner control points $\mathbf{q}_{ij}$, $3 \le i, j, \le 8$ are colored grey. These 36 inner control points are determined by the guiding M-patch as follows.

- The four corners of already defined control points (marked in Fig. 12 (3) by the dashed lines) are each rerepresented in bidegree $(9, 9)$ form;
- the degree-decreased corners are connected to form a part of a patch of bidegree $(9, 9)$; this part is displayed via black lines in Fig. 12 (4);
- the remaining 16 inner control points $\mathbf{q}_{i,j}^9$ (marked as grey circles in Fig. 12 (4)) are in $1 - 1$ corresponence to the 16 corner control points of one quarter $\tilde{G}$ of the patch; these corner control points are marked in Fig. 12 (4) as the black circles and in turn they are defined by the values of

$$\partial_u^i \partial_v^j \tilde{G}, \quad 0 \le i, j \le 3$$

at the corresponding corner;
- the just mentioned control points $\mathbf{q}_{i,j}^9$ are fixed using the M-patch:
  (a) the reparametrization γ^{R+1} from sect. 4.1 corresponding to the same sector as G_Q^s is selected;
  (b) the composition $F \circ \gamma^{R+1}$ is subdivided into four;
  (c) the control points $\mathbf{q}_{ij}^9$ are calculated to match the compositions;
- The auxiliary patch of bidegree $(9, 9)$ is degree-raised to bidegree $(11, 11)$; its 36 inner control points (with indices ij, $3 \le i, j \le 8$) are taken as the inner control points $\mathbf{q}_{ij}$ of G_Q^s.

Remark 7. The order 1 part of a tensor-border structure of an extraordinary patch is of degree 9. Hence no information is lost when an auxilary patch of bidegree $(9, 9)$ is built. The remaining tensor-border structure is defined either by the piecewise mapping PW_7 of degree 7 or by adjacent ordinary patches of bidegree $(5, 5)$. Thus the tensor-border of an extraordinary patch is controlled by the auxiliary patch. An auxiliary patch of bidegree $(9, 9)$ simplifies the implementation of the algorithm since only partial derivatives of lower order must be calculated for the guide surface.

5 Conclusion

We have presented the construction of a polynomial C^2 spline surface that approximates a rational multi-sided M-patch. The key of the approach is the construction of a C^2 reparametrization to transmit the data of a guiding M-patch to the spline surface.

The method can be applied to other surfaces, not necessarily rational and it can be used to construct an infinte sequence of annuli in the sprit of subdivision algorithms.

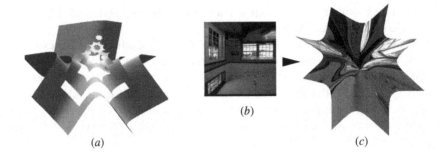

(b)

(a) (c)

Fig. 13. A sample surface based on an unsymmetric monkey saddle and valence $m = 8$. (a) shows the initial data, three annuli of bidegree $(5,5)$ and the cap of bidegree $(11,11)$. (b) is a bitmap placed as an environment map onto the surface (c); of course only interactive, detailed (sectional) curvature analysis and reflection lines can establish the claim of good shape which, in turn, depends on the assumption that the M-patch has good shape.

References

1. de Boor, C.: B-Form Basics. In Gerald Farin, editor, *Geometric Modeling : Algorithms and New Trends*, pages 131–148, Philadelphia, 1987. SIAM.
2. Catmull, E., Clark,J: Recursively generated B-spline surfaces on arbitrary topological meshes. *Computer Aided Design 10*, 350–355. (1978).
3. Farin, G.: 1997. Curves and Surfaces for Computer-Aided Geometric Design: A Practical Guide, 4th Edition. Academic Press, New York, NY, USA.
4. Gregory, J., Zhou, J.: Irregular C^2 surface construction using bi-polynomial rectangular patches. Computer Aided Geometric Design **16** (1999) 424–435
5. Karčiauskas, K.: Rational M-patches and tensor-border patches. Contemporary mathematics **334** (2003) 101–128
6. Karčiauskas, K., Peters, J., Reif, U.: Shape characterization of subdivision surfaces – Case studies. to appear in CAGD
7. Peters, J.: C^2 free-form surfaces of degree $(3,5)$. Computer Aided Geometric Design **19** (2002) 113–126
8. Peters, J., Reif, U.: Analysis of generalized B-spline subdivision algorithms. SIAM Journal on Numerical Analysis **35** (1998) 728–748
9. Peters, J., Reif, U.: Shape characterization of subdivision surfaces – Basic principles. to appear in CAGD
10. Peters, J.: Smoothness, Fairness and the need for better multi-sided patches. Contemporary mathematics **334** (2003) 55–64.
11. Prautzsch, H.: Freeform splines. *Comput. Aided Geom. Design*, 14(3):201–206, 1997.
12. Prautzsch H., Boehm, W., Paluszny, M.: Bézier and B-spline techniques. Springer Verlag, Berlin, Heidelberg. (2002)
13. Reif, U.: TURBS—topologically unrestricted rational B-splines. *Constr. Approx.*, 14(1):57–77, 1998.

A Recursive Taylor Method for Algebraic Curves and Surfaces

Huahao Shou[1,2], Ralph Martin[3], Guojin Wang[1], Adrian Bowyer[4], and Irina Voiculescu[5]

[1] State Key Laboratory of CAD & CG, Zhejiang University, Hangzhou, China
[2] Department of Applied Mathematics, Zhejiang University of Technology, Hangzhou, China
[3] School of Computer Science, Cardiff University, Cardiff, UK
[4] Department of Mechanical Engineering, University of Bath, Bath, UK
[5] Computing Laboratory, Oxford University, Oxford, UK

Abstract. This paper examines recursive Taylor methods for multivariate polynomial evaluation over an interval, in the context of algebraic curve and surface plotting as a particular application representative of similar problems in CAGD. The modified affine arithmetic method (MAA), previously shown to be one of the best methods for polynomial evaluation over an interval, is used as a benchmark; experimental results show that a second order recursive Taylor method (i) achieves the same or better graphical quality compared to MAA when used for plotting, and (ii) needs fewer arithmetic operations in many cases. Furthermore, this method is simple and very easy to implement. We also consider which order of Taylor method is best to use, and propose that second order Taylor expansion is generally best. Finally, we briefly examine theoretically the relation between the Taylor method and the MAA method.

1 Introduction

The aim of range analysis is to find the range of a function (usually a polynomial) in one or several variables over an input interval. In practice, finding an exact range is difficult, and it is more usual to find a range which includes the actual range. Information about the range of a function f, and related functions such as its partial derivatives, inverse, etc. are of considerable interest to people working in the fields of numerical and functional analysis, differential equations, linear algebra, approximation and optimization theory and other disciplines [7].

Range analysis has many important applications in CAGD and computer graphics, including the plotting and localisation of implicit curves and surfaces. Implicit surfaces are of direct use, for example, in CSG solid modelling, while implicit curves can be used to represent the intersection of two parametric surfaces, or the silhouette edges of a parametric surface with respect to a given view [10]. Many other geometric operations can also be performed by finding the simultaneous solution of a set of non-linear equations in several variables, and range analysis provides a means of localising such solutions [6]. Both as an interesting example in its own right, and as a *representative* problem, we thus consider in this paper the problem of solving $f(x, y) = 0$ in a rectangle or $f(x, y, z) = 0$ in a cuboid, and more particularly the problem of plotting this curve or surface into a set of pixels or voxels. Clearly, for other problems, e.g. finding the intersection of two surfaces, producing a pixel or voxel grid may not be appropriate, but our overall methodology and conclusions concerning localisation of implicit curves and surfaces remain valid.

Parametric curves or surfaces are very easy to plot. On the other hand, implicit curves or surfaces can not be plotted so readily. Implicit curve or surface plotting methods can be classified into two categories. The first are continuation methods [2–4], which are efficient. They find one or more seed cells (pixels or voxels) on a curve or a surface, and then trace the curve or surface continuously through appropriate adjacent cells—only cells containing the curve or surface are visited. However, continuation methods have one fundamental difficulty, that of finding a *complete* set of initial seed cells.

Subdivision methods [5, 8, 10–14] make up the second approach. These methods start with the whole plotting region itself as an initial cell. If a cell can be proven to be empty, it is discarded; otherwise it is subdivided into smaller cells, which are then visited recursively, until the cells reach pixel size. All pixels which contain the curve are thus guaranteed to be retained. In this way large portions of the plotting region can be discarded quickly and reliably at an early stage, leading to an efficient method. Such methods are generally based on ideas from interval arithmetic.

When $f(x, y)$ is a polynomial in two variables x and y, the curve is an *algebraic* curve. Similarly when $f(x, y, z)$ is a polynomial in three variables x, y and z, the surface is an *algebraic* surface. Algebraic curves or surfaces are a rich family, with several plotting methods [5, 8, 13] that exploit the properties of polynomials.

Taubin's method [13] is well known; we have shown in [5] that Taubin's method is equivalent to performing interval arithmetic on centered forms but without consideration of the even or odd properties of powers of polynomial terms. We have further shown that interval arithmetic on centered forms method is less accurate than a modified affine arithmetic method (MAA) which does take into consideration the even or odd properties.

In this paper we propose the use of a *recursive* Taylor method for function range evaluation and use it to plot algebraic curves and surfaces. We combine it with a point sampling technique and a subpixel (or subvoxel) technique to improve the results.

In our previous papers [5, 8] we showed that the modified affine arithmetic method is one of the best methods for polynomial evaluation over an interval, for use in recursive subdivision methods for plotting algebraic curves—we thus compare the Taylor method with that method. Our test results show that, when used for plotting algebraic curves and surfaces at a given resolution, the recursive Taylor method can give same or better graphical accuracy as the MAA method, and needs fewer arithmetic operations in most cases. Furthermore, this recursive Taylor method is simple and very easy to implement.

We also consider which order Taylor method to use, and show that 2nd order Taylor expansion seems to be best for general use.

Finally we examine theoretically the relation between the recursive Taylor method and the modified affine arithmetic method.

As noted above, the recursive Taylor technique presented in this paper is a general efficient method for computing bounds on a polynomial: its use here for algebraic curve and surface drawing is just an example application. The recursive Taylor method presented in this paper can be easily generalized to an arbitrary number of dimensions.

2   The Subdivision Algorithm

Subdivision algorithms for plotting implicit curves and implicit surfaces have much in common. We mainly focus on the case of plane implicit curves in this section.

In the following we use the standard notation that an interval A represents a range of real values between $\underline{a}$ and $\overline{a}$ such that $\underline{a} < \overline{a}$ and is written $[\underline{a}, \overline{a}]$.

The main idea of subdivision algorithms [11] for plotting implicit curves over a rectangular array of pixels is to consider various regions, initially the whole plotting region, $[\underline{x}, \overline{x}] \times [\underline{y}, \overline{y}]$, and to estimate bounds $[\underline{f}, \overline{f}]$ guaranteed to contain all values of $f(x, y)$ over this region. This is done using some range analysis method to estimate the range of the function. If $0 \notin [\underline{f}, \overline{f}]$, this means that the curve cannot pass through region, which therefore can be discarded. Otherwise the region is subdivided horizontally and vertically at its mid point into four sub-regions, and the pieces are considered in turn. The process stops when any region not yet discarded reaches pixel size.

In a basic version of the algorithm, we may just plot this pixel as if it did contain the curve. This can result in a "fat" curve if the bounds on the function obtained by range analysis method are too conservative, i.e. extra pixels which are actually not on the curve are plotted. Later, we will consider how to process the pixel-sized regions further to remove some, but not all, of the extraneous pixels. The basic procedure is summarized in fig. 1.

```
PROCEDURE Plot_Curve(x̲,x̄,y̲,ȳ):
[f̲,f̄] = Bound(f,x̲,x̄,y̲,ȳ);
IF f̲ ≤ 0 ≤ f̄ THEN
   IF x̄ - x̲ ≤ Pixel_size AND ȳ - y̲ ≤ Pixel_size THEN
      Plot_Pixel(x̲,x̄,y̲,ȳ)
   ELSE Subdivide(x̲,x̄,y̲,ȳ).

PROCEDURE Subdivide(x̲,x̄,y̲,ȳ):
x₀ = (x̲ + x̄)/2;
y₀ = (y̲ + ȳ)/2;
Plot_Curve(x̲,x₀,y̲,y₀);
Plot_Curve(x̲,x₀,y₀,ȳ);
Plot_Curve(x₀,x̄,y₀,ȳ);
Plot_Curve(x₀,x̄,y̲,y₀).
```

Fig. 1. Subdivision algorithm for curve plotting

The key step in subdivision algorithms of this type is to estimate the bounds $[\underline{f}, \overline{f}]$ on $f(x, y)$ over the region $[\underline{x}, \overline{x}] \times [\underline{y}, \overline{y}]$; this is done using some range analysis method. Different range analysis methods for computing the bounds have different effects on accuracy and efficiency of the plotting algorithm [5]. Generally, the more accurate the estimate is, the better the graphical result will be, and also less subdivision will be required. However, more accurate estimates usually need more arithmetic operations, which reduces the efficiency of the plotting algorithm. Obviously, accuracy and efficiency are to some extent trade-offs. In the next section we will present a Taylor method for computing these bounds.

In order to reduce the uncertainties associated with the regions remaining at pixel level, which may or may not contain the curve, as noted above, we use two further techniques. Point sampling [12] is done for regions of pixel size by evaluating the values of $f(x, y)$ at the four corner points of the pixel. If they do not all have the same sign (or zero), then the pixel must be include the curve (as f is a continuous function); otherwise, the pixel may or may not be on the curve. Thus, after point sampling, all pixels in the plotting region belong to one of three classes: (i) pixels discarded by the basic subdivision method, which are surely not on the curve, (ii) pixels accepted by the point sampling technique, which are surely on the curve, and (iii) pixels whose status is still not clear, and may or may not be on the curve. We now further attempt to discard as many pixels as possible in the third class. To this end we use a subpixel technique [14]. We subdivide pixels in the third category into four subpixels. If all four subpixels can be discarded by the range method, we discard this pixel, otherwise we keep the pixel.

A major advantage of the subdivision algorithm presented above is that it finds *all* points on the curve, and can handle singularities with no special processing. Thus, it can handle problems where continuation methods may typically fail, including curves with multiple components, cusps, self-intersections, touching components, and isolated points.

The subdivision algorithm for plotting implicit surfaces is a direct generalisation to three variables of the planar implicit curve algorithm. Plotting implicit space curve cases can also readily be done by finding regions *simultaneously* containing zeros of *two* implicit functions in three variables.

3   Taylor Method for Bounds

Constructing the natural inclusion function [10] giving the exact range of a function over an interval is often not easy, and may be impossible for general functions $f(x, y)$. Here we use a simple Taylor method [1] for computing bounds of $f(x, y)$ over $[\underline{x}, \overline{x}] \times [\underline{y}, \overline{y}]$, which can be combined with point sampling and subpixel techniques to solve the implicit curve plotting problem in a reliable, accurate and efficient way. For now, we assume the choice of a *second order* Taylor method, but we will return to the choice of order later. Suppose $f(x, y)$ has continuous second derivatives on $[\underline{x}, \overline{x}] \times [\underline{y}, \overline{y}]$. In many practical applications in CAGD and computer graphics, the functions encountered satisfy this condition, at least piecewise. To estimate the bound of $f(x, y)$ on $[\underline{x}, \overline{x}] \times [\underline{y}, \overline{y}]$, we expand $f(x, y)$ at the mid point (x_0, y_0) of the region $[\underline{x}, \overline{x}] \times [\underline{y}, \overline{y}]$ using

Taylor's formula:

$$f(x,y) = f(x_0, y_0) + h f_x(x_0, y_0) + k f_y(x_0, y_0) + \frac{1}{2}h^2 f_{xx}(x_0 + \theta h, y_0 + \theta k)$$
$$+ \frac{1}{2}k^2 f_{yy}(x_0 + \theta h, y_0 + \theta k) + hk f_{xy}(x_0 + \theta h, y_0 + \theta k),$$

where

$$(x,y) \in [\underline{x}, \overline{x}] \times [\underline{y}, \overline{y}], \quad x_0 = \frac{\underline{x} + \overline{x}}{2}, y_0 = \frac{\underline{y} + \overline{y}}{2}, \quad 0 < \theta < 1,$$

$$h = x - x_0 \in [-\frac{\overline{x} - \underline{x}}{2}, \frac{\overline{x} - \underline{x}}{2}] = \frac{\overline{x} - \underline{x}}{2}[-1, 1],$$

$$k = y - y_0 \in [-\frac{\overline{y} - \underline{y}}{2}, \frac{\overline{y} - \underline{y}}{2}] = \frac{\overline{y} - \underline{y}}{2}[-1, 1].$$

Suppose we know the interval bounds B_{xx}, B_{yy}, B_{xy} of the three second derivatives $f_{xx}(x, y), f_{yy}(x, y), f_{xy}(x, y)$ of the function $f(x, y)$ over the region $[\underline{x}, \overline{x}] \times [\underline{y}, \overline{y}]$ such that $f_{xx}(x, y) \in B_{xx}, f_{yy}(x, y) \in B_{yy}, f_{xy}(x, y) \in B_{xy}$. Let $x_1 = (\overline{x} - \underline{x})/2$, $y_1 = (\overline{y} - \underline{y})/2$. Then the bounds $[\underline{f}, \overline{f}]$ of $f(x, y)$ over the region $[\underline{x}, \overline{x}] \times [\underline{y}, \overline{y}]$ can be expressed as

$$[\underline{f}, \overline{f}] = f(x_0, y_0) + x_1 f_x(x_0, y_0)[-1, 1] + y_1 f_y(x_0, y_0)[-1, 1]$$
$$+ \frac{1}{2}x_1^2 B_{xx}[-1, 1] + \frac{1}{2}y_1^2 B_{yy}[-1, 1] + x_1 y_1 B_{xy}[-1, 1].$$

(To apply interval computation to the above formula, real numbers are converted where necessary to intervals with *equal* lower and upper bounds.)

The main potential limitation of this method is that we need estimates for the bounds B_{xx}, B_{yy}, B_{xy} of the three second derivatives $f_{xx}(x, y), f_{yy}(x, y), f_{xy}(x, y)$ of $f(x, y)$ on the region $[\underline{x}, \overline{x}] \times [\underline{y}, \overline{y}]$. (Note that the first derivatives required need only be computed at a specific point, and thus can readily be found.) For general implicit curves, finding bounds on the second derivatives is a difficult problem. However, as we show in the next section, they can be readily computed for algebraic curves.

Similarly, for surface plotting, to estimate the bound of $f(x, y, z)$ on $[\underline{x}, \overline{x}] \times [\underline{y}, \overline{y}] \times [\underline{z}, \overline{z}]$, we may expand $f(x, y, z)$ at the mid point (x_0, y_0, z_0) of the region $[\underline{x}, \overline{x}] \times [\underline{y}, \overline{y}] \times [\underline{z}, \overline{z}]$ using Taylor's formula:

$$f(x,y,z) = f(x_0, y_0, z_0) + h f_x(x_0, y_0, z_0) + k f_y(x_0, y_0, z_0) + l f_z(x_0, y_0, z_0)$$
$$+ \frac{1}{2}h^2 f_{xx}(x_0 + \theta h, y_0 + \theta k, z_0 + \theta l) + \frac{1}{2}k^2 f_{yy}(x_0 + \theta h, y_0 + \theta k, z_0 + \theta l)$$
$$+ \frac{1}{2}l^2 f_{zz}(x_0 + \theta h, y_0 + \theta k, z_0 + \theta l) + hk f_{xy}(x_0 + \theta h, y_0 + \theta k, z_0 + \theta l)$$
$$+ hl f_{xz}(x_0 + \theta h, y_0 + \theta k, z_0 + \theta l) + kl f_{yz}(x_0 + \theta h, y_0 + \theta k, z_0 + \theta l)$$

where

$$(x,y,z) \in [\underline{x}, \overline{x}] \times [\underline{y}, \overline{y}] \times [\underline{z}, \overline{z}], \quad x_0 = \frac{\underline{x} + \overline{x}}{2}, y_0 = \frac{\underline{y} + \overline{y}}{2}, z_0 = \frac{\underline{z} + \overline{z}}{2}, \quad 0 < \theta < 1,$$

$$h = x - x_0 \in [-\frac{\overline{x} - \underline{x}}{2}, \frac{\overline{x} - \underline{x}}{2}] = \frac{\overline{x} - \underline{x}}{2}[-1, 1],$$

$$k = y - y_0 \in [-\frac{\overline{y} - \underline{y}}{2}, \frac{\overline{y} - \underline{y}}{2}] = \frac{\overline{y} - \underline{y}}{2}[-1, 1].$$

$$l = z - z_0 \in [-\frac{\overline{z} - \underline{z}}{2}, \frac{\overline{z} - \underline{z}}{2}] = \frac{\overline{z} - \underline{z}}{2}[-1, 1].$$

Suppose we know the interval bounds $B_{xx}, B_{yy}, B_{zz}, B_{xy}, B_{xz}, B_{yz}$ of the six second derivatives $f_{xx}(x, y, z)$, $f_{yy}(x, y, z)$, $f_{zz}(x, y, z)$, $f_{xy}(x, y, z)$, $f_{xz}(x, y, z)$, $f_{yz}(x, y, z)$ of the function $f(x, y, z)$ over the region $[\underline{x}, \overline{x}] \times [\underline{y}, \overline{y}] \times [\underline{z}, \overline{z}]$ such that $f_{xx}(x, y, z) \in B_{xx}$, $f_{yy}(x, y, z) \in B_{yy}$, $f_{zz}(x, y, z) \in B_{zz}$, $f_{xy}(x, y, z) \in B_{xy}$, $f_{xz}(x, y, z) \in B_{xz}$, $f_{yz}(x, y, z) \in B_{yz}$. Let $x_1 = (\overline{x} - \underline{x})/2$, $y_1 = (\overline{y} - \underline{y})/2$,, $z_1 = (\overline{z} - \underline{z})/2$,. Then the bounds $[\underline{F}, \overline{F}]$ of $f(x, y, z)$ over the region $[\underline{x}, \overline{x}] \times [\underline{y}, \overline{y}] \times [\underline{z}, \overline{z}]$ can be expressed as

$$[\underline{F}, \overline{F}] = f(x_0, y_0, z_0) + x_1 f_x(x_0, y_0, z_0)[-1, 1] + y_1 f_y(x_0, y_0, z_0)[-1, 1]$$

$$+ z_1 f_z(x_0, y_0, z_0)[-1, 1] + \frac{1}{2}x_1^2 B_{xx}[-1, 1] + \frac{1}{2}y_1^2 B_{yy}[-1, 1] + \frac{1}{2}z_1^2 B_{zz}[-1, 1]$$

$$+ x_1 y_1 B_{xy}[-1, 1] + x_1 z_1 B_{xz}[-1, 1] + y_1 z_1 B_{yz}[-1, 1].$$

As above, again we need estimates for the bounds $B_{xx}, B_{yy}, B_{zz}, B_{xy}, B_{xz}, B_{yz}$.

4   Finding Bounds on Derivatives

When $f(x, y) = 0$ represents an *algebraic* curve, $f(x, y)$ is a *polynomial* function of two variables. In this case the three second derivatives $f_{xx}(x, y)$, $f_{yy}(x, y)$, $f_{xy}(x, y)$ are themselves also polynomials in two variables with lower degrees in x or y or both. Therefore we can use a recursive technique to estimate the bounds of the second derivatives, as given by the algorithm in fig. 2. Here, "IF $f \equiv c$ RETURN Interval$[c, c]$"

```
Bound (f, x, x̄, y, ȳ) :
IF  f ≡ c RETURN Interval[c, c]
ELSE
     x₀ = (x + x̄)/2;   y₀ = (y + ȳ)/2;   x₁ = (x̄ - x)/2;   y₁ = (ȳ - y)/2;
     [f, f̄] = f(x₀, y₀) + x₁fₓ(x₀, y₀)[-1, 1] + y₁f_y(x₀, y₀)[-1, 1]
          + ½x₁²[0, 1]Bound(fₓₓ, x, x̄, y, ȳ) + ½y₁²[0, 1]Bound(f_yy, x, x̄, y, ȳ)
          + x₁y₁[-1, 1]Bound(fₓy, x, x̄, y, ȳ);
     RETURN Interval[f, f̄].
```

Fig. 2. Recursive Taylor algorithm for polynomial bounding

tests if f is a constant, and if so terminates the recursion—the bound on a constant can be trivially computed. (Recursion could also be stopped one step earlier, as it is easy to compute exact bounds for linear functions.)

Note that only in the case that f is a polynomial can we guarantee that such recursion will terminate. For polynomials, successive differentiation must eventually result in a constant, which is not true for other functions.

A similar recursive technique can be used for trivariate polynomials.

5 Examples

In the above sections we proposed a recursive Taylor method combined with point sampling and a subpixel technique for plotting algebraic curves and surfaces. In this section we give some examples demonstrating the accuracy and efficiency of these methods.

Most of the examples we give involve low degree polynomials. While it is conceivable that somewhat different conclusions might be drawn for the cases of higher degree polynomials, other tests we have done on further higher degree polynomials support the conclusions here. Furthermore, in most CAGD applications, the polynomials used are generally of a low degree, justifying our choice of low degree test cases.

5.1 Algebraic Curves

Examples 1 to 10 are the same examples for plotting algebraic curves given in a recent survey of methods [5], with plotting region $[0, 1] \times [0, 1]$ and resolution 256×256 pixels. They were designed to test the efficiency and accuracy of range evaluation methods on a variety of problem cases, including curves with cusps, self-intersections, closely adjacent loops, and so on.

The corresponding figures produced by the new recursive Taylor (RT) method (including the use of point sampling and subpixel techniques, denoted RT++) are shown in Figures 3 to 12. The survey [5] showed that the modified affine arithmetic method (MAA) is one of the best methods for plotting algebraic curves. Therefore we have compared the recursive Taylor method with the MAA method. A detailed quantitative comparison of the MAA and RT methods, and also their variants MAA++ and RT++ which include point sampling and subpixel techniques, is given in Table 1 (*left*) for these examples.

Table 1 (*left*) shows, for each example, how many pixels are plotted by the different methods (the fewer, the more accurately the method has found the curve), the number of subdivisions used in the computation (the fewer, the better, as less stack operation overheads result), and the number of addition and multiplication operations used overall (the lower, the better).

The recorded number of additions and multiplications in Table 1 (*left*) does not include the arithmetic operations used to differentiate the polynomial. An implementation of the recursive Taylor method should calculate all necessary coefficients of the derivatives of the polynomial just once at the beginning, and store them in an array, to avoid differentiation of the polynomial during the subdivision process every time a derivative is needed. The number of arithmetic operations used to differentiate the polynomial once only is relatively small and can be neglected.

From Table 1 (*left*) we can see that in one case out of ten (Example 4), the recursive Taylor method produced better graphical quality than the modified affine arithmetic method (fewer pixels were plotted). The corresponding graphical output for the RT method is shown in fig. 14, where 801 pixels were plotted, and for the MAA method in fig. 13, where 816 pixels were plotted. (These two figures only differ in the lower left corner). In the other nine test cases the recursive Taylor method produced the same graphical quality as the modified affine arithmetic method.

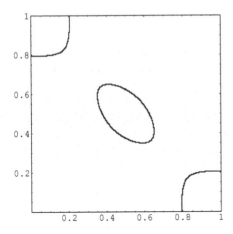

Fig. 3. Example 1: $\frac{15}{4} + 8x - 16x^2 + 8y - 112xy + 128x^2y - 16y^2 + 128xy^2 - 128x^2y^2 = 0$, plotted by RT++ method (522 pixels)

Fig. 4. Example 2: $20160x^5 - 30176x^4 + 14156x^3 - 2344x^2 + 151x + 237 - 480y = 0$, plotted by RT++ method (432 pixels)

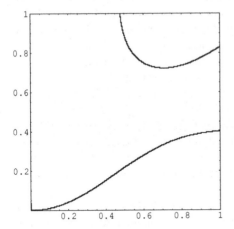

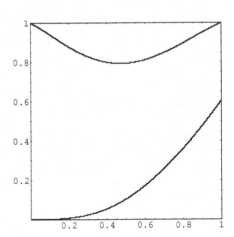

Fig. 5. Example 3: $0.945xy - 9.43214x^2y^3 + 7.4554x^3y^2 + y^4 - x^3 = 0$, plotted by RT++ method (601 pixels)

Fig. 6. Example 4: $x^9 - x^7y + 3x^2y^6 - y^3 + y^5 + y^4x - 4y^4x^3 = 0$, plotted by RT++ method (774 pixels)

In seven out of ten cases, the recursive Taylor method needed fewer arithmetic operations in total (the number of additions plus the number of multiplications) than the modified affine arithmetic method (Examples 2,4,6,7,8,9,10). In Examples 2,6,9,10 the number of arithmetic operations needed by the recursive Taylor method was much fewer than (less than half of) those for the modified affine arithmetic method. Although the recursive Taylor method needed more arithmetic operations than the modified affine

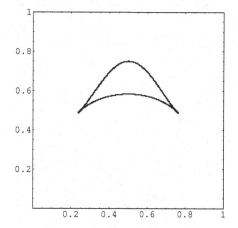

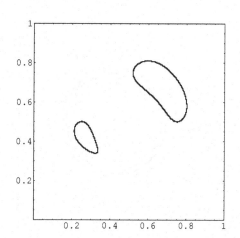

Fig. 7. Example 5. $-\frac{1801}{50} + 280x - 816x^2 + 1056x^3 - 512x^4 + \frac{1601}{25}y - 512xy + 1536x^2y - 2048x^3y + 1024x^4y = 0$, plotted by RT++ method (456 pixels)

Fig. 8. Example 6: $\frac{601}{9} - \frac{872}{3}x + 544x^2 - 512x^3 + 256x^4 - \frac{2728}{9}y + \frac{2384}{3}xy - 768x^2y + \frac{5104}{9}y^2 - \frac{2432}{3}xy^2 + 768x^2y^2 - 512y^3 + 256y^4 = 0$, plotted by RT++ method (456 pixels)

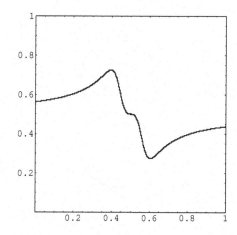

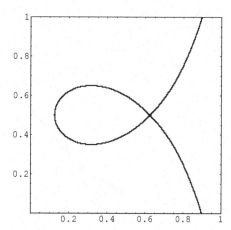

Fig. 9. Example 7: $-13 + 32x - 288x^2 + 512x^3 - 256x^4 + 64y - 112y^2 + 256xy^2 - 256x^2y^2 = 0$, plotted by RT++ method (460 pixels)

Fig. 10. Example 8: $-\frac{169}{64} + \frac{51}{8}x - 11x^2 + 8x^3 + 9y - 8xy - 9y^2 + 8xy^2 = 0$, plotted by RT++ method (808 pixels)

arithmetic method for Examples 1,3,5, we note that the numbers of arithmetic operations needed by both methods for these examples were very similar.

One minor disadvantage of the recursive Taylor method is that it often needs a few more recursive operations than MAA.

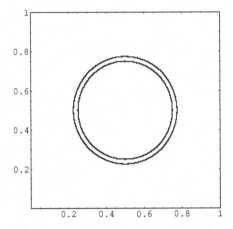

Fig. 11. Example 9: $47.6 - 220.8x + 476.8x^2 - 512x^3 + 256x^4 - 220.8y + 512xy - 512x^2y + 476.8y^2 - 512xy^2 + 512x^2y^2 - 512y^3 + 256y^4 = 0$, plotted by RT++ method (1088 pixels)

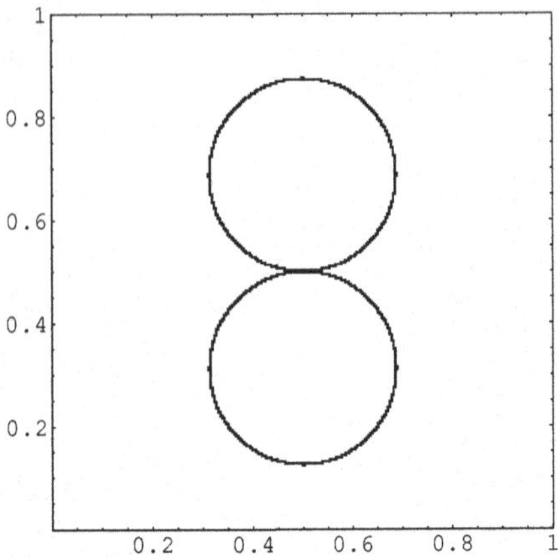

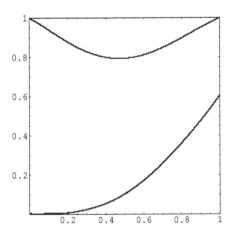

Fig. 12. Example 10: $\frac{55}{256} - x + 2x^2 - 2x^3 + x^4 - \frac{55}{64}y + 2xy - 2x^2y + \frac{119}{64}y^2 - 2xy^2 + 2x^2y^2 - 2y^3 + y^4 = 0$, plotted by RT++ method (772 pixels)

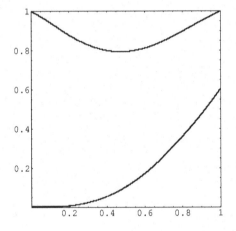

Fig. 13. Example 4, plotted by the MAA method (816 pixels)

Fig. 14. Example 4, plotted by the RT method (801 pixels)

Point sampling and subpixel techniques further improved the graphical quality of RT and MAA methods, especially for Examples 4,7,9 where the improvements are significant. However, for Examples 1,2,3,5,6,8,10 the improvements only affected a few pixels and insignificant. Of course, the price to pay for these improvements is an increase in arithmetic operations: every pixel which cannot be discarded by the basic subdivision process needs to be examined further. We can however see from Table 1 (*left*) that the increased number of arithmetic operations is not greatly significant. This is

Table 1. Comparison of 2D (*left*) and 3D (*right*) RT, MAA, RT++, MAA++ methods

Ex.	Method	Pixels plotted	Subdivisions	Additions	Multiplications
1	RT	526	571	415688	343892
1	MAA	526	563	404262	171226
1	RT++	522	575	436316	385080
1	MAA++	522	567	421448	207820
2	RT	433	461	241581	205717
2	MAA	433	459	601510	407812
2	RT++	432	462	253193	234577
2	MAA++	432	460	611148	434354
3	RT	608	637	1116344	936757
3	MAA	608	634	1178329	646933
3	RT++	601	653	1143206	992682
3	MAA++	601	650	1202312	694836
4	RT	801	845	4662221	4461229
4	MAA	816	857	6773822	6302500
4	RT++	774	876	4844054	4748416
4	MAA++	774	903	7139018	6757864
5	RT	464	627	664231	575815
5	MAA	464	611	599656	339853
5	RT++	456	635	690161	630353
5	MAA++	456	619	621248	387781
6	RT	460	567	442025	414092
6	MAA	460	560	1329630	788830
6	RT++	456	573	469450	478064
6	MAA++	456	566	1362826	853306
7	RT	512	629	445039	386359
7	MAA	512	627	873923	476708
7	RT++	460	719	512886	472534
7	MAA++	460	717	986288	569061
8	RT	818	829	563844	422917
8	MAA	818	827	855337	397078
8	RT++	808	843	595997	476088
8	MAA++	808	841	886530	444873
9	RT	1144	1281	998825	935312
9	MAA	1144	1269	3012696	1787102
9	RT++	1088	1351	1106039	1131219
9	MAA++	1088	1339	3214325	2018571
10	RT	784	849	662153	609761
10	MAA	784	845	2006376	1190110
10	RT++	772	861	710484	710732
10	MAA++	772	857	2068693	1294219

Ex.	Method	Voxels plotted	Subdivisions	Additions	Multiplications
11	RT	1791	592	397403	229152
11	MAA	1791	592	326348	110727
11	RT++	1791	592	432100	278177
11	MAA++	1791	592	361045	159752
12	RT	3992	1353	918367	588609
12	MAA	3992	1353	3289042	1476259
12	RT++	3952	1401	1163930	953372
12	MAA++	3944	1401	3513741	1733406
13	RT	3712	1433	958102	589014
13	MAA	3712	1433	1084217	692199
13	RT++	3712	1433	1023606	713910
13	MAA++	3712	1433	1149721	817095
14	RT	3272	1129	756950	491169
14	MAA	3272	1129	2735177	1231875
14	RT++	3176	1249	1145079	1038878
14	MAA++	3192	1249	3048966	1515888
15	RT	2192	985	4455080	3265689
15	MAA	2144	985	13108130	11792931
15	RT++	1904	1337	5603358	4948007
15	MAA++	1920	1289	16804088	15499281
16	RT	2376	1153	5232146	3831834
16	MAA	2344	1121	14953054	13456863
16	RT++	2148	1497	6291409	5435377
16	MAA++	2104	1433	18908261	17434612
17	RT	5276	1841	7081323	4483139
17	MAA	5256	1837	6948311	5854917
17	RT++	4896	2265	8662097	6658461
17	MAA++	4976	2241	8576707	7662017
18	RT	9424	2865	12975392	9497889
18	MAA	9376	2769	36866234	33149195
18	RT++	7236	5313	21000451	20340451
18	MAA++	7792	5169	64451180	59649509
19	RT	1832	961	4290881	3139995
19	MAA	1816	961	12656417	11259579
19	RT++	1572	1249	5248699	4536725
19	MAA++	1624	1233	15851779	14407101
20	RT	3428	1197	3139078	2100954
20	MAA	3416	1169	3739482	4352652
20	RT++	3288	1425	3913112	3195292
20	MAA++	3288	1385	4474204	5400382

because the RT and MAA methods already provide close to the best possible graphical quality at the given resolution, and thus the numbers of pixels left to be examined further by point sampling and subpixel techniques are relatively small.

5.2 Algebraic Surfaces

We have also experimented with algebraic surface plotting, as outlined below. Note that our main purpose in this paper is to compare our new range analysis method with existing methods, in this case for *localising* the surface to specific regions (voxels). We only use voxel plotting as a *representative* application; the graphical results of surface plotting shown at a resolution of $32 \times 32 \times 32$ are clearly crude. Such an approach is not meant to be a useful surface rendering algorithm in itself. A realistic surface plotting

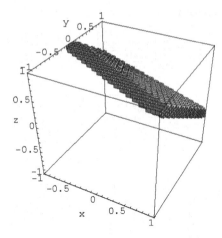

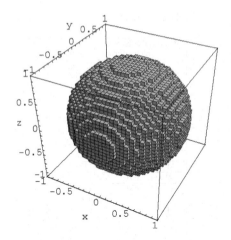

Fig. 15. Example 11: The plane plotted by 3D RT++ method (1791 voxels)

Fig. 16. Example 12: The sphere plotted by 3D RT++ method (3952 voxels)

algorithm would, for example, attempt to find a linear fit to the surface and estimate its normal in each region where the surface has been localised.

Example 11: this plots the plane $f(x, y, z) = x + 2y + 3z - 2$ inside the box box $[-1, 1] \times [-1, 1] \times [-1, 1]$, with resolution $32 \times 32 \times 32$ voxels. Figure 15 shows the plane plotted by the 3D recursive Taylor method using point sampling and subpixel techniques. A total of 1791 voxels were plotted.

Example 12: this plots the sphere $f(x, y, z) = 100x^2 + 100y^2 + 100z^2 - 81$ inside the box $[-1, 1] \times [-1, 1] \times [-1, 1]$, with resolution $32 \times 32 \times 32$ voxels. Figure 16 shows the sphere plotted by the recursive Taylor method using point sampling and subpixel techniques. A total of 3952 voxels were plotted.

Example 13: this plots the cylinder $f(x, y, z) = 100x^2 + 100y^2 - 81$ inside the box $[-1, 1] \times [-1, 1] \times [-1, 1]$, with resolution $32 \times 32 \times 32$ voxels. Figure 17 shows the cylinder plotted by the recursive Taylor method using point sampling and subpixel techniques. A total of 3712 voxels were plotted.

Example 14: this plots the cone $f(x, y, z) = 100x^2 + 100y^2 - 81z^2$ inside the box $[-1, 1] \times [-1, 1] \times [-1, 1]$, with resolution $32 \times 32 \times 32$ voxels. Figure 18 is the cone plotted by the recursive Taylor method using point sampling and subpixel techniques. A total of 3176 voxels were plotted.

Example 15: this plots the torus $f(x, y, z) = 64 - 500x^2 + 625x^4 - 500y^2 + 1250x^2y^2 + 625y^4 + 400z^2 + 1250x^2z^2 + 1250y^2z^2 + 625z^4$ inside the box $[-1, 1] \times [-1, 1] \times [-1, 1]$ with resolution $32 \times 32 \times 32$ voxels. Figure 19 is the torus plotted by the recursive Taylor method using point sampling and subpixel techniques. A total of 1904 voxels were plotted.

Example 16: this plots the cyclide $f(x, y, z) = -459 + 15600x - 55000x^2 + 90000x^4 - 45000y^2 + 180000x^2y^2 + 90000y^4 + 12600z^2 + 180000x^2z^2 + 180000y^2z^2 + 90000z^4$ inside the box $[-1, 1] \times [-1, 1] \times [-1, 1]$, with resolution $32 \times 32 \times 32$ voxels.

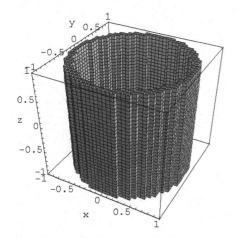

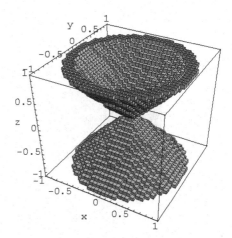

Fig. 17. Example 13: The cylinder plotted by 3D RT++ method (3712 voxels)

Fig. 18. Example 14: The cone plotted by 3D RT++ method (3176 voxels)

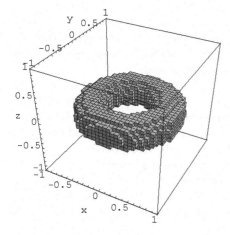

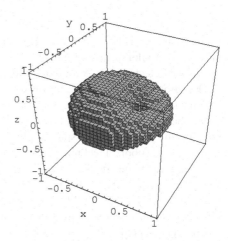

Fig. 19. Example 15: The torus plotted by 3D RT++ method (1904 voxels)

Fig. 20. Example 16: The cyclide plotted by 3D RT++ method (2148 voxels)

Figure 20 is the cyclide plotted by the recursive Taylor method using point sampling and subpixel techniques. A total of 2148 voxels were plotted.

Example 17: this plots a self-intersecting surface $f(x, y, z) = 16 - 32x - 25x^2 + 50x^3 - 25y^2 + 50xy^2 - 25z^2 + 50xz^2$ inside the box $[-1, 1] \times [-1, 1] \times [-1, 1]$, with resolution $32 \times 32 \times 32$ voxels. Figure 21 is the self-intersecting surface plotted by the recursive Taylor method using point sampling and subpixel techniques. A total of 4896 voxels were plotted.

Example 18: this plots a pair of parallel surfaces $f(x, y, z) = 1296 - 3625x^2 + 2500x^4 - 3625y^2 + 5000x^2y^2 + 2500y^4 - 3625z^2 + 5000x^2z^2 + 5000y^2z^2 + 2500z^4$

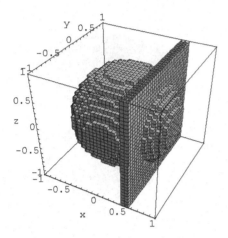
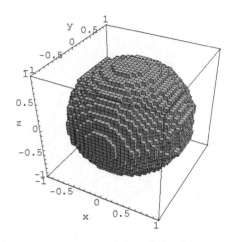

Fig. 21. Example 17: The self-intersecting surface plotted by 3D RT++ method (4896 voxels)

Fig. 22. Example 18: The pair of parallel surfaces plotted by 3D RT++ method (7236 voxels)

inside the box $[-1, 1] \times [-1, 1] \times [-1, 1]$ with resolution $32 \times 32 \times 32$ voxels. Figure 22 is the pair of parallel surfaces plotted by the recursive Taylor method using point sampling and subpixel techniques. A total of 7236 voxels were plotted.

Example 19: this plots a pair of just-touching surfaces (two spheres sharing a tangent plane) $f(x, y, z) = -16x^2 + 25x^4 + 50x^2y^2 + 25y^4 + 50x^2z^2 + 50y^2z^2 + 25z^4$ inside the box $[-1, 1] \times [-1, 1] \times [-1, 1]$, with resolution $32 \times 32 \times 32$ voxels. Figure 23 is the pair of tangent spheres plotted by the recursive Taylor method using point sampling and subpixel techniques. A total of 1572 voxels were plotted.

Example 20: this plots a cone-like surface with a line singularity $f(x, y, z) = -1 + 4x - 4x^2 + 2y^2 - 8xy^2 + 8x^2y^2 + 8z^2$ inside the box $[-1, 1] \times [-1, 1] \times [-1, 1]$, with resolution $32 \times 32 \times 32$ voxels. Figure 24 is the cone-like surface plotted by the recursive Taylor method using point sampling and subpixel techniques. A total of 3288 voxels were plotted.

Table 1 (*right*) gives a detailed quantitative comparison for these surface examples of the 3D MAA and 3D RT methods, and also of their improvements which include point sampling and subpixel techniques, 3D MAA++ and 3D RT++.

From Table 1 (*right*) we can see that:

- In 4 out of 10 cases (Examples 11–14) the RT method plotted the same number of voxels as the MAA method. In the other 6 cases (Examples 15–20) the RT method plotted slightly more voxels than the MAA method.
- In 9 out of 10 cases (all but Example 11) the RT method needed fewer arithmetic operations than the MAA method.
- In 5 out of 10 cases (Examples 14,15,17–19) the RT++ method plotted fewer voxels than the MAA++ method. In 3 cases (Examples 11,13,20) the RT++ method plotted the same number of voxels as the MAA++ method. In the other 2 cases (Examples 12,16) the RT++ method plotted slightly more voxels than the MAA++ method.

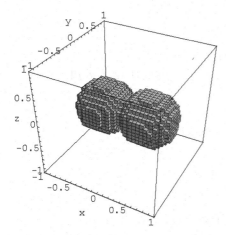

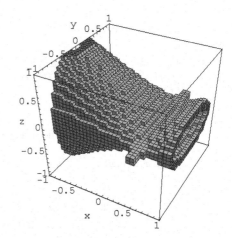

Fig. 23. Example 19: The pair of just touching surfaces plotted by 3D RT++ method (1572 voxels)

Fig. 24. Example 20: The cone like surface with a line singularity plotted by 3D RT++ method (3288 voxels)

- In 9 out of 10 cases (all but Example 11) the RT++ method needed fewer arithmetic operations than the MAA++ method.

Overall we may probably conclude that the 3D RT++ method is the best choice in terms of accuracy and efficiency.

6 Why use Order Two Taylor Expansion?

In sect. 4 we proposed an order 2 recursive Taylor method for finding the bound of a polynomial, and in sect. 5 we gave some examples to show that this method works well. Clearly, however, we could have chosen to use some other order for our Taylor expansion, so we will now justify why we use a second order expansion rather than some other order, particularly order 1, 3 or 4. To do so we give an experimental comparison between recursive Taylor methods of orders 1–4.

We first begin by explicitly stating order 1, 3 and 4 recursive Taylor algorithms for evaluating a bivariate polynomial $f(x, y)$. An order 1 recursive Taylor algorithm is given in fig. 25, while an order 3 recursive Taylor algorithm is given in fig. 26, and an order 4 recursive Taylor algorithm is given in fig. 27.

Using the same curves from Examples 1–10 as before, we compared the accuracy and efficiency of order 1, 2, 3 and 4 recursive Taylor methods, using the same criteria of assessment as before. The test results are shown in Table 2 (*left*).

From Table 2 (*left*) we can see that:

- The order 1 recursive Taylor method is less accurate than order 2, 3 and 4 recursive Taylor methods.
- Usually, but not always, the order 1 method needs more arithmetic operations than order 2, 3 and 4 methods (Example 2 is a counterexample).

```
Bound (f, x̲, x̄, y̲, ȳ) :
IF  f ≡ c RETURN Interval[c, c],
ELSE
```

$$x_0 = (x̲ + x̄)/2; \quad y_0 = (y̲ + ȳ)/2; \quad x_1 = (x̄ - x̲)/2, \quad y_1 = (ȳ - y̲)/2;$$

$$[f̲, f̄] \quad = \quad f(x_0, y_0) \quad + \quad x_1 \text{Bound}(f_x, x̲, x̄, y̲, ȳ)[-1, 1] \quad +$$

$$y_1 \text{Bound}(f_y, x̲, x̄, y̲, ȳ)[-1, 1];$$

```
RETURN Interval[f̲, f̄].
```

Fig. 25. Order 1 recursive Taylor algorithm

```
Bound (f, x̲, x̄, y̲, ȳ) :
IF  f ≡ c RETURN Interval[c, c],
ELSE
```

$$x_0 = (x̲ + x̄)/2; \quad y_0 = (y̲ + ȳ)/2; \quad x_1 = (x̄ - x̲)/2, \quad y_1 = (ȳ - y̲)/2;$$

$$[f̲, f̄] = f(x_0, y_0) + x_1 f_x(x_0, y_0)[-1, 1] + y_1 f_y(x_0, y_0)[-1, 1]$$

$$+ \tfrac{1}{2} x_1^2[0, 1] f_{xx}(x_0, y_0) + \tfrac{1}{2} y_1^2[0, 1] f_{yy}(x_0, y_0) + x_1 y_1[-1, 1] f_{xy}(x_0, y_0)$$

$$+ \tfrac{1}{6} x_1^3[-1, 1] \text{Bound}(f_{xxx}, x̲, x̄, y̲, ȳ) + \tfrac{1}{6} y_1^3[-1, 1] \text{Bound}(f_{yyy}, x̲, x̄, y̲, ȳ)$$

$$+ \tfrac{1}{2} x_1^2 y_1[-1, 1] \text{Bound}(f_{xxy}, x̲, x̄, y̲, ȳ) + \tfrac{1}{2} x_1 y_1^2[-1, 1] \text{Bound}(f_{xyy}, x̲, x̄, y̲, ȳ);$$

```
RETURN Interval[f̲, f̄].
```

Fig. 26. Order 3 recursive Taylor algorithm

```
Bound (f, x̲, x̄, y̲, ȳ) :
IF  f ≡ c RETURN Interval[c, c],
ELSE
```

$$x_0 = (x̲ + x̄)/2; \quad y_0 = (y̲ + ȳ)/2; \quad x_1 = (x̄ - x̲)/2, \quad y_1 = (ȳ - y̲)/2;$$

$$[f̲, f̄] = f(x_0, y_0) + x_1 f_x(x_0, y_0)[-1, 1] + y_1 f_y(x_0, y_0)[-1, 1]$$

$$+ \tfrac{1}{2} x_1^2[0, 1] f_{xx}(x_0, y_0) + \tfrac{1}{2} y_1^2[0, 1] f_{yy}(x_0, y_0) + x_1 y_1[-1, 1] f_{xy}(x_0, y_0)$$

$$+ \tfrac{1}{6} x_1^3[-1, 1] f_{xxx}(x_0, y_0) + \tfrac{1}{6} y_1^3[-1, 1] f_{yyy}(x_0, y_0)$$

$$+ \tfrac{1}{2} x_1^2 y_1[-1, 1] f_{xxy}(x_0, y_0) + \tfrac{1}{2} x_1 y_1^2[-1, 1] f_{xyy}(x_0, y_0)$$

$$+ \tfrac{1}{24} x_1^4[0, 1] \text{Bound}(f_{xxxx}, x̲, x̄, y̲, ȳ) + \tfrac{1}{24} y_1^4[0, 1] \text{Bound}(f_{yyyy}, x̲, x̄, y̲, ȳ)$$

$$+ \tfrac{1}{6} x_1^3 y_1[-1, 1] \text{Bound}(f_{xxxy}, x̲, x̄, y̲, ȳ)$$

$$+ \tfrac{1}{6} x_1 y_1^3[-1, 1] \text{Bound}(f_{xyyy}, x̲, x̄, y̲, ȳ)$$

$$+ \tfrac{1}{4} x_1^2 y_1^2[0, 1] \text{Bound}(f_{xxyy}, x̲, x̄, y̲, ȳ);$$

```
RETURN Interval[f̲, f̄].
```

Fig. 27. Order 4 recursive Taylor algorithm

- In 9 out of 10 cases the order 2 recursive Taylor method has the same accuracy as order 3 and 4 methods. In the other case (Example 4) the order 2 method is more accurate than the order 3 and 4 methods.
- In 6 out of 10 cases, the order 2 recursive Taylor method needs fewer arithmetic operations than the order 3 method (Examples 1,2,6,7,9,10).
- In all cases the order 4 recursive Taylor methods has the same accuracy as the order 3 method.
- In 6 out of 10 cases order 4 recursive Taylor method needs fewer arithmetic operations than the order 3 method (Examples 1,4,6,7,9,10).

Table 2. Comparison of order 1, 2, 3 and 4 RT methods under resolution 256 × 256 (*left*) and resolution 16 × 16 (*right*)

Ex.	Method	Pixels plotted	Subdi-visions	Addi-tions	Multipli-cations
1	1	550	631	795049	536562
1	2	526	571	415688	343892
1	3	526	567	460441	429975
1	4	526	563	252186	287257
2	1	438	497	248387	191938
2	2	433	461	241581	205717
2	3	433	460	246584	240250
2	4	433	459	334228	357296
3	1	619	681	1771000	1265762
3	2	608	637	1116344	936757
3	3	608	636	793926	808037
3	4	608	634	887844	1000846
4	1	843	952	12534981	9330145
4	2	801	845	4662221	4461229
4	3	816	860	3767717	4179094
4	4	816	857	2149817	3043237
5	1	484	803	1171467	869116
5	2	464	627	664231	575815
5	3	464	615	518665	535267
5	4	464	611	691345	764062
6	1	492	710	1053137	762808
6	2	460	567	442025	414092
6	3	460	560	743610	707035
6	4	460	560	281964	357439
7	1	562	755	990114	684256
7	2	512	629	445039	386359
7	3	512	627	644351	600905
7	4	512	627	273019	327424
8	1	846	895	612153	402862
8	2	818	829	563844	422917
8	3	818	827	258064	246520
8	4	818	827	337480	352408
9	1	1336	1625	2410713	1745518
9	2	1144	1281	998825	935312
9	3	1144	1269	1685062	1601793
9	4	1144	1269	639200	809781
10	1	844	997	1479305	1059079
10	2	784	849	662153	609761
10	3	784	845	1122246	1056562
10	4	784	845	425760	529126

Ex.	Method	Voxels plotted	Subdi-visions	Addi-tions	Multipli-cations
1	1	58	57	72137	48662
1	2	48	49	35848	29648
1	3	44	49	39969	37331
1	4	44	45	20266	23077
2	1	36	52	26059	20168
2	2	32	36	18977	16167
2	3	32	34	18348	17878
2	4	32	33	24200	25868
3	1	55	57	148840	106370
3	2	43	48	84512	70927
3	3	43	47	58950	60007
3	4	43	45	63340	71404
4	1	76	68	898473	668713
4	2	63	53	293765	281053
4	3	63	52	228897	253830
4	4	62	50	126073	178387
5	1	156	85	124747	92240
5	2	88	77	81927	70915
5	3	84	77	65225	67207
5	4	82	77	87465	96562
6	1	100	84	125089	90484
6	2	57	74	57845	54202
6	3	55	72	95878	91179
6	4	55	72	36344	46095
7	1	80	67	88282	60928
7	2	58	51	36311	31467
7	3	58	49	50663	47181
7	4	58	49	21507	25708
8	1	64	59	40545	26662
8	2	58	51	34876	26137
8	3	58	49	15400	14676
8	4	58	49	20128	20980
9	1	108	85	126601	91558
9	2	88	73	57193	53472
9	3	88	69	92038	87393
9	4	88	69	34976	44181
10	1	92	85	126537	90535
10	2	68	65	50905	46849
10	3	64	65	86646	81562
10	4	64	65	32880	40846

Obviously the order 1 recursive Taylor method is not as good as the order 2, 3 or 4 methods in accuracy or speed. On the other hand, we note that the order 3 and 4 recursive Taylor methods are not always at least as accurate as the order 2 method (see Example 4), or as efficient (see Example 2). While it is clear that the order 1 method can be rejected on grounds of poor performance, choice amongst the higher order methods is less clear-cut. Unsurprisingly, in most cases, using higher-order recursive Taylor methods leads to fewer recursive operations, but the decrease between using orders 1 and 2 is much greater than between using orders 2 and 3, and between higher orders. Overall the above results suggest using an order 2 recursive Taylor method as the best compromise between accuracy and efficiency, and ease of implementation.

However, a word of warning is necessary. This judgement strictly applies only to 256×256 resolution. If we reduce the resolution to 16×16 we get the results shown in Table 2 (*right*). This Table shows that under these conditions, a second order expansion need neither be most accurate (see Examples 1,4,5,6,10), nor most efficient. (The accuracy of the second order method is still quite close to that of the third and fourth order methods in all cases, however). Clearly, these results show that a theoretical proof that *any* particular order expansion is the best choice is not possible.

7 Theoretical Connection Between Taylor Method and MAA

In this section we briefly consider a theoretical relation between the Taylor method and the modified affine arithmetic method. It only concerns the intervals output by a direct (i.e. non-recursive) Taylor method and the MAA method; furthermore, it does not say how many operations are needed by each method.

Theorem 1 *Given a degree n polynomial, suppose $m > n$, and we perform an order m Taylor method. The output interval is equivalent to that produced by the modified affine arithmetic method.*

Proof We only prove the theorem here in the univariate case. The proofs for multivariate cases are similar.

Let $f(x) = \sum_{i=0}^{n} a_i x^i$ be the degree n polynomial in one variable whose range we wish to estimate over $[\underline{x}, \overline{x}]$. Let $x_0 = (\underline{x} + \overline{x})/2$, and $x_1 = (\overline{x} - \underline{x})/2 > 0$. Then the centered form of $f(x)$ on $[\underline{x}, \overline{x}]$ is

$$f(x) = f(x_0) + \sum_{i=1}^{n} \frac{f^{(i)}(x_0)}{i!}(x - x_0)^i. \tag{1}$$

It is known that the modified affine arithmetic method produces the same results as carrying out interval arithmetic on the centred form method with proper consideration of even and odd properties of polynomial terms [9]. If we evaluate $f(x)$ on $[\underline{x}, \overline{x}]$ using the modified affine arithmetic method we get

$$f_{MAA}[\underline{x}, \overline{x}] = f(x_0) + \sum_{i=1}^{n} \frac{f^{(i)}(x_0)}{i!} x_1^i \times \left\{ \begin{array}{l} [0,1], \text{ if } i \text{ is even} \\ [-1,1], \text{ if } i \text{ is odd} \end{array} \right\} \tag{2}$$

On the other hand, when $m > n$, the degree m Taylor form of $f(x)$ on $[\underline{x}, \overline{x}]$ is the same as eq. 1, because for any integer $i > n$, $f^{(i)}(x) = 0$ when $f(x)$ is a degree n polynomial. Therefore if we evaluate $f(x)$ on $[\underline{x}, \overline{x}]$ using a degree m Taylor method, we get the same interval as in eq. 2.

More work is needed to compare theoretically the intervals produced by the recursive Taylor method with those from MAA, and also to compare the numbers of operations. We intend to study these issues in the near future.

8   Conclusions

From the above experiments we can see that recursive Taylor methods can produce at least as good graphical results as the modified affine arithmetic method, and often need fewer arithmetic operations. Furthermore, the recursive Taylor method is simple and very easy to implement. One minor disadvantage of the recursive Taylor methods are that they often need a few more recursive operations than MAA. Repeating our earlier conclusions, overall we suggest using the *second order* recursive Taylor method as the best compromise (in terms of order) between accuracy and efficiency, and ease of implementation.

Acknowledgements

We wish to thank Dr. Stephen Cameron of Oxford University for insightful comments. This work was supported jointly by the National Natural Science Foundation of China (Grant No.60173034), the National Natural Science Foundation for Innovative Research Groups (No.60021201) and the Foundation of State Key Basic Research 973 Item (Grant No.2002CB312101).

References

1. Berz, M., Hoffstatter, G., Computation and application of Taylor polynomial with interval remainder bounds, *Reliable Computing*, 1998, 4: 83–97.
2. Cai, Y. Z., *Positive-Negative Method for Numerical Control Plot*, Zhejiang University Publishing House, Hangzhou, 1990 (in Chinese).
3. Chandler, R. E., A Tracking Algorithm for Implicitly Defined Curves, *IEEE Computer Graphics & Applications*, 1988, 8(2): 83–89.
4. Jin, T. G., T-N method for curve approximation, *Journal of Zhejiang University, Proceedings of Computational Geometry Conference*, 1982, 150–176 (in Chinese).
5. Martin R., Shou H., Voiculescu I., Bowyer A., Wang G., Comparison of Interval Methods for Plotting Algebraic Curves, *Computer Aided Geometric Design*, 2002, 19(7): 553-587.
6. Patrikalakis, N. M., Maekawa T., *Shape Interrogation for Computer Aided Design and Manufacturing*, Springer Verlag, 2002.
7. Ratschek, H., Rokne, J., *Computer Methods for the Range of Functions*, Ellis Horwood Ltd., 1984.
8. Shou H., Martin R., Voiculescu I., Bowyer A., Wang G., Affine arithmetic in matrix form for polynomial evaluation and algebraic curve drawing, *Progress in Natural Science*, 2002, 12(1): 77–80.
9. Shou H., Lin H., Martin R., Wang G., Modified affine arithmetic is more accurate than centered interval arithmetic or affine arithmetic, In: Michael J. Wilson, Ralph R. Martin (Eds.), Lecture Notes in Computer Science 2768, *Mathematics of Surfaces*, Springer-Verlag Berlin Heidelberg New York 2003, 355-365.
10. Snyder, J. M., Interval Analysis for Computer Graphics, *Computer Graphics (SIGGRAPH'92 Proceedings)*, 1992, 26(2): 121–130.
11. Suffern, K. G., Quadtree Algorithms for Contouring Functions of Two Variables, *The Computer Journal*, 1990, 33: 402–407.

12. Taubin, G., Distance Approximations for Rasterizing Implicit Curves, *ACM Transactions on Graphics*, 1994, 13(1): 3–42.
13. Taubin, G., Rasterizing Algebraic Curves and Surfaces, *IEEE Computer Graphics and Applications*, 1994, 14: 14–23.
14. Tupper, J., Realiable two-dimensional graphing methods for mathematical formulae with two free variables, *Proceedings of SIGGRAPH'2001*.

Self-Intersection Problems and Approximate Implicitization [*]

Jan B. Thomassen[1,2]

[1] Centre of Mathematics for Applications, University of Oslo
URL: http://www.cma.uio.no
janbth@math.uio.no
[2] SINTEF Applied Mathematics
URL: http://www.sintef.no

Abstract. We discuss how approximate implicit representations of parametric curves and surfaces may be used in algorithms for finding self-intersections. We first recall how approximate implicitization can be formulated as a linear algebra problem, which may be solved by an SVD. We then sketch a self-intersection algorithm, and discuss two important problems we are faced with in implementing this algorithm: What algebraic degree to choose for the approximate implicit representation, and – for surfaces – how to find self-intersection *curves*, as opposed to just points.

1 Introduction

Self-intersection algorithms are important for many applications within CAD/ CAM. Suppose we try to create an offset surface within a CAD system with an offset distance larger than the radius of curvature at some points on the surface. Offset surfaces are not in general rational, and it is therefore necessary to make a NURBS approximation. In a typical CAD-system, surfaces are often represented in terms of degree $(3,3)$ NURBS surfaces. A NURBS approximation of such an offset surface is shown in fig. 1. We would like to have parts of this surface trimmed away, or maybe even to have it rejected as an illegal surface. This requires an algorithm to find the self-intersection curves, or at least to detect that there are self-intersections.

Implicitization is the process of converting a parametric representation of a curve or surface into an implicit one. Figure 2 shows the implicit representation of a cubic curve with a node, given by the zero contour of the implicit function. Implicit representations have many uses. One example is finding intersections between different objects. Ray tracing is a simple case of this, where the parametric form of the ray – a straight line – is inserted into the implicit equation for a surface in order to get an equation in the parameter of the ray. This can then be solved to find the point, or points, of intersection. Note that self-intersections cannot be treated in this relatively simple way because inserting the parametric form into the implicit form for the same object gives identically zero. Thus, finding true self-intersections is a much harder problem.

[*] This work was supported by the EU project GAIA II, IST-2001-35512

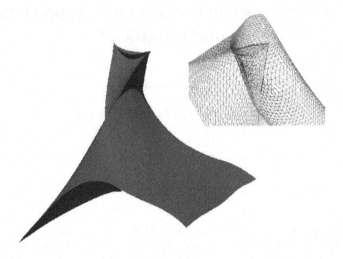

Fig. 1. NURBS approximation of an offset surface. Made within the CAD-system 'thinkdesign' from CAD-vendor think3.

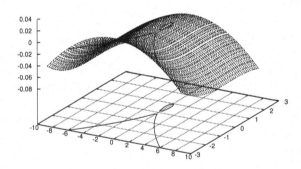

Fig. 2. Implicit representation of a nodal cubic

In this paper, we discuss an algorithm for finding self-intersections of NURBS curves and surfaces. This algorithm has in part grown out of the long-time experience collected at SINTEF [4, 12], and is still under development. The intersection problem has been studied a lot during the last few decades; for a partial list of references, see [12]. The present algorithm is optimized towards giving certifiable answers with respect to the self-intersections, rather than for speed. Therefore, central to the algorithm is an implicitization step and the use of implicit functions in detecting the self-intersections.

This will give us more information about the situation and secures the quality of the results compared to many brute force algorithms. Tangential (self-)intersections are in general difficult and will not be considered in this paper. We will therefore assume that all intersections are transversal.

Implicitization in CAGD has for a long time meant using classical algebraic techniques like resultants [10, 8] or Gröbner bases [3]. However, it is desirable to use approximate implicit representations of low algebraic degrees, which requires new methods. One recent method is the scattered data fitting method developed by Jüttler et al. [13]. This is very flexible but not accurate or fast. The method of choice for our algorithm is the one developed by Dokken [5, 6], which performs better on accuracy and speed [14]. This method essentially transforms the implicitization problem into the language of linear algebra. As far as we know, this is the simplest way to formulate the problem.

In the following, planar Bézier curves $\mathbf{p}(t)$ of degree n are defined in terms of $n+1$ control points $\mathbf{c}_i$ by

$$\mathbf{p}(t) = \sum_{i=0}^{n} \mathbf{c}_i B_{i,n}(t), \qquad t \in [0,1], \tag{1}$$

where the basis functions $B_{i,n}$ are Bernstein polynomials:

$$B_{i,n}(t) = \binom{n}{i} (1-t)^{n-i} t^i. \tag{2}$$

The unnormalized normal $\mathbf{n}$ is defined by

$$\mathbf{n}(t) = R_{\pi/2} \frac{d\mathbf{p}}{dt}(t), \tag{3}$$

where $R_{\pi/2} = \begin{pmatrix} 0 & -1 \\ 1 & 0 \end{pmatrix}$ is a matrix that rotates a vector by an angle of $\frac{\pi}{2}$. Implicit curves of degree d are defined in terms of a polynomial function $q(\mathbf{x})$. In the actual implementation of the algorithm we use barycentric coordinates $\mathbf{x} = (u, v, w)$ with respect to a triangle enclosing the curve. This improves numerical stability. Thus, for the implicit function we have

$$q(\mathbf{x}) = \sum_{i+j+k=d} b_{ijk} B_{ijk,d}(\mathbf{x}), \qquad u + v + w = 1, \tag{4}$$

where

$$B_{ijk,d}(\mathbf{x}) = \frac{d!}{i!j!k!} u^i v^j w^k, \qquad i + j + k = d, \tag{5}$$

are the Bernstein polynomials over this triangle. These $\frac{1}{2}(d+1)(d+2)$ monomials constitute a basis for the polynomials of degree d. In barycentric coordinates q is a homogeneous polynomial.

We may now sketch the self-intersection algorithm for curves. We are given a NURBS curve of degree n in the plane and we want to find the parameter values that correspond to self-intersections. Let us assume for simplicity that we are dealing with a non-rational[3] parametric curve – that is, the NURBS is in fact a B-spline.

1. Split the curve into Bézier segments.
2. For each segment $\mathbf{p}$, find candidates to self-intersection parameters. To do this,
   - find an implicit representation q,
   - form the quantity $\nabla q(\mathbf{p}(t)) \cdot \mathbf{n}(t)$ and find the roots.
3. For each pair of segments $\mathbf{p}_1$ and $\mathbf{p}_2$, find candidates to parameters for ordinary intersections. To do this,
   - find the implicit representations q_1 and q_2,
   - form the quantities $q_1(\mathbf{p}_2(t))$ and $q_2(\mathbf{p}_1(t))$ and find the roots.
4. Identify parameters for true self-intersections from the list of candidates. To do this, find pairs of parameters from this list whose corresponding 2D points match.

Although we have omitted the details, the structure of this algorithm should be straightforward. However, two things we will discuss further is the implicitization and the use of the quantity $\nabla q(\mathbf{p}(t)) \cdot \mathbf{n}(t)$. Thus, in the next section we will discuss exact implicitization. We will show that this is a linear algebra problem – in particular, it is an SVD problem. In sect. 3 we explain why $\nabla q(\mathbf{p}(t)) \cdot \mathbf{n}(t)$ is useful for finding self-intersections.

In sect. 4 we start considering surfaces, in particular tensor product B-splines. Unlike for curves, exact implicitization will typically require too high degrees, so if we want an algorithm similar to the one for curves, we need to find an approximate implicit representation. Approximate implicitization is the subject of sect. 5. Finally, in sect. 6 we discuss what are probably the two most important problems to solve before we can get an industrial strength surface intersection algorithm: What degree to choose for the approximate implicitization, and how to find self-intersection curves, as opposed to just points.

2 Implicitization

We are given a Bézier curve $\mathbf{p}(t)$ of degree n. We want to find a polynomial $q(\mathbf{x})$ such that the zero-set $\{\mathbf{x}|q(\mathbf{x}) = 0\}$ contains $\mathbf{p}$.

First of all, we want the degree d of q to be as small as possible for achieving exact implicitization. If $\mathbf{p}$ is sufficiently general, then $d = n$. If the true degree of $\mathbf{p}$ is smaller than n, for example if $\mathbf{p}$ is obtained by degree elevating a curve of lower degree, we still get an exact implicit representation, but with possible additional branches. This will typically produce phantom candidates for self-intersections in our algorithm, but that is not a problem because of the processing of the candidates at the end of the algorithm.

Thus we choose $d = n$, and we must find a nontrivial $\frac{1}{2}(d+1)(d+2)$-dimensional vector $\mathbf{b}$ of coefficients of q such that

$$q(\mathbf{p}(t)) = 0. \tag{6}$$

---

[3] In this paper, 'non-rational' means 'polynomial'. This is standard terminology in CAGD, see *e.g.* Farin's book [7].

This turns out to give a matrix equation in $\mathbf{b}$ [5,6]. To see this we calculate $q(\mathbf{p}(t))$ by using the product rule for Bernstein basis functions:

$$B_{i,m}(t)B_{j,n}(t) = \frac{\binom{m}{i}\binom{n}{j}}{\binom{m+n}{i+j}}B_{i+j,m+n}(t). \tag{7}$$

From this we have that $q(\mathbf{p}(t))$ can be expressed in a Bernstein basis of degree nd. We organize the basis functions in an $(nd+1)$-dimensional vector $\mathbf{B}(t)^T = (B_{0,nd}(t),\ldots,B_{nd,nd}(t))$. Thus we find that inserting $\mathbf{p}$ in q yields the factorization

$$q(\mathbf{p}(t)) = \mathbf{B}(t)^T\mathbf{D}\mathbf{b}, \tag{8}$$

where $\mathbf{D}$ is an $M \times N$ matrix with $M = nd + 1$ and $N = \frac{1}{2}(d+1)(d+2)$. Since $\mathbf{B}$ is a basis, we see that the problem we need to solve is the matrix equation

$$\mathbf{D}\mathbf{b} = 0. \tag{9}$$

The standard method for solving a matrix equation like (9) is to use SVD [9]. The theory of the SVD states that we can find the decomposition $\mathbf{D} = \mathbf{U}\mathbf{W}\mathbf{V}^T$, where $\mathbf{U}$ is an $M \times N$ column-orthogonal, $\mathbf{W}$ an $N \times N$ diagonal, and $\mathbf{V}$ an $N \times N$ orthogonal matrix. The singular values of $\mathbf{D}$ are the numbers on the diagonal of $\mathbf{W}$. The solution to $\mathbf{D}\mathbf{b} = 0$ is an eigenvector corresponding to the vanishing singular values in $\mathbf{W}$. If there is exactly one singular value equal to zero, we can find the associated eigenvector from the corresponding column in $\mathbf{V}$. If there are more than one, any linear combination of the corresponding columns will solve (9). Note that we avoid problems with the trivial solution $\mathbf{b} = 0$ to (9). We may always normalize $\mathbf{b}$ and thereby $q(\mathbf{x})$.

An example of this method of implicitization is the nodal cubic Bézier curve in fig. 3. The result we have already seen in fig. 2. This implicitization is numerically very exact, and a plot of $q(\mathbf{p}(t))$ as a function of t would show zero to within machine precision.

Rational Bézier curves can be treated in essentially the same way. In barycentric coordinates a rational Bézier curve is given by

$$\mathbf{p}(t) = \frac{\sum_{i=0}^{n} w_i\mathbf{c}_iB_{i,n}(t)}{\sum_{i=0}^{n} w_iB_{i,n}(t)}, \tag{10}$$

where $\mathbf{c}_i$ are the control points and w_i are the weights. Since the implicit function q is homogeneous in barycentric coordinates, the denominator $\sum_i w_iB_{i,n}(t)$ leads to a polynomial that factors out in the expression for $q(\mathbf{p}(t))$. This means that the non-rational procedure can be applied with just the numerator $\sum_i w_i\mathbf{c}_iB_{i,n}(t)$ instead of $\mathbf{p}$ itself.

3   Finding Self-Intersections

The other issue we need to discuss is how to find the self-intersections of a Bézier curve $\mathbf{p}(t)$ once an implicit representation $q(\mathbf{x})$ is provided. For each self-intersection point we want to find the two corresponding parameters t_1 and t_2 such that

$$\mathbf{p}(t_1) = \mathbf{p}(t_2), \qquad t_1 \neq t_2. \tag{11}$$

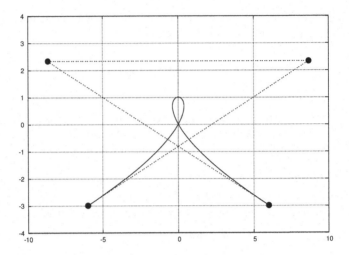

Fig. 3. Nodal cubic Bézier curve

This does not include cusps, which are "degenerate" self-intersections. We will briefly mention cusps later.

At a self-intersection point $\mathbf{p}_0 = \mathbf{p}(t_1) = \mathbf{p}(t_2)$ in the plane the gradient of the implicit function q vanishes: $\nabla q(\mathbf{p}_0) = 0$. In fact $\mathbf{p}_0$ is a saddle point of q, which looks like a hyperbolic paraboloid locally around $\mathbf{p}_0$. One way to find the self-intersection parameter values is to form $[\nabla q(\mathbf{p}(t))]^2$ and find the roots of this expression. However, this is not the optimal quantity for this purpose, because the polynomial degree is $2n(d-1)$, which is unnecessarily high, and all roots are located at minima. A better choice is the quantity

$$\nabla q(\mathbf{p}(t)) \cdot \mathbf{n}(t), \tag{12}$$

where $\mathbf{n}$ is the normal. The polynomial degree of (12) is $nd - 1$. This quantity is proportional to the projection of ∇q on the normal vector along the curve, and has the advantage that it changes sign at roots corresponding to transverse – i.e. non-tangential – self-intersections.

For the nodal cubic of fig. 3, the gradient dotted with the normal is plotted in fig. 4. The two roots occur at $t_1 = 0.25$ and $t_2 = 0.75$, which are the correct parameters for the self-intersection.

Roots of $\nabla q \cdot \mathbf{n}$ should be regarded as *candidates* for self-intersections. Indeed, ∇q may vanish along the curve $\mathbf{p}(t)$ if q is reducible so that the additional factors represent extra branches, or if $\mathbf{p}(t)$ reaches the endpoint, as we increase t, before the curve has time to complete the loop and self-intersect. This is why the last step of the algorithm is needed.

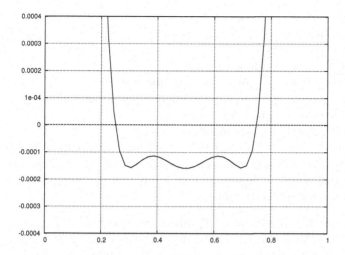

Fig. 4. $\nabla q \cdot \mathbf{n}$ for the nodal cubic of fig. 3

4   Surfaces

We now turn to surfaces. In this case we are given a tensor product NURBS surface. Let us assume this to be non-rational. Thus the surface consists of tensor product Bézier patches:

$$\mathbf{p}(u,v) = \sum_{i,j} \mathbf{c}_{ij} B_{i,n_1}(u) B_{j,n_2}(v), \tag{13}$$

where the degree is (n_1, n_2) and $\mathbf{c}_{ij}$ are the $(n_1 + 1)(n_2 + 1)$ control points. The unnormalized normal vector $\mathbf{n}(u,v)$ is defined by

$$\mathbf{n}(u,v) = \partial_u \mathbf{p}(u,v) \times \partial_v \mathbf{p}(u,v). \tag{14}$$

Like for curves we use barycentric coordinates $\mathbf{x} = (u,v,w,x)$ in our implementation, this time defined over a tetrahedron enclosing the surface. Then algebraic functions of degree d can be written

$$q(\mathbf{x}) = \sum_{i+j+k+l=d} b_{ijkl} B_{ijkl,d}(\mathbf{x}), \qquad u+v+w+x = 1, \tag{15}$$

with

$$B_{ijkl,d}(\mathbf{x}) = \frac{d!}{i!j!k!l!} u^i v^j w^k x^l, \qquad i+j+k+l = d, \tag{16}$$

the tetrahedral Bernstein basis. The polynomials $B_{ijkl,d}$ are now the $\frac{1}{6}(d+1)(d+2)(d+3)$-dimensional basis of degree d functions. Again, in barycentric coordinates $q(\mathbf{x})$ is a homogeneous function.

What we want now is an algorithm for finding self-intersections like the one for curves. *Ideally* this would look like:

1. Split the surface into Bézier patches.
2. For each patch $\mathbf{p}(u, v)$, find candidates to self-intersection curves in the (u, v)-plane. To do this,
   - find an implicit representation q,
   - form $\nabla q(\mathbf{p}(u, v)) \cdot \mathbf{n}(u, v)$ and find the roots.
3. For each pair of patches $\mathbf{p_1}(u, v)$ and $\mathbf{p_2}(u, v)$, find candidates to parameter curves for ordinary intersections. To do this,
   - find the implicit representations q_1 and q_2,
   - form $q_1(\mathbf{p_2}(u, v))$ and $q_2(\mathbf{p_1}(u, v))$ and find the zero curves in the (u, v)-plane.
4. Identify the curves in the parameter plane for true self-intersection curves from the list of candidates. To do this, find pairs of parameter curves from this list whose corresponding 3D curves match.

Unfortunately it is difficult to implement the algorithm in this way. There are essentially two problems. First, the necessary degree for exact implicitization is too high, both with respect to numerical stability and to speed. Hence the need to find an approximate implicit representation. For a chosen degree, the next section provides a review of how this can be done along the lines of the exact implicitization of sect. 2. However, this leads to the problem of what degree to choose. Second, self-intersection curves are difficult to handle.

In the current implementation of the algorithm, we have settled for a compromise to both these problems. We choose $d = 4$ for the approximate implicitization, based on the expectation that this degree is high enough to capture even complicated topologies. Our experience with this choice is good. Furthermore, instead of working with full self-intersection curves in the parameter plane, we work with points sampled on them. A sampling density must then be chosen, and a density of roughly 100×100 in the parameter plane for each patch gives acceptable results.

These two compromises lead to a change in the last step of the algorithm, since now also an iteration procedure is required for matching the points in 3D. This matching works roughly as follows. We loop over every pair of points from the list of candidates. Each pair of points in the (u, v)-plane corresponds to two points on the surface. We consider the squared Euclidean distance in 3D space between these two surface points. This is a four-variate function depending on the u and v parameter for the first and second points. Now we look for a minimum of this function. This is where the iterations come in – in our current implementation, a conjugate gradient method is used. If the iteration converges, and if the found minimum is a zero of the distance function, then we have a match and we record the corresponding parameters as points on a self-intersection curve. Then we go on with the next pair, etc. The points found in this way will in general not coincide exactly with the candidates we started with, but they still have the property that they lie on self-intersection curves.

An example is given by the offset surface we have already seen in fig. 1. As mentioned in the introduction, this is a realistic industrial example. It consists of many (rational) polynomial surface patches and has a very complex self-intersection structure. The result of running our algorithm on this surface is shown in fig. 5, where we have plotted points in the parameter plane that corresponds to points on the self-intersection curves in 3D space.

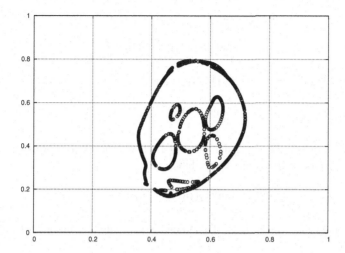

Fig. 5. Self-intersection points in the parameter plane of the offset surface in fig. 1

5   Approximate Implicitization

Although approximate implicitization is mostly relevant for surfaces, we will use curve language for simplicity in the following discussion. However, the results can easily be transferred to the surface case.

An implicitly defined curve $q(\mathbf{x}) = 0$ approximates a parametric curve $\mathbf{p}(t)$ within the tolerance ϵ if we can find a vector-valued function $\Delta\mathbf{p}(t)$ such that

$$q(\mathbf{p}(t) + \Delta\mathbf{p}(t)) = 0 \tag{17}$$

and

$$\max_t |\Delta\mathbf{p}(t)| \le \epsilon. \tag{18}$$

$\Delta\mathbf{p}$ does not need to be continuous. In fact, for nontrivial topologies like self-intersections, $\mathbf{p} + \Delta\mathbf{p}$ may jump from one branch of q to another as we increase t.

We may Taylor expand (17) in $\Delta\mathbf{p}$ around $\mathbf{p}$:

$$q(\mathbf{p}(t)) + \nabla q(\mathbf{p}(t)) \cdot \Delta\mathbf{p} + \cdots = 0. \tag{19}$$

Thus, provided $\Delta\mathbf{p}$ is small and the coefficients of q are normalized in some way, the quantity $\max_t |q(\mathbf{p}(t))|$ is a meaningful measure of how well q approximates $\mathbf{p}$. This quantity is the "algebraic distance".

How do we find an approximate implicit function q such that the algebraic distance is as small as possible? Again we get an answer from linear algebra. If we insert the Bézier segment $\mathbf{p}(t)$ into a function q with unknown coefficients $\mathbf{b}$ we obtain the factorization in eq. (8),

$$q(\mathbf{p}(t)) = \mathbf{B}(t)^T \mathbf{D} \mathbf{b}. \tag{20}$$

Since $\mathbf{B}(t)$ is a Bernstein basis, $\|\mathbf{B}(t)\| \leq 1$ for all $t \in [0,1]$. Thus we get the inequality

$$\max_t |q(\mathbf{p}(t))| \leq \|\mathbf{Db}\|. \tag{21}$$

Furthermore, the theory of the SVD tells us that

$$\min_{\|\mathbf{b}\|=1} \|\mathbf{Db}\| = \sigma_{\min}, \tag{22}$$

where $\sigma_{\min}$ is the smallest singular value of $\mathbf{D}$. If we choose the corresponding eigen-vector $\mathbf{b}_{\min}$ as the coefficients of q we have

$$\max_t |q(\mathbf{p}(t))| \leq \sigma_{\min}. \tag{23}$$

Thus, choosing the vector $\mathbf{b}_{\min}$ of coefficients gives us the implicit function we are looking for.

If we choose an algebraic degree d for q much lower than the required degree for exact implicitization, it may be that even the smallest singular value $\sigma_{\min}$ is greater than a given tolerance ϵ. It is then possible to improve the situation by subdividing. This leads us to consider convergence rates for approximate implicitization. Let us consider a curve $\mathbf{p}(t)$ and pick out an interval $[a,b] \subset [0,1]$ from the parameter domain with length h, that is, $b - a = h$. If we approximate the curve on this interval by an implicit function $q(\mathbf{x})$ of degree d we will have

$$|q(\mathbf{p}(t))| = O(h^{k+1}), \qquad t \in [a,b], \tag{24}$$

as h goes to zero, where the integer $k+1$ is the convergence rate. Dokken [5] proved that $k = \frac{1}{2}(d+1)(d+2) - 2$. Likewise for a surface $\mathbf{p}(u,v)$ we get a convergence rate from

$$|q(\mathbf{p}(u,v))| = O(h^{k+1}) \tag{25}$$

with $k = \left\lfloor \frac{1}{6}\sqrt{9 + 132d + 72d^2 + 12d^3} - \frac{3}{2} \right\rfloor$ (see [5]). The notation $\lfloor \cdots \rfloor$ means that '$\cdots$' is rounded downwards to the nearest integer. Thus, for surfaces an approximate implicitization of degree 4 will have convergence rate $O(h^7)$.

If we are not satisfied with the algebraic distance we get from implicitizing a surface, we may dramatically improve the situation with a few subdivisions of the parameter plane. It is therefore a useful strategy to include subdivision in a self-intersection algorithm that is based on implicitization.

6   Two Open Problems

Finally, in this section we address the two problems mentioned previously: What degree d should we choose for the implicit representation of a surface in our algorithm? And, also for surfaces, how do we describe and find self-intersection *curves*, as opposed to just points?

6.1   What Degree d Should we Choose?

For NURBS curves there is no problem with using the required degree for exact implicitization in our algorithm. That is, for a curve of degree n, we may use $d = n$. For all realistic test cases we have tried, this gives a fast and reliable algorithm.

For a NURBS surface of degree (m, n), the required degree for exact implicitization is in general $d = 2mn$. Thus the realistic case of degree $(3, 3)$ surfaces needs $d = 18$. A polynomial in 3D of this degree will in general have 1330 terms, which gives very poor precision and long evaluation times. Furthermore, the matrix $\mathbf{D}$ in the implicitization becomes so large that the SVD is very slow and has problems with converging.

There are several possibilities to overcome this. One is to choose some fixed low degree d and hardcode this in the algorithm. Another is to use some adaptive procedure where we start with a low value for d and then increase it until some requirement is met. We have experimented with both of these possibilities. The requirement in the adaptive procedure was that the smallest singular value in the SVD of $\mathbf{D}$ should be smaller than some number. However, it turned out to be very time consuming and not noticeably better than the hardcoding possibility. As already mentioned, $d = 4$ leads to acceptable results in all test cases we have tried.

Could we just as well manage with $d = 3$, which might lead to a faster algorithm? After all, implicitly defined surfaces of degree 3 are the simplest ones that may exhibit self-intersection curves. However, for $d = 3$ it turns out that the quality of the candidates to self-intersection points obtained from sampling the zeros of $\nabla q \cdot \mathbf{n}$ are very poor. This means that the iteration procedure in the matching step leads either to discarding most of the candidates or to produce a large number of redundant matches which is time consuming. On the other hand, for $d = 4$ iteration leads to matching for most of the candidates without too much redundancy.

An example is the surface shown in fig. 6, which has a closed self-intersection curve. The degree of the exact implicit representation is 6. Candidates and the result after iteration for $d = 3$ is shown in fig. 7, while the result for $d = 4$ is shown in 8.

Nevertheless, it is a suboptimal solution to hardcode the degree once and for all. It would be better to be able to determine for each surface patch which degree would be most suitable. But for the moment it is an open problem how to do that efficiently inside the algorithm.

6.2   How do we Find Self-Intersection *Curves* (as Opposed to *Points*)?

A possible strategy to finding self-intersection curves is to first find points on these curves and then use a marching technique to "march them out". The goal here would be to obtain a description of the intersection curves in terms of B-spline (or NURBS) curves in the parameter domain. We have already said a lot about finding such points, so let us discuss marching.

Marching is an iterative procedure, which can be used to trace out intersection curves [1, 2] (see [12] for more references). We start with a point on or nearby the curve, and then proceed by taking small steps at a time in a cleverly chosen direction until some stopping criterion is met. For instance, we stop when we reach the edge of the parameter domain or when the curve closes on itself.

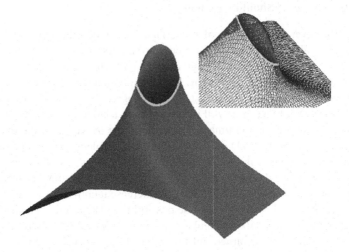

Fig. 6. Surface with points on self-intersection curve. Supplied by think3.

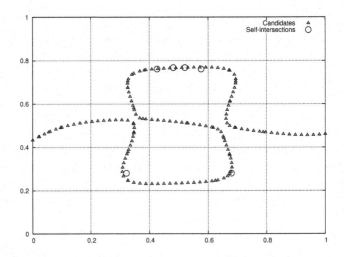

Fig. 7. Parameter plane with candidates and self-intersections obtained after iteration for the surface in fig. 6 for Degree 3

One problem we are faced with when marching out a self-intersection curve is the marching onto a singularity, which in this case means a point on the surface where the unnormalized normal vanishes. The normal is used in order to find the direction of the next step in the marching. A singularity of this kind is a cusp. For example, the surface in fig. 6 has two singularities, plotted in fig. 9. Closed self-intersection curves typically have singularities on them.

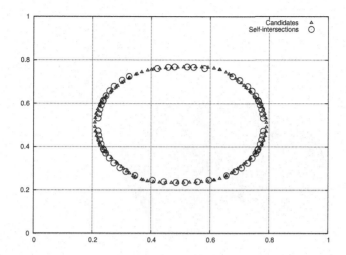

Fig. 8. Parameter plane with candidates and self-intersections obtained after iteration for the surface in fig. 6 for Degree 4

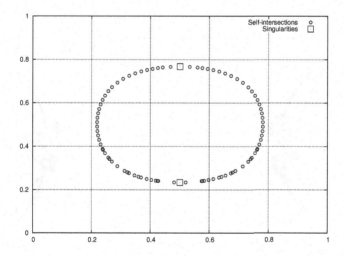

Fig. 9. The parameter plane of the surface in fig. 6, with self-intersection points and two cusp-like singularities

Thus, the solution is to identify all the isolated singularities and start the marching *from* these points. It is not difficult to find isolated cusp singularities. For instance, we may look for the zeros of $[\mathbf{n}(u, v)]^2$, which are also minima.

Unfortunately in real life things are not that easy. It may often happen that entire curves of cusp singularities exist. An example of this is the surface in fig. 10. Points sampled from the cusp curves and self-intersection curves in the parameter plane are plotted in fig. 11. As we may see from the plot, in order to start marching we would need

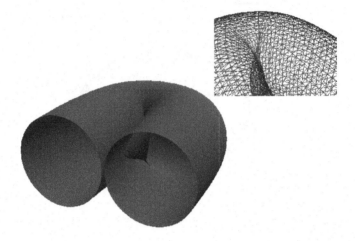

Fig. 10. Bent pipe surface with "supersingularities". Surface provided by think3.

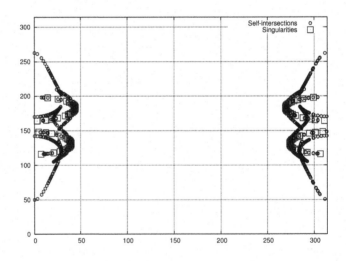

Fig. 11. Parameter plane of the pipe surface with self-intersection points and points on the cusp curves

to locate an intersection point between a self-intersection curve and a cusp curve. This is a very special kind of singularity, let us call it a "supersingularity". At the moment we do not know how to properly characterize and find such supersingular points. Until we do, the treatment of self-intersection curves in such cases is another open problem.

7 Conclusion

We have described how we may use implicit representations of curves and surfaces in an algorithm for finding self-intersections. We have also described how to find the implicit representation given a NURBS curve or surface. We do this by formulating it as a linear algebra problem. For curves we may use an exact implicitization, while for surfaces it is necessary to use an approximate implicitization.

By using the implicit representation q and the normal vector $\mathbf{n}$ we are able to find candidates for the parameters of self-intersections by looking for the roots of $\nabla q \cdot \mathbf{n}$. For curves, where we use exact implicitization, the self-intersection parameters are included in the list of candidates. For surfaces, it is necessary to process the candidates with an iterative procedure to match them in order to get points on the self-intersection curves.

Surfaces are much more challenging than curves, and we have described two open problems: How do we choose the degree for the approximate implicitization, and how do we deal with the fact that self-intersections are in general curves?

Acknowledgements I thank Tor Dokken for discussions, and for reading and commenting on the manuscript.

References

1. C.L. Bajaj, C.M. Hoffmann, R.E. Lynch, and J.E.H. Hopcroft, Tracing surface intersections, *Computer Aided Geometric Design* 5 (1988) 285–307.
2. R.E. Barnhill, and S.N. Kersey, A marching method for parametric surface/surface intersection, *Computer Aided Geometric Design* 7 (1990) 257–280.
3. D. Cox, J. Little, and D. O'Shea, *Ideals, Varieties, and Algorithms*, Second Edition, Springer-Verlag, New York, 1997.
4. T. Dokken, V. Skytt, and A.-M. Ytrehus, Recursive Subdivision and Iteration in Intersections and Related Problems, in Mathematical Methods in Computer Aided Geometric Design, T. Lyche and L. Schumaker (eds.), Academic Press, 1989, 207–214.
5. T. Dokken, *Aspects of Intersection Algorithms and Applications*, Ph.D. thesis, University of Oslo, July 1997.
6. T. Dokken, and J.B. Thomassen, Overview of Approximate Implicitization, in *Topics in Algebraic Geometry and Geometric Modeling*, AMS Cont. Math. 334 (2003), 169–184.
7. G. Farin, *Curves and surfaces for CAGD: A practical guide*, Fourth Edition, Academic Press, 1997.
8. R.N. Goldman, T.W. Sederberg, and D.C. Anderson, Vector elimination: A technique for the implicitization, inversion, and intersection of planar rational polynomial curves, *Computer Aided Geometric Design* 1 (1984) 327–356.
9. W.H. Press, S.A. Teukolsky, W.T. Vetterling, and B.P. Flannery, *Numerical Recipes in C*, Second Edition, Cambridge University Press, 1992.
10. T.W. Sederberg, D.C. Anderson, and R.N. Goldman, Implicit Representation of Parametric Curves and Surfaces, *Comp. Vision, Graphics, and Image Processing* **28**, 72–84 (1984).
11. T.W. Sederberg, J. Zheng, K. Klimaszewski, and T. Dokken, Approximate Implicitization Using Monoid Curves and Surfaces, *Graph. Models and Image Proc.* **61**, 177–198 (1999).
12. V. Skytt, Challenges in surface-surface intersections, these proceedings.

13. E. Wurm, and B. Jüttler, Approximate implicitization via curve fitting, in L. Kobbelt, P. Schröder, H. Hoppe (eds.), *Symposium on Geometry Processing*, Eurographics, ACM Siggraph, New York 2003, 240–247.
14. E. Wurm, and J.B. Thomassen, Deliverable 3.2.1 – Benchmarking of the different methods for approximate implicitization, Internal report in the GAIA II project, October 2003.

Singularities of
Some Projective Rational Surfaces

Ragni Piene

Centre of Mathematics for Applications & Department of Mathematics,
University of Oslo, P.O. Box 1053 Blindern, NO-0316 Oslo, Norway
`ragnip@math.uio.no`

Abstract. We discuss the singularities of some rational algebraic surfaces in complex projective space. In particular, we give formulas for the degrees of the various types of singular loci, in terms of invariants of the surface. These enumerative results can be used, on the one hand, to show the existence of singularities in the complex case, and, on the other hand, as an "upper bound" for the singularities that can occur on a *real* rational surface.

1 Introduction

The simplest rational surfaces spanning projective 3-space are the quadric and cubic surfaces. The classification of real and complex quadric surfaces is very old, and the nonsingular and singular complex and real cubic surfaces were classified in the 19th century by Cayley, Salmon, Schläfli, Klein, Zeuthen (see [5, 13, 7, 19, 15]). Affine real 2- and 3-dimensional geometry was the object of intense study among 19th century mathematicians, but eventually the focus shifted to projective complex algebraic geometry, which from many points of view is much simpler. The classical work in algebraic geometry remains a rich source of inspiration for people working in Computer Aided Geometric Design (CAGD).

In this article, the objects of study are surfaces that admit a rational parameterization; because of the methods used, we choose to work in the complex projective setting. The two types of rational parameterizations most commonly used in CAGD are the "triangular" and "tensor product" parameterizations. These are patches of what in algebraic geometry are called Veronese and Segre surfaces. Some times these parameterizations have base points; we shall look at some examples, namely Del Pezzo surfaces and monoid surfaces.

In the GAIA II project "Intersection algorithms for geometry based IT-applications using approximate algebraic methods" we realized the need for a better understanding of the singularities of algebraic curves and surfaces that are used in geometric modeling. Though the work presented here is carried out in the complex projective setting, the enumerative results can be interpreted as an "upper bound" for the singularities that may appear in the real affine case. There are of course a lot of hard and interesting problems that can not be addressed via these methods, such as determining the number of connected components and positions of a real curve or surface (cf. Hilbert's 16th problem), questions of boundedness and of equidimensionality, and the problem of determining which real forms a given complex singularity may have.

2   Numerical Characters of a Projected Surface

Let $\mathbb{P}^N_{\mathbb{C}}$, or $\mathbb{P}^N$ for short, denote the complex projective N-space. Let S be a nonsingular surface and let $S \hookrightarrow \mathbb{P}^N$ be any embedding of S. One can show that for most linear projections $p \colon S \to \mathbb{P}^3$, the induced map $p \colon S \to X := p(S)$ is finite and birational, and X has only ordinary singularities. By this we mean the following: Let $\Gamma := \mathrm{Sing}\, X$ denote the singular locus of X, and let $P \in \Gamma$. Assume $P = (1 : 0 : 0 : 0)$ and that $f(x_1, x_2, x_3)$ is the equation of X in the affine space where $x_0 \neq 0$. Consider the Taylor series expansion of f, which, by abuse of notation, we also denote by f. Then, up to change of coordinates, there are only three possibilities [9, p. 202]:

1. The point P is a point where the surface X intersects itself transversally, i.e., we have $f = x_1 x_2 +$ terms of higher degree, and $\#p^{-1}(P) = 2$. All but finitely many points of Γ are of this type.
2. The point P is a point where three sheets of the surface intersect transversally, i.e., we have $f = x_1 x_2 x_3 +$ terms of higher degree, and $\#p^{-1}(P) = 3$. This point is a triple point of the curve Γ and of the surface X.
3. The point P is a point of selfintersection of the surface where the projection map is ramified. Here we have $f = x_2{}^2 - x_1{}^2 x_3 +$ terms of higher degree, and $\#p^{-1}(P) = 1$. Such points are called pinch points. (The real part of the surface at such a point is the Whitney umbrella.)

The curve Γ is called the *nodal curve* of X; its only singular points are the triple points (the pinch points are not singular on Γ).

Introduce the following numerical characters:

$$n := \deg X$$

$$m := \deg \Gamma$$

$$t := \#\text{ triple points of } X \text{ and of } \Gamma$$

$$\nu_2 := \#\text{ pinch points of } X$$

$$c_1^2 := K_S^2 = \text{the self intersection of the canonical divisor of } S$$

$c_2 :=$ the second Chern class of S ($=$ the topological Euler characteristic of S).

Then we have the following formulas [9, Prop. 1, p. 211]:

$$c_1^2 = (n-4)^2 n - (3n-16)m + 3t - \nu_2 \tag{1}$$

$$c_2 = n(n^2 - 4n + 6) - (3n-8)m + 3t - 2\nu_2 \tag{2}$$

Subtracting (1) from (2) gives

$$\nu_2 = 2n(2n-5) - 8m + c_1^2 - c_2 \tag{3}$$

and substituting (3) in (1) gives

$$3t = 3(n-8)m - n(n^2 - 12n + 26) + 2c_1^2 - c_2 \tag{4}$$

The *sectional genus* π of S is the geometric genus of a general hyperplane section of S. By the adjunction formula it is given by $2\pi - 2 = \deg K_S + n$.

Among the other characters considered classically are:

$$\mu_2 := \text{class of } X$$

$$\mu_1 := \text{rank of } X$$

$$\epsilon_1 := \text{rank of } \Gamma$$

Recall that the class of X is the number of tangent planes containing a given line, hence it is the same as the degree of the dual surface of X; the rank of X is the number of tangent lines in a given plane through a given point in that plane (hence is equal to the class of a plane section of X); the rank of the curve Γ is the number of tangent lines to Γ that meet a given line. From the formulas in [9, Thm. 2, p. 204; Prop. 3, p. 212] we deduce the following formulas for the characters of the projected surface:

$$\mu_1 = n(n-1) - 2m \tag{5}$$

$$\mu_2 = n(2n-5) - 4m + c_2 \tag{6}$$

$$\epsilon_1 = n(n^2 - 14n + 31) - 2m(n-13) + \tfrac{1}{2}(3c_2 - 5c_1^2) \tag{7}$$

3 The Veronese Surfaces

The dth Veronese embedding of the plane $\mathbb{P}^2$ is the map

$$v_d \colon \mathbb{P}^2 \to \mathbb{P}^N,$$

where $N := \binom{d+2}{2} - 1$, given by

$$v_d(t_0 : t_1 : t_2) = (t_0^{i_0} t_1^{i_1} t_2^{i_2})_{i_0+i_1+i_2=d}.$$

In affine coordinates, this is just the triangular parameterization

$$(t_1, t_2) \mapsto (t_1, t_2, t_1^2, t_1 t_2, t_2^2, \ldots, t_2^d)$$

We want to study the surface $X \subset \mathbb{P}^3$ obtained by projecting the nonsingular surface $S = v_d(\mathbb{P}^2)$ to $\mathbb{P}^3$ from a linear subspace $L \subset \mathbb{P}^N$ of dimension $N - 4$. The surface X has a rational parameterization by polynomials of degree d, and, conversely, any surface in $\mathbb{P}^3$ that has such a parameterization is obtained in this way.

Assume $d \geq 2$, $L \cap v_d(\mathbb{P}^2) = \emptyset$, and that the induced morphism $p \colon \mathbb{P}^2 \to X$ is birational. In this case, the degree of X is $n = d^2$. Since the morphism p is finite (because it is a projection), it is equal to the normalization map of X. It is well known that therefore, X can have no isolated singular points, so that $\operatorname{Sing} X$ is either empty or purely one-dimensional.

The following proposition shows that X must be singular, and gives a bound for the degree of the singular locus.

Proposition 1. *In the situation above, set* $\Gamma := \text{Sing } X$. *Then* Γ *is a curve, of degree*

$$\deg \Gamma \leq \tfrac{1}{2}d(d-1)(d^2+d-3).$$

Moreover, equality holds if and only if the components of Γ *are ordinary double curves or cuspidal edges on* X.

Proof. Let $H \subset \mathbb{P}^3$ be a general plane, and set $C := X \cap H$. Then C is a plane curve of the same degree, d^2, as X. Let D denote the inverse image of C in $\mathbb{P}^2$; this is a plane curve of degree d, birationally equivalent to C. It follows from Bertini's theorem that D is nonsingular, and therefore D is equal to the desingularization of C. Hence, the sectional genus π of S, by definition equal to the geometric genus $g(C)$ of C, is equal to the geometric genus of D, namely

$$\pi = g(C) = g(D) = \binom{d-1}{2}.$$

By the genus formula for the plane curve C, we have

$$\binom{d-1}{2} = \binom{d^2-1}{2} - \sum_{x \in \text{Sing } C} \delta_x,$$

where δ_x denotes the δ-invariant of the point $x \in \text{Sing } C$. Note that $\delta_x \geq 1$, with equality if and only if x is an ordinary double point (a node) or an ordinary cusp, i.e., if and only if the components of Γ are ordinary double curves or cuspidal edges on X. Since $d \geq 2$ by assumption, we get $\sum_{x \in \text{Sing } C} \delta_x > 0$, hence C is singular. Using Bertini's theorem again, we have $\text{Sing } C = H \cap \text{Sing } X = H \cap \Gamma$, therefore $\#\text{Sing } C = \#H \cap \Gamma = \deg \Gamma$. The last equality holds because, since H is general, we may assume that H and Γ intersect transversally.

If we consider a general projection of the dth Veronese surface $S = v_d(\mathbb{P}^2)$, we have $n = d^2$, $c_1^2 = 9$, and $c_2 = 3$. Moreover, from Proposition 1, we have

$$m = \tfrac{1}{2}d(d-1)(d^2+d-3).$$

Then (3) and (4) give the following formulas:

$$\nu_2 = 6(d-1)^2$$

$$t = \tfrac{1}{6}(d-1)(d^5 + d^4 - 11d^3 - 2d^2 + 42d - 30)$$

From (5), (6), and (7), we get

$$\mu_1 = 3d(d-1)$$

$$\mu_2 = 3(d-1)^2$$

$$\epsilon_1 = 3(d-1)^2(d-2)(d+3)$$

Example 2. Consider the case $d = 2$. From Proposition 1, we get $\deg \Gamma \leq 3$. Since in this case the rank ϵ_1 of Γ is 0, Γ must be a union of lines. If the lines are of multiplicity two on X, the formula for the number t of triple points of Γ applies, so that we have $t = 1$. One can show that there are three possibilities: (i) Γ is the union of three lines meeting in a non-planar triple point, and each line contains two pinch points; (ii) Γ is the union of two intersecting lines, one of these is an ordinary double line on X and the other is a "tacnodal" line (meaning that a plane intersects X in a curve with a tacnode, or A_3 singularity[1]); (iii) Γ is a "higher order tacnodal" line on X, meaning a plane intersection has an A_5 singularity at the intersection with Γ.

A general plane section $C = X \cap H$ is a plane rational quartic curve, with three nodes in case (i), one node and one tacnode (an A_3 singularity) in case (ii), and one A_5 singularity in case (iii). In all cases, the rank of X is $\mu_1 = 6$, and its class is $\mu_2 = 3$. So the dual surface $X^\vee \subset (\mathbb{P}^3)^\vee$ is a cubic surface, of rank equal to the rank of X, namely 6, and of class 4, equal to the degree of X. From this, it follows that the cubic surface $X^\vee$ must be singular (otherwise it would have class 12), but that it cannot have a one-dimensional singular locus, since the rank is the maximum possible for a cubic, by (5). In fact, it is known classically that in case (i), $X^\vee$ is equal to a cubic surface with 4 nodes and containing 9 lines (see [17], p. 153).

To determine the type of the dual surface in the cases (ii) and (iii), since, by reciprocity, the dual surface of $X^\vee$ is X, one could try to use the classification of cubic surfaces: there are in all five types of cubic surfaces that have a dual surface of degree 4 [4, p. 255]. Instead we used Singular [18] to compute the equation of the dual of a typical surface in each of the cases (ii) and (iii).

Case (ii): We may assume the surface has equation

$$w^4 - 2w^2 x(y+z) + x^2(x-z)^2 = 0.$$

The dual surface is the cubic

$$4stu - v^2(s+t) = 0.$$

Using the methods of [4, Lemma 3, p. 249] one finds that this cubic surface has two A_1 singularities and one A_3 singularity,[2] and it contains five lines.

Case (iii): We may assume the surface has equation

$$(wy - z^2)^2 - xy^3 = 0.$$

The dual surface is the cubic

$$s^2 u - 4tuv + v^3 = 0.$$

This cubic surface has one A_1 and one A_5 singularity, and it contains two lines.

Note that the deformations (iii) $\to$ (ii) $\to$ (i), obtained by changing the center of projection, correspond to deformations of the dual surfaces, which fits with the deformations of simple singularities

$$A_1 A_5 \to 2A_1 A_3 \to 4A_1$$

---

[1] A plane curve has an A_k singularity at the origin if the "local normal form" of its defining polynomial is $x^2 + y^{k+1}$ [2, pp. 20–28]. An ordinary node is A_1, an ordinary cusp is A_2.

[2] A surface in 3-space has an A_k singularity at the origin if the "local normal form" of its defining polynomial is $x^2 + y^2 + z^{k+1}$ [2, pp. 20–28].

coming from the deformations

$$A_5 \to A_1 A_3 \to 3A_1$$

corresponding to removing each time one inner vertex of the Dynkin diagram and its adjoining edges (as in [4, pp. 255–256] for $\hat{E}_6 \to A_1 A_5$ and $\hat{E}_6 \to 3A_2$).

One of the real versions in case (i) of this example is Steiner's Roman surface. There are three different real types in case (i), depending on the reality of the components of Γ and of the pinch points, two different real types in case (ii), and one in case (iii). This is all explained, with pictures, in [1, 6].

4   The Segre Surfaces

The Segre embedding of bidegree (a, b) of the product $\mathbb{P}^1 \times \mathbb{P}^1$ is the map

$$\sigma_{a,b} \colon \mathbb{P}^1 \times \mathbb{P}^1 \to \mathbb{P}^N,$$

where $N := (a + 1)(b + 1) - 1$, given by

$$\sigma_{a,b}((s_0 : s_1), (t_0 : t_1)) = (s_0^i s_1^{a-i} t_0^j t_1^{b-j})_{i=0,\ldots,a, j=0,\ldots,b}.$$

In affine coordinates, this is the tensor product parameterization

$$(s, t) \mapsto (s, t, st, s^2, t^2, s^2 t, st^2, \ldots, s^a t^b).$$

We want to study the surface $X \subset \mathbb{P}^3$ obtained by projecting $\sigma_{a,b}(\mathbb{P}^1 \times \mathbb{P}^1)$ to $\mathbb{P}^3$ from a linear subspace $L \subset \mathbb{P}^N$ of dimension $N - 4$.

If $a = b = 1$, then $X = \sigma_{1,1}(\mathbb{P}^1 \times \mathbb{P}^1)$ is a nonsingular quadric surface in $\mathbb{P}^3$. So assume $a > 1$ or $b > 1$, that $L \cap \sigma_{a,b}(\mathbb{P}^1 \times \mathbb{P}^1) = \emptyset$, and that the induced morphism $\mathbb{P}^1 \times \mathbb{P}^1 \to X$ is birational. In this case, the degree of X is $n = 2ab$. As in the case of the Veronese surfaces, X can have no isolated singularities.

Proposition 3. *In the situation above, set $\Gamma := \operatorname{Sing} X$. Assume $a > 1$ or $b > 1$. Then Γ is a curve, of degree*

$$\deg \Gamma \le 2a^2 b^2 - 4ab + a + b.$$

Moreover, equality holds if and only if the components of Γ are ordinary double curves or cuspidal edges on X.

Proof. Let $H \subset \mathbb{P}^3$ be a general plane, and set $C := X \cap H$. Then C is a plane curve of the same degree, $2ab$, as X. Let D denote the inverse image of C in $\mathbb{P}^1 \times \mathbb{P}^1$; this is a curve of bidegree (a, b), birationally equivalent to C. It follows from Bertini's theorem that D is nonsingular, and therefore D is equal to the desingularization of C. Hence, the sectional genus π of S, equal to the geometric genus $g(C)$ of C, is equal to the geometric genus of D:

$$\pi = g(C) = g(D) = ab - a - b + 1.$$

By the genus formula for the plane curve C, we have

$$ab - a - b + 1 = \binom{2ab-1}{2} - \sum_{x \in \text{Sing } C} \delta_x,$$

so that $\sum_{x \in \text{Sing } C} \delta_x > 0$, and hence C is singular. Using Bertini's theorem again, we have $\text{Sing } C = H \cap \text{Sing } X = H \cap \Gamma$, therefore $\#\text{Sing } C = \#H \cap \Gamma = \deg \Gamma$. The last equality holds because, since H is general, we may assume that H and Γ intersect transversally.

As in the previous section, if the projection is general, we get formulas for the degree of the double curve of X and the number of its triple points and pinch points. For $S = \sigma_{a,b}(\mathbb{P}^1 \times \mathbb{P}^1)$, we have $c_1^2 = 8$ and $c_2 = 4$, so Prop. 3 gives:

$$m = 2a^2b^2 - 4ab + a + b. \tag{8}$$

This gives the following formulas, using (3), (4), (5), (6), and (7):

$$\nu_2 = 4(3ab - 2a - 2b + 1) \tag{9}$$

$$t = \tfrac{4}{3}ab(a^2b^2 + 11) - 8a^2b^2 + 2ab(a + b) - 8(a + b) + 4. \tag{10}$$

$$\mu_1 = 2(3ab - a - b)$$

$$\mu_2 = 2(3ab - 2a - 2b + 2)$$

$$\epsilon_1 = 12a^2b^2 - 4ab(a + b) - 42ab + 26(a + b) - 14$$

Example 4 (The biquadric surface). Consider the case $a = b = 2$. The surface X has degree 8. If the projection is general, X has a double curve of degree 20 and rank $\epsilon_1 = 50$, with 20 triple points and 36 pinch points. The surface has rank $\mu_1 = 16$ and class $\mu_2 = 12$. Its sectional genus is $\pi = 1$.

Example 5 (The bicubic surface). Consider the case $a = b = 3$. Then X has degree 18. If the projection is general, X has a double curve of degree 132 and rank $\epsilon_1 = 540$, with 520 triple points and 64 pinch points. The surface has rank $\mu_1 = 42$ and class $\mu_2 = 34$. Its sectional genus is $\pi = 4$. In spite of the presence of all these singularities in the global situation, patches of these surfaces are used extensively in CAGD as bicubic splines.

5 Del Pezzo Surfaces

Consider the 3rd Veronese map $v_3 : \mathbb{P}^2 \to \mathbb{P}^9$. For $r \leq 6$, take r points $P_1, \ldots, P_r$ in general position in $\mathbb{P}^2$ and let $L \subset \mathbb{P}^9$ denote the linear space spanned by the points $v_3(P_1), \ldots, v_3(P_r)$. The (closure of the) image of the projection $p : v_3(\mathbb{P}^2) \dashrightarrow \mathbb{P}^{9-r}$ with center L, is a nonsingular surface S_{9-r}, called a Del Pezzo surface; it is isomorphic to $\mathbb{P}^2$ blown up in the r points P_i. For $r = 6$, the resulting surface S_3 is a nonsingular cubic surface in $\mathbb{P}^3$, and this is a standard way of representing cubic surfaces. (For more on Del Pezzo surfaces, see Schicho's article in these proceedings [12].) If $r \leq 5$, we may project such a surface S_{9-r} to $\mathbb{P}^3$; the projection is a rational surface X_n of degree $n = 9 - r$ and with characters $c_1^2 = n = 9 - r$ and $c_2 = 12 - n = 3 + r$.

The inverse image in S_n of a general plane section of X_n is a curve isomorphic to a nonsingular cubic plane curve, hence the surfaces X_n all have sectional genus 1. If the projection to $\mathbb{P}^3$ is general, X_n will have a double curve of degree

$$m = \binom{n-1}{2} - 1 = \tfrac{1}{2}(9 - r)(6 - r).$$

Using again the formulas of sect. 2, we get

$$\nu_2 = 4(n - 3) = 4(6 - r)$$

$$t = \binom{n-2}{3} = \tfrac{1}{6}(7 - r)(6 - r)(5 - r).$$

$$\mu_1 = 2(9 - r) = 2n$$

$$\mu_2 = 12$$

$$\epsilon_1 = 2(n - 3)^2 = 2(6 - r)^2$$

Example 6 (r = 5). For $r = 5$, we get a quartic surface S_4 in $\mathbb{P}^4$. It is well known that S_4 is the complete intersection of two quadric hypersurfaces, and, conversely, all such complete intersections are obtained in this way [3, Prop. IV.16, p. 67]. The cubic surface $(r = 6)$ contains 27 lines; more generally, one can show that S_{9-r} contains precisely $r + \binom{r}{2} + \binom{r}{5}$ lines [3, Prop. IV.12, p. 63], and hence there are 16 lines on S_4.

A general projection $X_4 \subset \mathbb{P}^3$ will have a double curve of degree 2 and rank 2. There are 4 pinch points and no triple points. A thorough study of the surface S_4 and all possible projections of it can be found in [16].

By considering points $P_1, \ldots, P_r$ that are not in general position, and/or projections that are not general, we can of course produce many more types of singular Del Pezzo surfaces.

6   Rational Scrolls

A rational normal scroll of type (d_1, d_2) is a surface $S \subset \mathbb{P}^{d_1+d_2+1}$ obtained as follows: Choose disjoint subspaces $\mathbb{P}^{d_i} \subset \mathbb{P}^{d_1+d_2+1}$, $i = 1, 2$, and rational normal curves

$$v_{d_i} : \mathbb{P}^1 \to \mathbb{P}^{d_i} \subset \mathbb{P}^{d_1+d_2+1}.$$

Let S be the ruled surface swept out by the lines joining corresponding points $v_{d_1}(P)$ and $v_{d_2}(P)$. If $t \mapsto (1, t, t^2, \ldots, t^{d_i})$ are affine parameterizations of the curves, then an affine parameterization of S is given by

$$(t, \lambda) \mapsto (t, t^2, \ldots, t^{d_1}, \lambda t, \lambda t^2, \ldots, \lambda t^{d_2}).$$

(For a more precise description, with homogeneous coordinates, see [11].)

The degree of the surface S is $n = d_1 + d_2$, so a hyperplane section will be a curve of degree $d_1 + d_2$ spanning a $\mathbb{P}^{d_1+d_2}$, hence this curve is necessarily rational. So the sectional genus of any rational scroll is $\pi = 0$.

We may assume $d_1 \leq d_2$. If $d_1 = d_2 = b > 0$, we say that S is *balanced*. In this case, $S \cong \mathbb{P}^1 \times \mathbb{P}^1$, and $S = \sigma_{1,b}(\mathbb{P}^1 \times \mathbb{P}^1)$ is equal to the Segre surface of bidegree $(1, b)$. If $0 \leq d_1 < d_2$, then S is isomorphic to the Hirzebruch surface $\mathbb{F}_e$, where $e := d_2 - d_1$.

For all S we have $c_1^2 = 8$ and $c_2 = 4$. If $d_1 > 0$ and the projection $S \to \mathbb{P}^3$ is general, X will have a nodal curve of degree m, with t triple points and ν_2 pinch points. As in the previous section, these numbers can be computed, in terms of the degree $n = d_1 + d_2$ of X. We find:

$$m = \tfrac{1}{2}(n-1)(n-2) = \tbinom{n-1}{2} \tag{11}$$

$$\nu_2 = 2(n-2) \tag{12}$$

$$t = \tfrac{1}{6}(n-2)(n-3)(n-4) = \tbinom{n-2}{3} \tag{13}$$

$$\mu_1 = 2(n-1)$$

$$\mu_2 = n$$

$$\epsilon_1 = 2(n-2)(n-3)$$

In particular, as is of course well known, the only nonsingular rational scroll $X \subset \mathbb{P}^3$ occurs when $d_1 = d_2 = 1$, i.e., when $S = X = \sigma_{1,1}(\mathbb{P}^1 \times \mathbb{P}^1) \subset \mathbb{P}^3$ is a quadric surface. Note that the fact that the degree μ_2 of the dual surface $X^\vee$ is equal to the degree of X, holds for all (nondevelopable) scrolls (see [10, Prop. 7, p. 341]).

Example 7 (Balanced scrolls). Consider the case of a balanced scroll, $d_1 = d_2 = b$. Then we saw above that S is equal to the Segre surface $\sigma_{1,b}(\mathbb{P}^1 \times \mathbb{P}^1)$. So, both (8) and (11) give that a general projection X has a nodal curve of degree $m = (2b-1)(b-1)$; from (9) or (12), X has $\nu_2 = 4(b-1)$ pinch points, from (10) or (13), it has $t = \tbinom{2b-2}{3}$ triple points. For example, if $d_1 = d_2 = 2$, X has degree 4 and its nodal curve has degree 3; it has 4 pinch points and no triple points.

Thus we see that in order to get new examples from these rational scrolls, we must take $d_2 > d_1$. For example, with $d_1 = 2$ and $d_2 = 3$ we obtain a quintic scroll $X \subset \mathbb{P}^3$ with a nodal curve of degree 6.

Example 8 ($d_1 = 1$). Consider the case $d_1 = 1$, $d_2 = 2$. Then $S \subset \mathbb{P}^4$ is a cubic normal scroll, and a general projection $X \subset \mathbb{P}^3$ is a cubic scroll with a nodal line, and no other singularities. This is in fact the only example of an irreducible cubic surface in $\mathbb{P}^3$ which has a singular locus of dimension 1 and is not a cone.

If $d_1 = 1$ and $d_2 = 3$, then a general projection X has degree 4 and a nodal curve of degree 3, hence a general plane section is a trinodal quartic. Further, there are 4 pinch points and no triple points in this case.

Example 9 ($d_1 = 0$). When $d_1 = 0$, $d_2 = 1$, then $S = \mathbb{P}^2$ is a plane. When $d_1 = 0, d_2 = 2$, $S \subset \mathbb{P}^3$ is a cone over a conic and has the vertex as the only singular point. When $d_1 = 0$, $d_2 \geq 3$, $S \subset \mathbb{P}^{d_2+1}$ is a cone over a rational normal curve D_2 of degree d_2. Though the formulas above do not apply, one sees that a general projection X will have singular lines corresponding to the $\tbinom{d_2-1}{2}$ nodes of the plane projection of the curve D_2 from the vertex of S.

7 Monoid Surfaces

Another type of rational algebraic surface treated classically, and of potential interest in CAGD (see [14, 8]), are the monoid surfaces. These are irreducible surfaces in 3-space with a singular point of multiplicity one less than the degree of the surface. Monoid surfaces are rational, as can easily be seen by exhibiting a rational parameterization; we shall do this below. We have already seen some examples of such surfaces: any singular irreducible cubic surface in $\mathbb{P}^3$ is a monoid, and any general projection of the Veronese surface $v_2(\mathbb{P}^2)$ is a monoid. We note that general projections of rational scrolls of degree 4 have no triple points, hence are not monoids.

If $X \subset \mathbb{P}^3$ is a monoid surface of degree n, we may assume that the point $P := (0 : 0 : 0 : 1)$ is a point of multiplicity $n - 1$. The defining polynomial of the surface can then be written as

$$F(x_0, x_1, x_2, x_3) = x_3 f_{n-1}(x_0, x_1, x_2) + f_n(x_0, x_1, x_2),$$

where f_i is homogeneous of degree i. Since F is irreducible, f_n is not identically 0, and f_{n-1} and f_n have no nonconstant common factors. If $f_{n-1} = 0$, then X is a cone over a plane curve of degree n; in what follows, we shall only consider monoids that are not cones.

To find the explicit rational parameterization of X, consider the projection $p \colon X \dashrightarrow \mathbb{P}^2$ from the point P, so that $p(a_0 : a_1 : a_2 : a_3) = (a_0 : a_1 : a_2)$, for $(a_0 : a_1 : a_2 : a_3) \in X \setminus P$. The inverse map $p^{-1} \colon \mathbb{P}^2 \dashrightarrow X$ is given by

$$p^{-1}(x_0 : x_1 : x_2) = \big(x_0 : x_1 : x_2, -f_n(x_0, x_1, x_2)/f_{n-1}(x_0, x_1, x_2)\big)$$
$$= \big(x_0 f_{n-1}(x_0, x_1, x_2) : x_1 f_{n-1}(x_0, x_1, x_2) : x_2 f_{n-1}(x_0, x_1, x_2) : -f_n(x_0, x_1, x_2)\big).$$

This rational map is defined outside the common zeros of f_{n-1} and f_n, i.e., outside the intersections of the two plane curves C_{n-1} and C_n defined by these polynomials. As in the case of Del Pezzo surfaces, this map can be viewed as the projection of the nth Veronese embedding; the projection center is a linear space that intersects $v_n(\mathbb{P}^2)$ in the points $v_d(C_{n-1} \cap C_n)$. (Computing the degree of the projected surface from this point of view gives $n = n^2 - n(n-1)$, as expected.)

Assume Q is a singular point on X different from P. Then the line through P and Q will intersect X in at least $n - 1 + 2 = n + 1$ points (counted with multiplicities). So Bezout's theorem implies that the line must be contained in X. If a point $(a_0 : a_1 : a_2) \in C_{n-1} \cap C_n$, then the whole line in $\mathbb{P}^3$ projecting to this point will lie on X. Moreover, it is easy to see, by looking at the partial derivatives of F, that if $(a_0 : a_1 : a_2)$ is singular on both C_{n-1} and C_n, then every point on this line will be singular on X.

Example 10 (Quartic monoids). In [4], cubic monoids were classified according to the shapes and intersections of the plane curves f_2 and f_3. The same approach can be used for monoids of higher degree.

Consider a quartic monoid X, given by $F = x_3 f_3 + f_4 = 0$. Depending on how the curves $f_3 = 0$ and $f_4 = 0$ are, and how they intersect, we get various types of monoids X. We can use SINGULAR [18] to compute the Milnor number and the type and normal form of the singularities of X. The "simplest" singularities come from taking $f_3 = 0$ to be a nonsingular curve. For example, if we take

$$F = (x_0^3 + x_1^3 + x_2^3 - x_0 x_1 x_2) x_3 + x_3^4,$$

then X will have a singularity with Milnor number 8 and of type T[3,3,3] (this type is also called P_8, see [2, p. 33]). If we let $f_3 = 0$ have a node, we get something else: for example, taking

$$F = (x_1^2 x_2 - x_0^2 x_2 - x_0^3) x_3 + x_3^4$$

we obtain a singularity with Milnor number 9 and of type T[3,3,4]. Here is a worse example, also starting with a nodal $f_3 = 0$:

$$F = (x_0^3 - x_0^2 x_1 + x_1^3 - x_0 x_1 x_2) x_3 + x_1 x_2^3 - x_0^2 x_2^2 + x_0^3 x_2 - x_0^4$$

defines a surface with a singularity with Milnor number 15 and of type T[3,3,10]. The real type of the singularities will of course depend on the real types of the plane curves.[3]

References

1. Apéry, F.; Models of the real projective plane. Computer graphics of Steiner and Boy surfaces. With a preface by Egbert Brieskorn. Friedr. Vieweg & Sohn, Braunschweig, 1987.
2. Arnold, V. I., Gorynov, V. V., Lyashko, O. V., Vasil'ev, V. A.: Singularity Theory, Springer 1998.
3. Beauville, A.: *Surfaces algébriques complexes*. Astérisque **54** (1978).
4. Bruce, J. W., Wall, C. T. C.; *On the classification of cubic surfaces*. J. London Math. Soc. (2), **19** (1979), 245–256.
5. Cayley, A.: *A memoir on cubic surfaces*. Phil. Trans. Roy. Soc. London Ser. A **159** (1869), 231–326.
6. Coffman, A.; Schwartz, A. J.; Stanton, C.: *The algebra and geometry of Steiner and other quadratically parametrizable surfaces*. Comput. Aided Geom. Design **13** (1996), no. 3, 257–286.
7. Klein, F.: *Über Flächen dritter Ordnung*. Math. Ann. **6** (1873), 551–581.
8. Pérez-Díaz, S., Sendra, J., Sendra, J. R.: *Parametrizations of approximate algebraic surfaces by lines*. Submitted.
9. Piene, R.: *Some formulas for a surface in $\mathbb{P}^3$*. In *Algebraic Geometry (Proc. Sympos., Tromsoe, Norway 1977)*, ed. L. Olson, LNM **687**, Springer-Verlag, 1978, 196–235.
10. Piene, R.: *On higher order dual varieties, with an application to scrolls*. Proc. Symp. Pure Math. **40** (Part 2), AMS 1983, 335–342.
11. Piene, R., Sacchiero, G.: *Duality for rational normal scrolls*. Comm. Algebra **12**(9–10) (1984) 1041–1066.
12. Schicho, J.: *Elementary theory of Del Pezzo surfaces*, these proceedings.
13. Schläfli, L.: *On the distribution of surfaces of the third order into species, in reference to the absence or presence of singular points, and the reality of their lines*. Phil. Trans. Roy. Soc. London Ser. A **153** (1863), 195–241 = L. Schläfli, Ges. Math. Abh., Birkhäuser 1953, Vol. II.

---

[3] A classification of quartic monoid surfaces along these lines is the topic of the forthcoming Master thesis of Magnus Loeberg at the University of Oslo.

14. Sederberg, T., Zheng, J., Klimaszewski, K., Dokken, T.: *Approximate implicitization using monoid curves and surfaces.* Graphical Models and Image Processing **61** (1999), 177–198.
15. Segre, B.: The nonsingular cubic surface. Oxford University Press, 1942.
16. Segre, C.: *Étude des différentes surfaces du 4e ordre à conique double ou cuspidale (générale ou décomposé) considérées comme des projections de l'intersection de deux variétés quadratiques de l'espace à quatre dimensions.* Math. Ann. **24** (1884), 313–442.
17. Semple, J. G., Roth, L.: Introduction to algebraic geometry. Oxford University Press, 1949.
18. SINGULAR (Greuel, G.-M., Pfister, G., Schönemann, H.): A Computer Algebra System for Polynomial Computations. Centre for Computer Algebra, University of Kaiserslautern (2001). http://www.singular.uni-kl.de
19. Zeuthen, H. G.: *Études des propriétés de situation des surfaces cubiques.* Math. Ann. **8** (1875), 1–30

On the Shape Effect of a Control Point: Experimenting with *NURBS* Surfaces

Panagiotis Kaklis[1] and Spyridon Dellas

Ship Design Laboratory, School of Naval Architecture & Marine Engineering,
National Technical University of Athens, Greece
Zografos 157 73, Athens
`kaklis@deslab.ntua.gr`,
WWW home page: `http//www.naval.ntua.gr/kaklis.html`

Abstract. In this paper we further the theoretical investigation, initiated in [1], on the shape effect of a single control point, measured in terms of the sign of Gaussian and mean curvatures of the underlying parametric surface. The so far obtained theoretical results are illustrated by experimenting with three typical *NURBS* surfaces, namely a cylinder, an ellipsoid and a torus.

1 Introduction and Preliminaries

Creating and processing the *shape* of curves and surfaces remains and will remain to be a central issue in *Computer Aided Geometric Design (CAGD)*. This is grounded on the fact that, besides its intrinsic interest, the request for robust and efficient handling of the shape is more or less present in every application that involves geometric information. As an indication, a web-search via the keyword pairs: {*shape, curves/surfaces*} it is likely to provide several hundred thousands of results; e.g., the search engine [4] yielded $519/744 \times 10^3$ outcomes, respectively.

In this context, Koras and Kaklis [1] initiated a study on the link between the shape of smooth parametric surfaces, quantified on a first level via the sign of their Gaussian and mean curvatures, and one of the most popular, among users and researchers, surface representation approaches, the so-called *control-point paradigm*. The purpose of this paper is to present additional pertinent theoretical results (see §§3, 4) and illustrate the so far gained understanding on the shape effect of a control point, by experimenting with *NURBS* surfaces (§5), more specifically with two typical 2nd-order algebraic surfaces: a cylinder (§5.1) and an ellipsoid (§5.2) and a 4th-order surface: a torus (§5.3).

As previously noted, the theoretical basis of the present paper is provided in [1], where the following problem is being investigated: *For those parametric surface representations that adopt the control point paradigm, determine all admissible loci of a specific control point for which the Gaussian or mean curvatures have prescribed signs at a specific surface point.* The basic outcome of this investigation is that, if our aim is to preserve local convexity over a finite set of specific surface points, then the selected control point is permitted to move within a convex polyhedron.

To be more analytic, we have to introduce some notation and terminology. Let $\mathbf{s}(u, v)$, be a parametric surface defined as

$$\mathbf{s}(u, v) = \sum_{\rho \in I} \mathbf{d}_\rho \, N_\rho(u, v), \, (u, v) \in \Omega, \tag{1}$$

where $\mathbf{d}_\rho$ are the so-called control points and $\{N_\rho(u, v)\}_{\rho \in I}$ is an appropriate set of weight functions, defined on a compactum Ω of $\mathbb{R}^2$, while I denotes the finite range of the multi-index ρ. Next, we introduce a series of notions, namely the notions of *parabolic/minimal loci, elliptic/hyperbolic domains, mean-positive/mean-negative domains* (see Def. 1) and the notions of *convex/concave domains* (see Def. 2) for a control point $\mathbf{d}_l$, that will be also referred to as the *free control point*.

Definition 1. *a) The parabolic (minimal) loci of the control point $\mathbf{d}_l$, $l \in I$, with respect to a point $\widetilde{\mathbf{w}} = (\widetilde{u}, \widetilde{v})$ of the parametric domain of definition, Ω, of the surface $\mathbf{s}(u, v)$, consists of all possible locations of $\mathbf{d}_l$ for which the Gaussian curvature $K(\widetilde{\mathbf{w}})$ (the mean curvature $H(\widetilde{\mathbf{w}})$) vanishes.*
b) The parabolic loci separate the elliptic, D_{K_+}, from the hyperbolic domain, D_{K_-}, of $\mathbf{d}_l$, where the Gaussian curvature is positive and negative, respectively.
c) Analogously, the minimal loci separate the positive-mean, D_{H_+}, from the negative-mean domain, D_{H_-}, of $\mathbf{d}_l$, where mean curvature is positive and negative, respectively.

Definition 2. *Let $D_{K^o_{+/-}} = D_{K_{+/-}} \cup \{\mathbf{p}(\mathbf{d}_l; \widetilde{\mathbf{w}}) = 0\}$ and $D_{H^o_{+/-}} = D_{H_{+/-}} \cup \{\mathbf{m}(\mathbf{d}_l; \widetilde{\mathbf{w}}) = 0\}$, where $\mathbf{p}(\mathbf{d}_l; \widetilde{\mathbf{w}}) = 0$, $\mathbf{m}(\mathbf{d}_l; \widetilde{\mathbf{w}}) = 0$ represent the parabolic- and minimal-loci surface, respectively. Then $D_{K^o_+} \cap D_{H^o_{+/-}}$ will be correspondingly referred to as the convex/concave domains of the control point $\mathbf{d}_l$, $l \in I$, with respect to the parametric point $\widetilde{\mathbf{w}} \in \Omega$.*

2 Parabolic Loci

Let us now recall from [1] the following auxiliary result: the parabolic-loci surface is a quadric surface, defined by

$$p(\mathbf{d}_l; \widetilde{\mathbf{w}}) = [\mathbf{d}_l^T \ 1] \, \mathbf{Q}(\widetilde{\mathbf{w}}) \begin{bmatrix} \mathbf{d}_l \\ 1 \end{bmatrix} = 0, \tag{2}$$

with $rank(\mathbf{Q}) \leq 2$ and $rank(\mathbf{Q}_{44}) \leq 2$, $\mathbf{Q}_{44}$ denoting the upper-left 3×3 submatrix of $\mathbf{Q}$. Matrix $\mathbf{Q}$ depends on the first- and second-order partial derivatives of the function-set $\{N_\rho(u, v)\}_{\rho \in I}$, namely:

$$\mathbf{Q}(\widetilde{\mathbf{w}}) = \frac{1}{2} \begin{bmatrix} \mathbf{b}^{uu} (\mathbf{b}^{vv})^T + \mathbf{b}^{vv} (\mathbf{b}^{uu})^T - 2\mathbf{b}^{uv} (\mathbf{b}^{uv})^T & a^{uu}\mathbf{b}^{vv} + a^{vv}\mathbf{b}^{uu} - 2a^{uv}\mathbf{b}^{uv} \\ (a^{uu}\mathbf{b}^{vv} + a^{vv}\mathbf{b}^{uu} - 2a^{uv}\mathbf{b}^{uv})^T & 2 \left(a^{uu}a^{vv} - (a^{uv})^2 \right) \end{bmatrix} \tag{3}$$

where

$$\mathbf{b}^\square = \left(\mathbf{s}_v^{(0)} N_{lu} - \mathbf{s}_u^{(0)} N_{lv} \right) \times \mathbf{s}_\square^{(0)} + N_{l\square} \left(\mathbf{s}_u^{(0)} \times \mathbf{s}_v^{(0)} \right), \quad \square \in \{uu, uv, vv\}, \tag{4}$$

and

$$a^\square = [\mathbf{s}_u^{(0)}, \mathbf{s}_v^{(0)}, \mathbf{s}_\square^{(0)}]. \tag{5}$$

Here $[\mathbf{a}, \mathbf{b}, \mathbf{c}]$ signifies the standard triple scalar product of vectors $\mathbf{a}$, $\mathbf{b}$ and $\mathbf{c}$, $\mathbf{s}^{(0)}(u, v)$ is the remaining part of $\mathbf{s}(u, v)$ after subtracting $\mathbf{d}_l N_l(u, v)$, while the subscript $u(v)$ denotes partial differentiation with respect to the parameter $u(v)$, respectively.

Combining this result with a standard classification theorem for quadric surfaces, we are led to the following basic theorem:

Theorem 3. *Let* $\{\lambda_1, \lambda_2, 0\}$ *be the eigenvalues of the symmetric matrix* $\mathbf{Q}_{44}$*. The parabolic loci* $\mathbf{p}(\mathbf{d}_l; \widetilde{\mathbf{w}}) = 0$ *of the control point* $\mathbf{d}_l$ *with respect to the parametric point* $\widetilde{\mathbf{w}} \in \Omega$*, of a surface defined as in (1), is a degenerate quadric, consisting of:*
(i) A pair of real intersecting planes if $\lambda_1 \lambda_2 < 0$ *(*$rank(\mathbf{Q}) = rank(\mathbf{Q}_{44}) = 2$*),*
(ii) A pair of complex planes if $\lambda_1 \lambda_2 > 0$ *(*$rank(\mathbf{Q}) = rank(\mathbf{Q}_{44}) = 2$*),*
(iii) A pair of parallel (real or imaginary) planes if $\lambda_1 \lambda_2 = 0$*,* $rank(\mathbf{Q}) = 2$ *and* $rank(\mathbf{Q}_{44}) = 1$ *or*
(iv) A pair of coincident planes if $\lambda_1 \lambda_2 = 0$*,* $rank(\mathbf{Q}) = 1$ *and* $rank(\mathbf{Q}_{44}) = 1$*.*

According to our numerical experience the case that occurs more frequently in practice is case (i) of Theorem 3, for which parabolic loci consists of a pair of intersecting planes; see planes $\mathcal{P}_1$ and $\mathcal{P}_2$ in the left of Figs 1. The elliptic domain of the free control point is then one of the two resulting wedge pairs, that share equal dihedral angles, the remaining wedge pair constituting the hyperbolic domain; see the elliptic wedge pair, where $K(\widetilde{\mathbf{w}}) > 0$, and the hyperbolic wedge pair, where $K(\widetilde{\mathbf{w}}) < 0$, in the afore-mentioned figure. Furthermore, recalling that local convexity (concavity) is equivalent to ellipticity enhanced with the positivity (negativity) of either of the triple products $[\mathbf{s}_u, \mathbf{s}_v, \mathbf{s}_{uu}]$ or $[\mathbf{s}_u, \mathbf{s}_v, \mathbf{s}_{vv}]$, the convex(concave) domain is the one of two elliptic wedges that lies in the positive(negative) halfspaces bounded by the planes

$$[\mathbf{s}_u, \mathbf{s}_v, \mathbf{s}_\square] = a^\square + \mathbf{d}_l^T \mathbf{b}^\square = 0, \quad \square = uu, vv; \tag{6}$$

the separating plane $\mathcal{P}_3$ in the *right* of Figs 1 is obtained by setting $\square = uu$ in (6).

3   Minimal Loci

The minimal-loci surface is a cubicoid, defined as

$$m(\mathbf{d}_l; \widetilde{\mathbf{w}}) = \begin{bmatrix} \mathbf{d}_l^T & 1 \end{bmatrix} \mathbf{H}(\widetilde{\mathbf{w}}) \begin{bmatrix} \mathbf{d}_l \\ 1 \end{bmatrix} + \mathbf{d}_l^T \mathbf{h}_3(\widetilde{\mathbf{w}})(x^2 + y^2 + z^2) = 0, \tag{7}$$

where x, y and z are the cartesian coordinates of the free control point $\mathbf{d}_l$, while

$$\mathbf{H}(\widetilde{\mathbf{w}}) = \begin{bmatrix} \frac{1}{2}(\mathbf{H}_2 + \mathbf{H}_2^T) & \frac{1}{2}\mathbf{h}_1 \\ \frac{1}{2}\mathbf{h}_1^T & h_0 \end{bmatrix} \tag{8}$$

is a symmetric matrix with its entries being defines as below:

$$\mathbf{H}_2(\widetilde{\mathbf{w}}) = b^{uu}(\mathbf{q}^{uu})^T - 2b^{uv}(\mathbf{q}^{uv})^T + (a^{uu}\beta^{vv} - 2a^{uv}\beta^{uv} + a^{vv}\beta^{uu})\mathbf{I}_3, \tag{9}$$

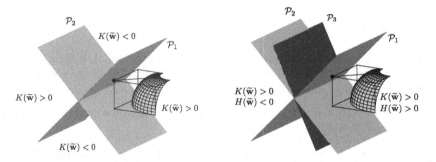

Fig. 1. For the control point (big bullet) of a spherical patch and an interior parametric point $\widetilde{\mathbf{w}}$: parabolic loci (planes $\mathcal{P}_i, i = 1, 2$), elliptic domain ($K(\widetilde{\mathbf{w}}) > 0$), hyperbolic domain ($K(\widetilde{\mathbf{w}}) < 0$), convex domain ($K(\widetilde{\mathbf{w}}) > 0$, $H(\widetilde{\mathbf{w}}) > 0$), concave domain ($K(\widetilde{\mathbf{w}}) > 0$, $H(\widetilde{\mathbf{w}}) < 0$) and their separating plane $\mathcal{P}_3$.

$$\mathbf{h}_1(\widetilde{\mathbf{w}}) = a^{vv}\mathbf{q}^{uu} + a^{vv}\mathbf{b}^{uu} - 2a^{uv}\mathbf{q}^{uv} - a^{uv}\mathbf{b}^{uv} + a^{uu}\mathbf{q}^{vv} + a^{uu}\mathbf{b}^{vv}, \qquad (10)$$

$$h_0(\widetilde{\mathbf{w}}) = a^{uu}a^{vv} - 2a^{uv}a^{uv} + a^{uu}a^{vv} \qquad (11)$$

$$\beta^{rs} = N_{lr}N_{ls}, \quad \mathbf{q}^{rs} = \mathbf{s}_r^{(0)}N_{ls} + \mathbf{s}_s^{(0)}N_{lr}, \quad r, s \in \{u, v\}, \qquad (12)$$

$\mathbf{I}_3$ denoting the $3{\times}3$ identity matrix, while $\mathbf{h}_3(\widetilde{\mathbf{w}})$ is a vector expressed as

$$\mathbf{h}_3(\widetilde{\mathbf{w}}) = \beta^{vv}\mathbf{b}^{uu} - 2\beta^{uv}\mathbf{b}^{uv} + \beta^{uu}\mathbf{b}^{vv}. \qquad (13)$$

As already noted, whenever the planes $\mathcal{P}_1$ and $\mathcal{P}_2$ intersect, the convex wedge is separated from its concave counterpart through the planes (6). Then, since the Gaussian curvature should also vanish along their intersection, we conclude that $[\mathbf{s}_u, \mathbf{s}_v, \mathbf{s}_{uv}] = 0$ as well. By virtue of the two previous remarks we conclude that, whenever it exists, the intersection line of $\mathcal{P}_1$ and $\mathcal{P}_2$ is a locus along which all three triple scalar products $[\mathbf{s}_u, \mathbf{s}_v, \mathbf{s}_\square]$, $\square = uu, uv, vv$, vanish. This in its turn implies that the minimal-loci surface passes through the intersection of the two parabolic planes; see Fig. 2. It is worth-noticing that, possessing aline on a nonsingular cubic surface, is a very useful knowledge for building up a parameterization; see, e.g., [[2], §7].

Furthermore, it is readily seen from the righthand side of (7) that, as $\|\mathbf{d}_l\| \to \infty$, the third-degree terms become dominant and the minimal-loci surface tends to a plane, namely

$$m(\mathbf{d}_l; \widetilde{\mathbf{w}}) = 0 \quad \to \quad \mathbf{h}_3^T \mathbf{d}_l = 0 \quad \text{as} \quad \|\mathbf{d}_l\| \to \infty. \qquad (14)$$

4   On the Influence of the Extrema of the Weight Function of the Free Control Point

Assuming that the weight functions $\{N_\rho(u, v)\}_{\rho \in I}$, appearing in the righthand side of (1), are sufficiently differentiable, the results presented in the previous two sections hold

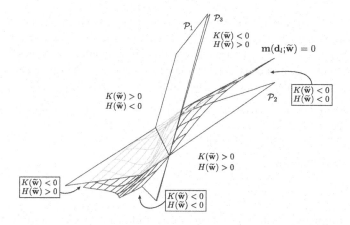

Fig. 2. For the free control point (bullet) of Fig. 1: Parabolic planes ($\mathcal{P}_i, i = 1, 2$), the plane ($\mathcal{P}_3$) that separates the convex from the concave wedge, the minimal-loci cubicoid surface ($\mathbf{m}(\mathbf{d}_l; \widetilde{\mathbf{w}}) = 0$), that passes through their common intersection line, and the partition of the space according to the sign of the Gaussian, $K(\mathbf{w})$, and mean, $H(\mathbf{w})$, curvatures at the specified parametric point $\widetilde{\mathbf{w}}$.

independently of the specific properties of the function set. We shall now reveal a case that links directly the properties of the parabolic and minimal loci with the behaviour of the weight function $N_l(u, v)$, that corresponds to the free control point $\mathbf{d}_l$, in the vicinity of the selected parametric point $\widetilde{\mathbf{w}}$. More specifically, let

$$N_{lu}(\widetilde{\mathbf{w}}) = N_{lv}(\widetilde{\mathbf{w}}) = 0, \tag{15}$$

which is a necessary condition for an extrema to arrive at $\widetilde{\mathbf{w}}$. Then $\mathbf{b}^{\square}$ (see (4)) can be written as

$$\mathbf{b}^{\square} = N_{l\square}(\mathbf{s}_u^{(0)} \times \mathbf{s}_v^{(0)}), \tag{16}$$

which implies that all three vectors $\mathbf{b}^{\square}$, $\square = uu, uv, vv$, are parallel. Thus, a pair of real constants κ and λ, depending solely on the second-order partial derivatives of N_l at $\mathbf{w} = \widetilde{\mathbf{w}}$, exists, such that

$$\mathbf{b}^{vv} = \kappa \mathbf{b}^{uu}, \quad \mathbf{b}^{uv} = \lambda \mathbf{b}^{uu}. \tag{17}$$

If, e.g., $N_{luu}(\widetilde{\mathbf{w}}) \neq 0$ then $\kappa = N_{lvv}(\widetilde{\mathbf{w}})/N_{luu}(\widetilde{\mathbf{w}})$ and $\lambda = N_{luv}(\widetilde{\mathbf{w}})/N_{luu}(\widetilde{\mathbf{w}})$. Substituting now (17) into the expression for the upper-left submatrix $\mathbf{Q}_{44}$ of $\mathbf{Q}$ (see eq. (3)), we finally arrive at

$$\mathbf{Q}_{44} = (\kappa - \lambda^2) \mathbf{b}^{uu} (\mathbf{b}^{uu})^T. \tag{18}$$

Combining (18) with the remark that, for any pair of non-zero vectors $\mathbf{v}_i, i = 1, 2$, the rank of matrix $\mathbf{v}_1(\mathbf{v}_2)^T$ is equal to 1, we can state

Lemma 4. *If $N_{lu}(\widetilde{\mathbf{w}}) = N_{lv}(\widetilde{\mathbf{w}}) = 0$ and $N_{luu}(\widetilde{\mathbf{w}})N_{lvv}(\widetilde{\mathbf{w}}) - N_{luv}^2(\widetilde{\mathbf{w}}) \neq 0$ then* $rank(\mathbf{Q}_{44}(\widetilde{\mathbf{w}})) = 1$.

In view of parts (iii) and (iv) of Theorem 3, the above lemma simply means that, if it is possible to change the sign of the Gaussian curvature at a parametric point $\mathbf{w}) = \widetilde{\mathbf{w}}$, where the hypotheses of the lemma hold true, then the parabolic-loci planes $\mathcal{P}_i, i = 1, 2$, are (real) planes and, as a consequence, the convex(concave) domain becomes either a halfspace or the infinitely extended strip that lies in between $\mathcal{P}_1$ and $\mathcal{P}_2$.

As for the minimal-loci surface, it can be readily seen from (12) that, if equalities (15) hold true, then

$$\beta^{rs} = 0, \quad \mathbf{q}^{rs} = \mathbf{0}, \qquad r, s \in \{u, v\}, \tag{19}$$

which, in view of (9) and (13), leads to

$$\mathbf{H}_2 = \mathbf{0}, \quad \mathbf{h}_3 = \mathbf{0}. \tag{20}$$

Then, substituting (20) into (7) we arrive at the following conclusion: if the first-order partial derivatives of $N_l(\mathbf{w})$ vanish at $\mathbf{w} = \widetilde{\mathbf{w}}$, then the minimal loci surface degenerates to a plane, namely $m(\mathbf{d}_l; \widetilde{\mathbf{w}}) = h_0 + \mathbf{d}_l^T \mathbf{h}_1 = 0$.

5   Numerical Experimentation with *NURBS*

This section reports on our experimentation with *Waterloo Mapple*®, *version 8,* for illustrating the theoretical results, obtained in the preceding sections, in the case of a triplet of standard algebraic surfaces, more specifically a cylinder, an ellipsoid and a torus, represented as *NURBS* surfaces.

For this purpose we have developed two Maple modules, namely the *NURBS* and the *ConvexDomain* packages. The first package resides on the Maple's built-in implementation of $B-$spline basis functions. The second one makes use of a GNU-distributed Maple package on convex geometry, called *convex* and created by Mathhias Franz [3], that can create and manipulate polytopes and polyhedra. One of its most useful, for our purposes, features lies in its functionality to create a polyhedron by intersecting a sequence of halfspaces and polyhedra. Thus, the creation of the convex polyhedron that describes the convex domain of a control point $\mathbf{d}_l$ with respect to a finite set of parametric points $\widetilde{\mathbf{w}}_i, i = 1, ..., N$, becomes straightforward.

Furthermore, our *ConvexDomain* package provides the functionality of evaluating the Hausdorff distance between two convex polytopes, that is useful for monitoring the convergence of convex domain sequences that result as the density of the chosen parametric points increases. Note that the Hausdorff distance

$$d(K, L) := inf\{\sigma \mid K + \sigma B \supseteq L, L + \sigma B \supseteq K\} \tag{21}$$

between two three-dimensional convex bodies K and L, has been implemented by using the maximum norm for defining the unit ball B in (21). The approach we have adopted for "covering" the, in our test cases (see Figs. 3, 5 and 8) rectangular, influence area of a free control point, can then be outlined as follows: we construct a sequence of $u-$ and $v-$uniform grids over the parametric domain of interest, evaluating the Hausdorff distance between the convex domains corresponding to each grid and the previous (coarser) one. When the Hausdorff distance gets smaller than a user-defined percentage of the diameter of the tolerance box we are working with, then we consider that denseness of the grid is satisfactory for our purposes and the stopping criterion gets activated.

5.1   Cylinder

If we adopt the popular CAGD parameterization of a cylinder, namely as a *NURBS* surface of degree 1 with respect to the one of its variables, say u, then it is straightforward to prove that the matrix $\mathbf{Q}$ in (3) degenerates to

$$\mathbf{Q} = - \begin{bmatrix} \mathbf{b}^{uv} \\ a^{uv} \end{bmatrix} \begin{bmatrix} (\mathbf{b}^{uv})^T & a^{uv} \end{bmatrix}. \tag{22}$$

This implies that $rank(\mathbf{Q}) \leq 1$, $rank(\mathbf{Q}_{44}) \leq 1$, and thus, by virtue of case (iv) of Theorem 3, the parabolic loci of a control point $\mathbf{d}_l$ with respect to any point $\tilde{\mathbf{w}}$ in the parametric domain of definition of the cylinder surface will be a pair of coincident planes. In addition, recall that a *NURBS* surface, which is of degree 1 with respect to one of its variables, can be classified as ruled, implying that the Gaussian curvature of the surface will remain non-positive for every possible location of $\mathbf{d}_l$. Then, the non-zero eigenvalue matrix of $\mathbf{Q}_{44}$ will always be negative and, in conjunction with the fact that the parabolic planes coincide, it is deduced that, for the chosen parameterization of the cylinder, the convex domain will be constantly empty.

The parameterization dependency of the convex-domain notion, which is also true for the other notions introduced in §1, can be alternatively illustrated by degree elevating with respect to the u variable, which produces non empty convex domain; see Fig. 3.

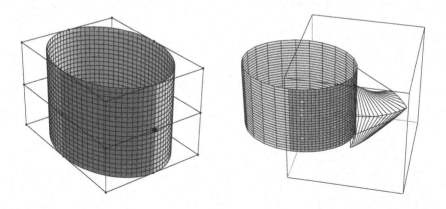

Fig. 3. *Left*: the cylinder with its control-point net after one-level degree elevation with respect to u. The darker area denotes the influence area of the free control point we have experimented with. *Right*: The free-form convex polyhedron for a dense set of parametric points, "covering" the whole influence area.

5.2   Ellipsoid

The ellipsoid, appearing in Fig. 4, is represented as a *NURBS* surface of degree 2. The surface consists of 8 patches, each containing a singularity at the pole. We have experimented with two different control points, which are not on the poles and their area of influence is limited within the boundaries of one and the same patch. On the basis of our numerical experience we dare asserting that, for any control point with the above characteristics, the convex domain is non-empty for any sample of parametric points and in fact converges to a free-form convex polytope as the density of the sample increases; see Fig. 5. Finally, the case of parallel parabolic planes (case (iii) of Th. 3), implying that the convex domain is either a halfspace or an infinite strip (see comments after Lemma 4),has been also observed; see Figs. 6.

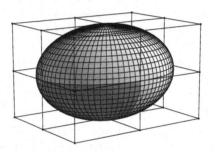

Fig. 4. The ellipsoid with its control-point net. The influence area of the two control points, we have experimented with, is depicted darker.

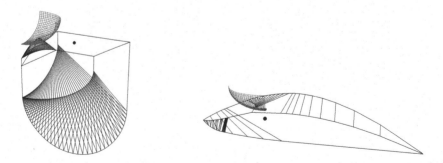

Fig. 5. Two instances of the free-form convex polytope for two different control points and a dense set of parametric "covering" the whole (common) influence area.

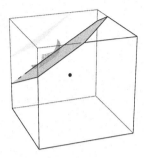 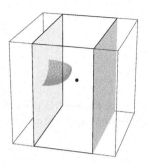

Fig. 6. *Left*: Parallel parabolic planes corresponding to a parametric point where $N_l(u, v)$ becomes maximum. The convex domain is here a halfspace, limited by the tolerance box. *Right*: Parallel parabolic planes for which the convex domain resides in the strip between them.

5.3   Torus

The torus, appearing in Fig. 7, is represented as a *NURBS* surface of degree 4, consisting of 16 patches with half of them belonging to the elliptic part of the surface. We have experimented with two control points, the one affecting an elliptic and the other a hyperbolic patch. In the first case the convex domain converges, as the set of parametric points covers increasingly the whole elliptic patch, to a zero thickness finite plate with free-form boundaries; see Fig. 8. For the second control point, that affects the hyperbolic patch of the torus (see in the *left* of Figs. 9, we find that the convex domain becomes eventually empty as we try to cover the whole parametric area of its influence. This is further illustrated in the *right* of Figs. 9, that has been obtained by evaluating, for a dense sampling over the parameter domain, the sign of the product of the two nonzero eigenvalues of submatrix Q_{44}. Recall that, according to cases (i) and (ii) of Th. 3, negative/positive sign of the eigenvalue product signifies existence/nonexistence of real intersecting parabolic planes, respectively.

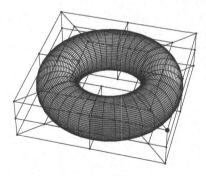

Fig. 7. The torus along with its control-point net. The darker elliptic part of the torus is the area of influence of the first free control point, we have experimented with.

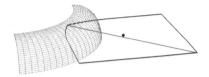

Fig. 8. The free-form convex polytope for the control point of Fig. 7 and a dense set of parametric points, "covering" its whole influence area.

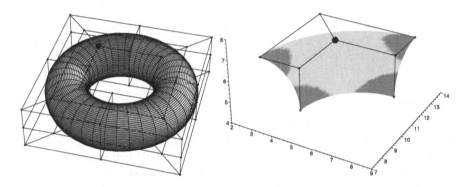

Fig. 9. *Left*: The torus along with its control point net. The darker (hyperbolic) part of the torus is the selected area of influence of the second free control point, we have experimented with. *Right*: The hyperbolic part colored according to the existence (light gray) or not (dark gray) of real intersecting planes.

References

1. G.D. Koras and P.D. Kaklis, On the Local Shape Effect of a Moving Control Point, Comp. Aided Geom. Design **20** (2003) 549-562.
2. T.G. Berry and R.D. Patterson, Implicitization and Parameterization of Nonsingular Cubic Surfaces, Comp. Aided Geom. Design **18** (2001) 723-738.
3. M. Franz: http://www-fourier.ujf-grenoble.fr/franz/
4. http://www.google.com

Third Order Invariants of Surfaces

Jens Gravesen

Department of Mathematics, Matematiktorvet Building 303,
Technical University of Denmark,
DK-2800 Kgs. Lyngby, Denmark
J.Gravesen@mat.dtu.dk

Abstract. The classical invariant theory from the 19th century is used to determine a complete system of 3rd order invariants on a surface in three-space. The invariant ring has 18 generators and the ideal of syzygies has 65 generators. The invariants are expressed as polynomials in the components of the first fundamental form, the second fundamental form and the covariant derivative of the latter, or in the case of an implicitly defined surface – $M = f^{-1}(0)$ – as polynomials in the partial derivatives of f up to order three.

As an application some commonly used fairings measures are written in invariant form. It is shown that the ridges and the subparabolic curve of a surface are the zero set of invariant functions and it is finally shown that the Darboux classification of umbilical points can be given in terms of two invariants.

1 Introduction

An *nth order invariant* on a surface M in $\mathbb{R}^3$ is a function $M \to \mathbb{R}$ whose value at a point $P \in M$ depends only on the nth order Taylor expansion of a parameterization of M around P, see Definition 2. E.g. the mean curvature H and the Gauss curvature K are second order invariants, in fact *any* second order invariant can be written as a function of H and K, so they form a *complete system*. In this paper a similar complete minimal system are found for the third order invariants, together with the complete system of relations.

The problem of finding such a system of generators and relations turns out to be a purely algebraic question that was much studied in the 19th century. The literature is immense, so we just refer to the books [1–6], and the survey [7]. Some of the classical algorithms from that time will be used but they will be phrased in the modern language of tensor analysis.

In Sect. 2 we give the precise definition of an invariant and we reduce the problem to a purely algebraic one. The main results are the list of invariants in Table 1 which forms a complete minimal system of generators and Theorem 8 which describe the surprisingly simple structure of the invariant ring.

The proof of Theorem 8 is in three stages. In Sect. 3 we use an algorithm from the 19th century, cf. [1], to determine a minimal set of generators. In Sect. 4 we find a set of relations – called syzygies – among these generators. Finally in Sect. 5 we show that we have found enough syzygies, i.e., they generate the whole ideal of relations. This is done by using Weyls character formula and the residue theorem to calculate the

Hilbert-Molien series which tells the dimension of the space of invariants of a fixed degree.

Section 6 is devoted to implicit surfaces, given by an equation $f(\mathbf{x}) = 0$. It is explained how the invariants can be expressed in terms of the function f.

Once the structure of the invariant ring is established it can be used without knowledge of the proof. So the reader interested in applications only can skip most of the paper and go directly to the examples. The way the theory is used is to perform a calculation using *principal coordinates* in which the first fundamental form, the second fundamental form and its covariant derivative is particular simple. The result is translated into an expression of invariants and then Table 1 can be used to tell what the expression is in an arbitrary parameterization. To demonstrate how this works we present some applications in Sect. 7. First we express some functions used as fairing measures in terms of our invariants. Then we characterize ridges, the subparabolic curve, and the Darboux classification of umbilical points using invariants.

2 Invariants on Surfaces

Let $\mathbf{x}_0$ be a point on a surface $M \subset \mathbb{R}^3$ and denote the unit normal vector and the tangent plane at $\mathbf{x}_0$ by $\mathbf{N}_{\mathbf{x}_0}$ and $T_{\mathbf{x}_0} M$ respectively. Let $\mathbf{r}_1, \mathbf{r}_2$ be a basis for the tangent space $T_{\mathbf{x}_0} M$ and let (x^1, x^2) denote the coordinates on $T_{\mathbf{x}_0} M$ with respect to this basis. Around the point $\mathbf{x}_0$ we can write the surface as a graph of a function on the tangent space. More precisely the map

$$(x^1, x^2) \mapsto \mathbf{x}(x^1, x^2) = \mathbf{x}_0 + x^1 \mathbf{r}_1 + x^2 \mathbf{r}_2 + h(x^1, x^2) \mathbf{N}_{\mathbf{x}_0} , \qquad (1)$$

is a local parameterization of M. The inverse map is simply the orthogonal projection $M \to T_{\mathbf{x}_0} M$, and h is the height of the surface over the tangent plane.

Normally the letter g is used for the first fundamental form, but we shall consider three different forms and it seems natural to use the letters a, b, c. So we let $a_{ij} = \mathbf{r}_i \cdot \mathbf{r}_j$ be the components of the first fundamental form. We can Taylor expand the function h to third order:

$$h(x^1, x^2) = \frac{1}{2} b_{ij} x^i x^j + \frac{1}{6} c_{ijk} x^i x^j x^k + \text{higher order terms} , \qquad (2)$$

where we use *Einsteins summation convention*, so if an index appears once as a subscript and once as a superscript, then it is tacitly understood that we sum over it. We may furthermore assume that b_{ij} and c_{ijk} are symmetric in the indices. We do not need an explicit expression of h in order to determine the coefficients b_{ij} and c_{ijk}. Indeed, we have

Proposition 1. *The coefficients b_{ij} and c_{ijk} are the components of the second fundamental form and the covariant derivative of the second fundamental form respectively, both with respect to the basis $\mathbf{r}_1, \mathbf{r}_2$.*

Proof. In the parameterization (1) we have $h(x^1, x^2) = (\mathbf{x}(x^1, x^2) - \mathbf{x}_0) \cdot \mathbf{N}_{\mathbf{x}_0}$, so

$$b_{ij} = \left. \frac{\partial^2 h}{\partial x^i \partial x^j} \right|_{(0,0)} = \left. \frac{\partial^2 \mathbf{x}}{\partial x^i \partial x^j} \right|_{(0,0)} \cdot \mathbf{N}_{\mathbf{x}_0} ,$$

and this is exactly the components of the second fundamental form at $\mathbf{x}_0$. To first order we have that $\frac{\partial \mathbf{x}}{\partial x^i} = \mathbf{r}_i + b_{ij}x^j\mathbf{N}_{\mathbf{x}_0}$ and as $\mathbf{N}_{\mathbf{x}_0} \perp \mathbf{r}_i$ the components of the first fundamental form are constant to first order: $\frac{\partial \mathbf{x}}{\partial x^i} \cdot \frac{\partial \mathbf{x}}{\partial x^j} \approx a_{ij}$. Likewise $\left| \frac{\partial \mathbf{x}}{\partial x^1} \times \frac{\partial \mathbf{x}}{\partial x^2} \right|^2 \approx |\mathbf{r}_1 \times \mathbf{r}_2|^2$. Finally

$$\frac{\partial^2 \mathbf{x}}{\partial x^i \partial x^j} \cdot \left(\frac{\partial \mathbf{x}}{\partial x^1} \times \frac{\partial \mathbf{x}}{\partial x^2} \right)$$
$$\approx (b_{ij} + c_{ijk}x^k)\mathbf{N}_{\mathbf{x}_0} \cdot \left((\mathbf{r}_1 + (b_{11}x^1 + b_{12}x^2)\mathbf{N}_{\mathbf{x}_0}) \times (\mathbf{r}_2 + (b_{12}x^1 + b_{22}x^2)\mathbf{N}_{\mathbf{x}_0}) \right)$$
$$= (b_{ij} + c_{ijk}x^k)\mathbf{N}_{\mathbf{x}_0} \cdot (\mathbf{r}_1 \times \mathbf{r}_2) = |\mathbf{r}_1 \times \mathbf{r}_2| (b_{ij} + c_{ijk}x^k) .$$

So the components of the second fundamental form are to first order $b_{ij} + c_{ijk}x^k$. Moreover, the ordinary derivative has components c_{ijk} and as the Christoffel symbols vanishes at $\mathbf{x}_0$ the covariant derivative at $\mathbf{x}_0$ agrees with the ordinary derivative and has components c_{ijk} too.   □

We have two interpretations of the quantities a_{ij}, b_{ij}, and c_{ijk}, as the coefficients of a homogeneous polynomial in two variables (called a *binary form*) and as the components of a k-form on M. We will also use a third interpretation, namely as the components of an element of the space $S^k\mathbb{R}^2$ of *symmetric k-tensors* on $\mathbb{R}^2$.

We have said that a third order invariant is a function that depends only on the third order behaviour of the surface. We can now make this precise:

Definition 2. *Let* $\mathbf{r} : U \to M \subset \mathbb{R}^3$ *be a parameterization of a surface. Let the components of the first fundamental form be* $\mathbf{a} = a_{11}, a_{12}, a_{22}$, *let the components of the second fundamental form be* $\mathbf{b} = b_{11}, b_{12}, b_{22}$, *and let the components of the covariant derivative of the second fundamental form be* $\mathbf{c} = c_{111}, c_{112}, c_{122}, c_{222}$. *A third order invariant is a function* $f : M \to \mathbb{R}$ *that can be written on the form* $f(\mathbf{r}(u,v)) = F(\mathbf{a}(u,v), \mathbf{b}(u,v), \mathbf{c}(u,v))$, *where* $F : S^2\mathbb{R}^2 \times S^2\mathbb{R}^2 \times S^3\mathbb{R}^2 \to \mathbb{R}$.

The function f is a function on the surface and is thus independent of the parameterization. On the other hand, if we change the parameterization of the surface we change the basis in the tangent plane and this in turn changes the components $\mathbf{a}, \mathbf{b}, \mathbf{c}$ of the three forms on the tangent plane. So F can not be arbitrary; it has to be invariant under the change of basis, i.e., under the action of $GL_2(\mathbb{R})$ on $S^2\mathbb{R}^2 \times S^2\mathbb{R}^2 \times S^3\mathbb{R}^2$.

We want to determine a finite set $F_1, \ldots, F_n$ of such invariant functions such that an arbitrary invariant function F can be written

$$F(\mathbf{a}, \mathbf{b}, \mathbf{c}) = \widehat{F}(F_1(\mathbf{a}, \mathbf{b}, \mathbf{c}), \ldots, F_n(\mathbf{a}, \mathbf{b}, \mathbf{c})).$$

We will in fact find a set of invariant *polynomials* such that any invariant polynomial can be written in the form above. Then the same is true for arbitrary invariant functions too, because such a set of polynomials separates orbits, see [3, Theorem 8.17]. The advantage is that the polynomial problem is a purely algebraic problem.

3   The Generators

We first consider a slightly different problem. We will consider forms or symmetric tensors over the complex numbers so we are given three binary forms $a_{ij}x^ix^j$, $b_{ij}x^ix^j$, and $c_{ijk}x^ix^jx^k$ where $(x^1, x^2) = \mathbf{x} \in \mathbb{C}^2$, and we ask for polynomials in the variables $\mathbf{a}, \mathbf{b}, \mathbf{c}, \mathbf{x}$ that are invariant under the action of $SL_2(\mathbb{C})$. More precisely we want to determine the structure of the *invariant ring* $\mathbb{C}[\mathbf{a}, \mathbf{b}, \mathbf{c}, \mathbf{x}]^{SL_2(\mathbb{C})}$. A polynomial is the sum of components, homogeneous in each set of variables $\mathbf{a}, \mathbf{b}, \mathbf{c}, \mathbf{x}$, and it is invariant if and only if each of its homogeneous components is invariant.

In the classical literature a *joint covariant* of *multi-degree* (d_1, d_2, d_3) and *order* k is a homogeneous invariant polynomial which has degree d_1 in $\mathbf{a}$, degree d_2 in $\mathbf{b}$, degree d_3 in $\mathbf{c}$, and degree k in $\mathbf{x}$. In particular, the forms themselves are covariants, and a *joint invariant* is a joint covariant of order 0.

An SL_2-invariant will in general not be invariant under the action of GL_2. Indeed, a diagonal matrix $\left(\begin{smallmatrix} t & 0 \\ 0 & t \end{smallmatrix}\right)$ acts on x^i by multiplication with t^{-1}, on $a_{i,j}, b_{i,j}$ by multiplication with t^2, and on $c_{i,j,k}$ by multiplication with t^3, so an $A \in GL_2$ acts on a joint covariant by multiplication with $\det A^\rho$, where $2\rho = 2d_1 + 2d_2 + 3d_3 - k$, we say that it is a *relative GL_2-covariant* of *weight* (or *index*) ρ.

The description of the invariant ring $\mathbb{C}[\mathbf{a}, \mathbf{b}, \mathbf{c}, \mathbf{x}]^{SL_2}$ is a classical problem that was studied intensely in the nineteenth century, and the two basic problems are the following

- Find a set of basic covariants $C_1, \ldots, C_p$, called a *complete system*, such that any covariant can be written as a polynomial in the basic covariants. I.e., such that the map $\phi : \mathbb{C}[X_1, \ldots, X_p] \rightarrow \mathbb{C}[\mathbf{a}, \mathbf{b}, \mathbf{c}, \mathbf{x}]$, given by $\phi(P) = P(C_1, \ldots, C_p)$ maps onto $\mathbb{C}[\mathbf{a}, \mathbf{b}, \mathbf{c}, \mathbf{x}]^{SL_2}$.
- Find all syzygies among the basic covariants, i.e., find the kernel $\mathcal{S}$ of ϕ. That is, all polynomials with $P(C_1, \ldots, C_p) = 0$.

In 1890 Hilbert showed that there always exists a finite system of generators and relations, see [8]. I.e., there exists generators $C_1, \ldots, C_p \in \mathbb{C}[\mathbf{a}, \mathbf{b}, \mathbf{c}, \mathbf{x}]$ and syzygies $S_1, \ldots, S_q \in \mathbb{C}[X_1, \ldots, X_p]$ such that the map $X_i \mapsto C_i$ gives an isomorphism

$$\mathbb{C}[X_1, \ldots, X_p] \,/\, (S_1, \ldots, S_q) \cong \mathbb{C}[\mathbf{a}, \mathbf{b}, \mathbf{c}, \mathbf{x}]^{SL_2} \,,$$

where $(S_1, \ldots, S_q)$ denotes the ideal generated by $S_1, \ldots, S_q$.

Before Hilbert the emphasis was on the explicit construction of covariants, often using *transvectants*. They can be defined symbolically, see [1, Chapter III] or [2, (20.18)] or they can be defined using differential operators, but we will use contraction of tensors to define them. We can assume that the components $f_{i_1, \ldots, i_n}$ of a polynomial $f = f_{i_1, \ldots, i_n} x^{i_1} \ldots x^{i_n}$ are symmetric in the indices so we may consider them as components of a symmetric tensor $f_{i_1, \ldots, i_n} x^{i_1} \otimes \cdots \otimes x^{i_n} \in S^n(\mathbb{C}^2)$. The *$r$th transvectant* of two symmetric tensors f and g is denoted $(f, g)^{(r)}$ and is defined by having components

$$(f, g)^{(r)}_{i_{r+1}, \ldots, i_n, j_{r+1}, \ldots, j_m} = S\big(\varepsilon^{i_1 j_1} \ldots \varepsilon^{i_r j_r} f_{i_1, \ldots, i_n} g_{j_1, \ldots, j_m}\big) \,, \tag{3}$$

where ε^{ij} is the completely anti symmetric symbol $\varepsilon^{11} = \varepsilon^{22} = 0$ and $\varepsilon^{12} = -\varepsilon^{21} = 1$. The symbol S stands for symmetrization of the free indices. Observe that the symmetry of f and g implies, that up to a sign, this is the only non zero contraction. As a contraction of a tensor is a new tensor we see that the transvectants of two covariants is a new covariant. Obviously

$$\deg(f, g)^{(r)} = \deg f + \deg g \ ,$$
$$\operatorname{order}(f, g)^{(r)} = \operatorname{order} f + \operatorname{order} g - 2r \ ,$$
$$\operatorname{weight}(f, g)^{(r)} = \operatorname{weight} f + \operatorname{weight} g + r \ .$$

It should be mentioned that all this is part of the representation theory of $SL_2\mathbb{C}$ (or $SU(2)$), see [9]. The irreducible representations are the spaces $S^n\mathbb{C}^2$ of symmetric tensors, and a tensor product of two of these has the following decomposition

$$S^n\mathbb{C}^2 \otimes S^m\mathbb{C}^2 = \bigoplus_{r=0}^{\lfloor \frac{n+m}{2} \rfloor} S^{n+m-2r}\mathbb{C}^2 \ .$$

The rth transvectant is exactly the projection from $S^n\mathbb{C}^2 \otimes S^m\mathbb{C}^2$ to $S^{n+m-2r}\mathbb{C}^2$.

The following theorem tells us how to get a complete system for a single binary form, see [1, § 86] or [2, Theorem 24.3].

Theorem 3. *Any covariant of a binary form f of degree d in its coefficients can be written as a linear combination of transvectants of the form itself and covariants of degree $d - 1$.*

As the only covariant of degree 1 is the form itself this shows that we can find a complete system of covariants of single form consisting of transvectants. E.g. a complete system of covariants for a single quadratic binary form a consists of the form itself and its discriminant:

$$a = a_{ij}x^i x^j \ , \qquad (a, a)^{(2)} = 2(a^{11}a^{22} - a^{12}a^{12}) \ , \tag{4}$$

see [1, 2, 4, 5].

A complete system of single cubic form c consists of the form itself, its Hessian, the discriminant of the Hessian and the Jacobian of the form with its Hessian:

$$c \ , \qquad H = (c, c)^{(2)} \ , \qquad D = (H, H)^{(2)} \ , \qquad T = (H, c)^{(1)} \ , \tag{5}$$

see [1, 2, 4, 5]. The joint covariants of a collection of forms can be created from complete subsystems, see [1, § 103].

Theorem 4. *If S_1 and S_2 are two finite and complete systems of forms, then there exists a finite and complete system consisting of transvectants of products of elements of S_1 and products of forms of S_2.*

The problem with this theorem is that we don't know how many products we need to form, before we take transvectants. But if one of the systems above is the complete system (4) of a single binary quadratic form then more is true, see [1, § 141].

Theorem 5. *If S_1 is the system (4) for a quadratic form a, and S_2 is an arbitrary system, then the irreducible transvectants belong to one of the three classes, $(C, a^r)^{(2r-1)}$, $(C, a^r)^{(2r)}$, and $(C_1C_2, a^r)^{(2r)}$, where C_1 and C_2 have odd order and the order of the product C_1C_2 is $2r$.*

By using this result twice we can find a complete system for two quadratic and one cubic binary form, i.e., a complete system for the third order invariants of a surface. The result is a rather large system and not all elements are needed. To get rid of the superfluous elements we use the following important result, see [6, Chapter VIII, §7], or [4, Chapter 4.3].

Theorem 6. *If $C, C_1, \ldots, C_n$ are covariants and $C = p_1 C_1 + \cdots + p_n C_n$ for some polynomials p_i, then we can assume that p_i are covariants too. In other words, if a covariant is contained in the ideal generated by some covariants then it is contained in the algebra generated by the same covariants:*

$$C \in (C_1, \ldots, C_n) \Rightarrow C \in \mathbb{C}[C_1, \ldots, C_n] .$$

So using the results above we first find a large set of generators. Then we sort them such that the partial ordering induced by the order and the multi degree is respected, i.e., $k \le k' \wedge d_1 \le d_1' \wedge d_2 \le d_2' \wedge d_3 \le d_3' \implies C \le C'$. Finally we take each covariant in turn and if it is contained in the ideal generated by the previous ones then we throw it away otherwise we keep it. All this was done using the algebra program "Singular" [10]. The result is a minimal system of generators consisting of 18 invariants, 13 linear covariants, 6 quadratic covariants, and 4 cubic covariants. The 18 invariants are

$$(a,a)^{(2)}, \quad (a,b)^{(2)}, \quad (b,b)^{(2)}, \quad ((c,c)^{(2)},a)^{(2)}, \quad ((c,c)^{(2)},b)^{(2)},$$

$$(((c,c)^{(2)},b)^{(1)},a)^{(2)}, \quad (c^2,a^3)^{(6)}, \quad (c(c,b)^{(2)},a^2)^{(4)}, \quad ((c,b)^{(2)^2},a)^{(2)},$$

$$(c^2,b^3)^{(6)}, \quad (c(c,b)^{(1)},a^3)^{(6)}, \quad (c(c,b^2)^{(3)},a^2)^{(4)}, \quad ((c,b)^{(2)}(c,b^2)^{(3)},a)^{(2)},$$

$$((c,c)^{(2)},(c,c)^{(2)})^{(2)}, \quad (c((c,c)^{(2)},c)^{(1)},a^3)^{(6)}, \quad (c(((c,c)^{(2)},c)^{(1)},b)^{(2)},a^2)^{(4)},$$

$$((c,b)^{(2)}(((c,c)^{(2)},c)^{(1)},b)^{(2)},a)^{(2)}, \quad (c((c,c)^{(2)},c)^{(1)},b^3)^{(6)} .$$

The symmetrization in (3) means that we in general will get a sum of different contractions. But each contraction in such a sum is an invariant and at least one of them is irreducible. So we can obtain a complete system where each generator is a single contraction. In Table 1 we have listed one possible choice of generators along with their multi-degree and weight. There are at first sight up to 2^{18} terms, but the symmetries reduces this number to 54, see [11] where the sums have been expanded.

4 The Syzygies

We now turn to the problem of finding all syzygies, i.e., all relations between the basic invariants in Table 1. In this section we present a set of syzygies and in the next section we prove that this set generates the ideal of syzygies.

Proposition 7. *There are 39 syzygies of the form*

$$J_i J_j = Q_{ij}^0 + \sum_{k=1}^{2} Q_{ij}^k J_k , \qquad 1 \le i \le j \le 2 \text{ or } 3 \le i \le j \le 10 , \qquad (6)$$

Table 1. The basic invariants, their multi-degree, and their weight. See [11] for expanded expressions.

Invariant	multi-degree	weight
$I_0 = \frac{1}{2}\varepsilon^{i_1j_1}\varepsilon^{i_2j_2}a_{i_1i_2}a_{j_1j_2}$	$(2,0,0)$	2
$I_1 = \varepsilon^{i_1j_1}\varepsilon^{i_2j_2}a_{i_1i_2}b_{j_1j_2}$	$(1,1,0)$	2
$I_2 = \frac{1}{2}\varepsilon^{i_1j_1}\varepsilon^{i_2j_2}b_{i_1i_2}b_{j_1j_2}$	$(0,2,0)$	2
$I_3 = \frac{1}{2}\varepsilon^{i_1j_3}\varepsilon^{i_2k_3}\varepsilon^{j_1k_1}\varepsilon^{j_2k_2}a_{i_1i_2}c_{j_1j_2j_3}c_{k_1k_2k_3}$	$(1,0,2)$	4
$I_4 = \frac{1}{2}\varepsilon^{i_1j_3}\varepsilon^{i_2k_3}\varepsilon^{j_1k_1}\varepsilon^{j_2k_2}b_{i_1i_2}c_{j_1j_2j_3}c_{k_1k_2k_3}$	$(0,1,2)$	4
$I_5 = \varepsilon^{i_1l_1}\varepsilon^{i_2m_1}\varepsilon^{j_1l_2}\varepsilon^{j_2m_2}\varepsilon^{k_1l_3}\varepsilon^{k_2m_3}a_{i_1i_2}a_{j_1j_2}a_{k_1k_2}c_{l_1l_2l_3}c_{m_1m_2m_3}$	$(3,0,2)$	6
$I_6 = \varepsilon^{i_1l_1}\varepsilon^{i_2m_1}\varepsilon^{j_1l_2}\varepsilon^{j_2m_2}\varepsilon^{k_1l_3}\varepsilon^{k_2m_3}a_{i_1i_2}b_{j_1j_2}b_{k_1k_2}c_{l_1l_2l_3}c_{m_1m_2m_3}$	$(0,3,2)$	6
$I_7 = \frac{1}{2}\varepsilon^{i_1j_1}\varepsilon^{i_2j_2}\varepsilon^{k_1l_1}\varepsilon^{k_2l_2}\varepsilon^{i_3k_3}\varepsilon^{j_3l_3}c_{i_1i_2i_3}c_{j_1j_2j_3}c_{k_1k_2k_3}c_{l_1l_2l_3}$	$(0,0,4)$	6
$J_1 = \varepsilon^{i_1l_1}\varepsilon^{i_2m_1}\varepsilon^{j_1l_2}\varepsilon^{j_2m_2}\varepsilon^{k_1l_3}\varepsilon^{k_2m_3}a_{i_1i_2}a_{j_1j_2}b_{k_1k_2}c_{l_1l_2l_3}c_{m_1m_2m_3}$	$(2,1,2)$	6
$J_2 = \varepsilon^{i_1l_1}\varepsilon^{i_2m_1}\varepsilon^{j_1l_2}\varepsilon^{j_2m_2}\varepsilon^{k_1l_3}\varepsilon^{k_2m_3}a_{i_1i_2}b_{j_1j_2}b_{k_1k_2}c_{l_1l_2l_3}c_{m_1m_2m_3}$	$(1,2,2)$	6
$J_3 = \varepsilon^{i_1j_1}\varepsilon^{i_2k_3}\varepsilon^{j_2l_3}\varepsilon^{k_1l_1}\varepsilon^{k_2l_2}a_{i_1i_2}b_{j_1j_2}c_{k_1k_2k_3}c_{l_1l_2l_3}$	$(1,1,2)$	5
$J_4 = \varepsilon^{i_1l_1}\varepsilon^{i_2m_1}\varepsilon^{j_1m_2}\varepsilon^{j_2n_1}\varepsilon^{k_1m_3}\varepsilon^{k_2n_2}\varepsilon^{l_2n_3}a_{i_1i_2}a_{j_1j_2}a_{k_1k_2}b_{l_1l_2}c_{m_1m_2m_3}c_{n_1n_2n_3}$	$(3,1,2)$	7
$J_5 = \varepsilon^{i_1l_1}\varepsilon^{i_2m_1}\varepsilon^{j_1m_2}\varepsilon^{j_2n_1}\varepsilon^{k_1m_3}\varepsilon^{k_2n_2}\varepsilon^{l_2n_3}a_{i_1i_2}a_{j_1j_2}b_{k_1k_2}b_{l_1l_2}c_{m_1m_2m_3}c_{n_1n_2n_3}$	$(2,2,2)$	7
$J_6 = \varepsilon^{i_1l_1}\varepsilon^{i_2m_1}\varepsilon^{j_1m_2}\varepsilon^{j_2n_1}\varepsilon^{k_1m_3}\varepsilon^{k_2n_2}\varepsilon^{l_2n_3}a_{i_1i_2}b_{j_1j_2}b_{k_1k_2}b_{l_1l_2}c_{m_1m_2m_3}c_{n_1n_2n_3}$	$(1,3,2)$	7
$J_7 = \varepsilon^{i_1l_1}\varepsilon^{i_2m_1}\varepsilon^{j_1l_2}\varepsilon^{j_2m_2}\varepsilon^{k_1l_3}\varepsilon^{k_2n_1}\varepsilon^{m_3p_1}\varepsilon^{n_2p_2}\varepsilon^{n_3p_3}a_{i_1i_2}a_{j_1j_2}a_{k_1k_2}c_{l_1l_2l_3}c_{m_1m_2m_3}c_{n_1n_2n_3}c_{p_1p_2p_3}$	$(3,0,4)$	9
$J_8 = \varepsilon^{i_1l_1}\varepsilon^{i_2m_1}\varepsilon^{j_1l_2}\varepsilon^{j_2m_2}\varepsilon^{k_1l_3}\varepsilon^{k_2n_1}\varepsilon^{m_3p_1}\varepsilon^{n_2p_2}\varepsilon^{n_3p_3}a_{i_1i_2}a_{j_1j_2}b_{k_1k_2}c_{l_1l_2l_3}c_{m_1m_2m_3}c_{n_1n_2n_3}c_{p_1p_2p_3}$	$(2,1,4)$	9
$J_9 = \varepsilon^{i_1l_1}\varepsilon^{i_2m_1}\varepsilon^{j_1l_2}\varepsilon^{j_2m_2}\varepsilon^{k_1l_3}\varepsilon^{k_2n_1}\varepsilon^{m_3p_1}\varepsilon^{n_2p_2}\varepsilon^{n_3p_3}a_{i_1i_2}b_{j_1j_2}b_{k_1k_2}c_{l_1l_2l_3}c_{m_1m_2m_3}c_{n_1n_2n_3}c_{p_1p_2p_3}$	$(1,2,4)$	9
$J_{10} = \varepsilon^{i_1l_1}\varepsilon^{i_2m_1}\varepsilon^{j_1l_2}\varepsilon^{j_2m_2}\varepsilon^{k_1l_3}\varepsilon^{k_2n_1}\varepsilon^{m_3p_1}\varepsilon^{n_2p_2}\varepsilon^{n_3p_3}b_{i_1i_2}b_{j_1j_2}b_{k_1k_2}c_{l_1l_2l_3}c_{m_1m_2m_3}c_{n_1n_2n_3}c_{p_1p_2p_3}$	$(0,3,4)$	9

and 16 of the form

$$J_i J_j = \sum_{k=3}^{10} Q_{ij}^k J_k \ , \qquad 1 \le i \le 2 \ and \ 3 \le j \le 10 \ , \tag{7}$$

where Q_{ij}^k is a polynomial in $I_0, \dots, I_7$ of the form

$$Q_{ij}^k = \sum_{\sum \deg I_{i_p} = \deg J_i + \deg J_j - \deg J_k} d_{i_1 \dots i_k} I_{i_1} \dots I_{i_k} \ ,$$

(deg $J_0 = 0$). We furthermore have two syzygies of the form

$$Q_i^1 J_1 + Q_i^2 J_2 = Q_i^0 \ , \qquad\qquad i = 1, 2 \ , \tag{8}$$

and eight of the form

$$Q_i^3 J_3 + \dots + Q_i^{10} J_{10} = 0 \ , \qquad\qquad i = 3, \dots, 10 \ , \tag{9}$$

where Q_i^k are polynomials of the form

$$Q_i^k = \sum_{\sum \deg I_{i_p} = (\gamma_1^i, \gamma_2^i, \gamma_3^i) - \deg J_k} d_{i_1 \dots i_k} I_{i_1} \dots I_{i_k} \ ,$$

(deg $J_0 = 0$), and the degrees $(\gamma_1^i, \gamma_2^i, \gamma_3^i)$ are

$$\begin{array}{ccccc}
(3,2,6) \ , & (2,3,6) \ , & (4,1,4) \ , & (3,2,4) \ , & (2,3,4) \ , \\
(1,4,4) \ , & (3,1,6) \ , & (2,2,6) \ , & (1,3,6) \ , & (3,3,6) \ .
\end{array} \tag{10}$$

Proof. We only sketch the proof. Equations (6) and (7) are finite dimensional inhomogeneous linear equations in the coefficients of the polynomials Q_{ij}^k. Using Maple or a similar system it is not hard to solve these equations, see [11].

When the degrees (10) are known then the existence of the polynomials Q_i^k in (8) and (9) reduces to a finite dimensional linear algebra problem, but we have to be careful. If we take syzygies of degree $(1,3,6)$, $(3,1,6)$, $(2,3,4)$, and $(3,2,4)$ and multiply with I_0, I_2, I_3, and I_4 respectively we obtain four syzygies of degree $(3,3,6)$. So when we solve (9) to find the polynomials Q_{10}^k, the space of solutions has dimension greater than one. We need to pick a solution that is not a $\mathbb{C}[I_0, \dots, I_7]$ linear combinations of syzygies of lower degree, but this is not hard to do using a computer algebra system, see [11]. In fact, this is how the 65 syzygies were found in the first place. Starting with low degree, Maple was used to determine syzygies of a fixed degree that can't be expressed as a $\mathbb{C}[I_0, \dots, I_7]$ linear combinations of the syzygies previously found. This process was iterated until no new syzygies emerged for some degrees. We might at this point believe we have all syzygies, but it is not proved – that will be done in the next section.

To simplify the calculations we can pick a good basis for the tangent plane, so we may assume that $a_{ij} = \delta_{ij}$ and b_{ij} is diagonal and obtain the expressions in Table 2, c.f. Sect. 7.   □

Note that if we eliminate $\mathbf{a}, \mathbf{b}, \mathbf{c}$ from the ideal $(X_0 - I_0, \ldots, X_7 - I_7, Y_1 - J_1, \ldots, Y_{10} - J_{10})$ in the ring $\mathbb{C}[\mathbf{X}, \mathbf{Y}, \mathbf{a}, \mathbf{b}, \mathbf{c}]$ then we obtain a system of generators for the ideal of syzygies. It is in principle possible to use Gröbner basis methods to do this, but the present problem is apparently too large to be solved this way. At least my implementation in Singular ran for months and never terminated.

Now consider the ring $\mathbb{C}[\mathbf{X}, \mathbf{Y}] = \mathbb{C}[X_0, \ldots, X_7, Y_1, \ldots, Y_{10}]$. We introduce a triple grading by letting

$$\deg(X_i) = \deg(I_i) = (\alpha_1^i, \alpha_2^i, \alpha_3^i), \quad \deg(Y_i) = \deg(J_i) = (\beta_1^i, \beta_2^i, \beta_3^i). \qquad (11)$$

The values of $(\alpha_1^i, \alpha_2^i, \alpha_3^i)$ and $(\beta_1^i, \beta_2^i, \beta_3^i)$ can be found in Table 1. We can in an obvious manner consider Q_{ij}^k and Q_i^k as polynomials in $\mathbf{X}$, i.e., as elements in $\mathbb{C}[\mathbf{X}]$ and we now put

$$S_{ij} = Y_i Y_j - \left(Q_{ij}^0(\mathbf{X}) + \sum_{k=1}^{10} Q_{ij}^k(\mathbf{X}) Y_k \right), \qquad 1 \leq i \leq j \leq 10,$$

$$S_i = Q_i^1(\mathbf{X}) Y_1 + Q_i^2(\mathbf{X}) Y_2 - Q_i^0(\mathbf{X}), \qquad i = 1, 2,$$

$$S_i = Q_i^3(\mathbf{X}) Y_3 + \cdots + Q_i^{10}(\mathbf{X}) Y_{10}, \qquad i = 3, \ldots, 10,$$

then $\deg S_i = (\gamma_1^i, \gamma_2^i, \gamma_3^i)$ is given by (10). Now let $\mathbb{C}[\mathbf{a}, \mathbf{b}, \mathbf{c}]$ denote the polynomial ring in the variable a_{ij}, b_{ij}, c_{ijk}, and consider the map $\phi : \mathbb{C}[\mathbf{X}, \mathbf{Y}] \rightarrow \mathbb{C}[\mathbf{a}, \mathbf{b}, \mathbf{c}]$ given by $\phi(X_i) = I_i$ and $\phi(Y_i) = J_i$. It preserves the grading, and maps onto the invariant ring. The polynomials S_{ij} and S_i are in the kernel of ϕ, so if $\mathcal{S}$ is the ideal generated by S_{ij} and S_i then we have a surjective map

$$\mathbb{C}[\mathbf{X}, \mathbf{Y}]/\mathcal{S} \rightarrow \mathbb{C}[\mathbf{a}, \mathbf{b}, \mathbf{c}]^{SL_2(\mathbb{C})}, \qquad (12)$$

and we want to show it is an isomorphism. If the degree is fixed then we have a linear map between finite dimensional vector spaces, so we need only to show that the two spaces have the same dimension. This is done in the next section.

5   The Structure of the Invariant Ring

Consider the subspace of invariants of multi-degree $\mathbf{d} = (d_1, d_2, d_3)$ and denote the dimension by $D_{\mathbf{d}}$. The *Hilbert-Molien series* is $H(\mathbf{z}) = \sum D_{\mathbf{d}} \mathbf{z}^{\mathbf{d}}$, where $\mathbf{z}^{\mathbf{d}} = z_1^{d_1} z_2^{d_2} z_3^{d_3}$. From the point of view of Lie group theory we have a representation of $SL_2(\mathbb{C})$ on $S^{\mathbf{d}}(S^2(\mathbb{C}^2) \times S^2(\mathbb{C}^2) \times S^3(\mathbb{C}^2))$ and the space of invariants of multi-degree $\mathbf{d}$ is exactly the subspace where $SL_2(\mathbb{C})$ acts trivially. We can split $S^{\mathbf{d}}$ in a direct sum of irreducible representations and the number of times the trivial representation occur is $D_{\mathbf{d}}$. This number can be computed by Weyls character formula, see [9]. Let $g_n : SL_2(\mathbb{C}) \rightarrow GL(S^n(\mathbb{C}^2))$ be the n'th symmetric representation of $SL_2(\mathbb{C})$, let T be a maximal torus in $SL_2(\mathbb{C})$ and let $\mathbf{dt}$ be a Haar measure on T, then with $(n_1, n_2, n_3) = (2, 2, 3)$, we have

$$H(z_1, z_2, z_3) = \int_{\mathbf{t} \in T} \frac{\prod_{1 \leq i < j \leq 2} \left(1 - \frac{t_i}{t_j} \right)}{\prod_{k=1}^{3} \det(1 - z_k g_{n_k}(\mathbf{t}))} \, \mathbf{dt}$$

$$= \frac{1}{2\pi} \int_0^{2\pi} \frac{1 - e^{-2i\theta}}{\prod_{k=1}^{3} \left(\prod_{l=0}^{n_k} (1 - z_k e^{(2l-n_k)i\theta}) \right)} \, d\theta$$

$$= \frac{1}{2\pi i} \int_{S^1} \frac{1 - \zeta^{-2}}{\prod_{k=1}^{3} \left(\prod_{l=0}^{n_k} (1 - z_k \zeta^{2l-n_k}) \right)} \zeta^{-1} \, d\zeta$$

$$= \frac{1}{2\pi i} \int_{S^1} \frac{\zeta^5 (\zeta^2 - 1)}{Q(\zeta)} \, d\zeta,$$

where

$$Q(\zeta) = (\zeta^2 - z_1)(1 - z_1)(1 - z_1 \zeta^2)(\zeta^2 - z_2)(1 - z_2)(1 - z_2 \zeta^2)$$
$$(\zeta^3 - z_3)(\zeta - z_3)(1 - z_3 \zeta)(1 - z_3 \zeta^3).$$

We will use the residue theorem to evaluate the last integral, and if we assume that $|z_1|, |z_2|, |z_3| < 1$ and that $\xi_1^2 = z_1$, $\xi_2^2 = z_2$, and $\eta^3 = z_3$ then the poles inside the unit circle are $\pm\xi_1, \pm\xi_2, \eta, e^{\pm 2\pi i/3}\eta, z_3$. We use Maple to calculate the residues for the eight poles and sum the results. The details can be found in [11] and the final result is

$$H(z_1, z_2, z_3) = \frac{1 + \sum_{j=1}^{10} z_1^{\beta_1^i} z_2^{\beta_2^i} z_3^{\beta_3^i} - \sum_{k=1}^{10} z_1^{\gamma_1^k} z_2^{\gamma_2^k} z_3^{\gamma_3^k} - z_1^4 z_2^4 z_3^8}{\prod_{i=0}^{7} \left(1 - z_1^{\alpha_1^i} z_2^{\alpha_2^i} z_3^{\alpha_3^i} \right)}, \qquad (13)$$

where the exponents α_i^j, β_i^j, and γ_i^j are given in (11) and (10).

We now consider the corresponding series – called the *Hilbert series* – for the ring at the left hand side of (12), but first we find a simple description of the ring. We define the ideals $\mathcal{S}_0 = (\ldots, S_{ij}, \ldots)$ and $\mathcal{S}_1 = (S_1, \ldots, S_{10})$ and the rings $R_0 = \mathbb{C}[\mathbf{X}] = \mathbb{C}[X_0, \ldots, X_7]$ and $R_1 = \mathbb{C}[\mathbf{X}, \mathbf{Y}]/\mathcal{S}_0 = R_0[\mathbf{Y}]/\mathcal{S}_0$. Then $\mathcal{S} = (\mathcal{S}_0 \cup \mathcal{S}_1)$ and $\mathbb{C}[\mathbf{X}, \mathbf{Y}]/\mathcal{S} = (\mathbb{C}[\mathbf{X}][\mathbf{Y}]/\mathcal{S}_0)/\mathcal{S}_1 = R_1/\mathcal{S}_1$. The syzygies S_{ij} tells us that in the ring R_1 any element can be uniquely written as $p = p_0 + \sum_{i=1}^{10} p_i Y_i$ where $p_i \in R_0$. Put an other way, as an R_0 module we have $R_1 = R_0 \oplus R_0 Y_1 \oplus \cdots \oplus R_0 Y_{10}$. We now proceed to look at $\mathcal{S}_1$ as an R_0 module. As an R_1 module it is generated by $S_1, \ldots, S_{10}$ so as an R_0 module it is generated by $S_1, \ldots, S_{10}$ and all products $S_i Y_j$. We put $S_0 = Y_1 S_2$, and using a computer algebra system we find that $Y_i S_j$ is contained in $\text{span}_{R_0}\{S_0, \ldots, S_{10}\}$ for all i, j, see [11]. Observe that $(\gamma_1^0, \gamma_2^0, \gamma_3^0) = \deg S_0 = \deg Y_1 + \deg S_2 = (4, 4, 8)$ which is the last exponent in the Hilbert-Molien series (13). So $\mathcal{S}_1$ is generated by $S_0, \ldots, S_{10}$ as an R_0 module. Furthermore, we can write

$$\begin{bmatrix} S_0 \\ S_1 \\ S_2 \end{bmatrix} = \mathbf{A}_1 \begin{bmatrix} 1 \\ Y_1 \\ Y_2 \end{bmatrix} \quad \text{and} \quad \begin{bmatrix} S_3 \\ \vdots \\ S_{10} \end{bmatrix} = \mathbf{A}_2 \begin{bmatrix} Y_3 \\ \vdots \\ Y_{10} \end{bmatrix},$$

where $\mathbf{A}_1$ and $\mathbf{A}_2$ are matrices with elements in R_0. We find that $\det \mathbf{A}_2 = 2 \det \mathbf{A}_1 \neq 0$ in R_0, see [11]. So $\mathcal{S}_1$ is a free R_0 module: $\mathcal{S}_1 = R_0 S_0 \oplus \cdots \oplus R_0 S_{10}$. The Hilbert series for the polynomial ring R_0 is

$$H_0(z_1, z_2, z_3) = \left(\prod_{i=0}^{7} (1 - z_1^{\alpha_1^i} z_2^{\alpha_2^i} z_3^{\alpha_3^i}) \right)^{-1}.$$

The ring R_1 is a free R_0 module and has the Hilbert series

$$H_1(z_1, z_2, z_3) = H_0(z_1, z_2, z_3) \left(1 + \sum_{i=1}^{10} z_1^{\beta_1^i} z_2^{\beta_2^i} z_3^{\beta_3^i} \right) .$$

The ideal $\mathcal{S}_1$ is a free R_0 module too and has the Hilbert series

$$H_2(z_1, z_2, z_3) = H_0(z_1, z_2, z_3) \left(\sum_{i=0}^{10} z_1^{\gamma_1^i} z_2^{\gamma_2^i} z_3^{\gamma_3^i} \right) .$$

So all in all we have that the Hilbert series for $R_1/\mathcal{S}_1$ is

$$H(z_1, z_2, z_3) = H_1(z_1, z_2, z_3) - H_2(z_1, z_2, z_3)$$

$$= \left(\prod_{i=0}^{7} (1 - z_1^{\alpha_1^i} z_2^{\alpha_2^i} z_3^{\alpha_3^i}) \right)^{-1} \left(1 + \sum_{i=1}^{10} z_1^{\beta_1^i} z_2^{\beta_2^i} z_3^{\beta_3^i} - \sum_{i=0}^{10} z_1^{\gamma_1^i} z_2^{\gamma_2^i} z_3^{\gamma_3^i} \right) .$$

This is exactly the same as the Hilbert-Molien series (13) for the ring of SL_2 invariants. As a consequence we have that the kernel of ϕ is $\mathcal{S}$. This is our main result and we formulate it as the following theorem.

Theorem 8. *The map ϕ given by $X_i \mapsto I_i$ and $Y_i \mapsto J_i$ induces an isomorphism*

$$\mathbb{C}[\mathbf{X}, \mathbf{Y}] \, / \, \mathcal{S} \cong \mathbb{C}[\mathbf{a}, \mathbf{b}, \mathbf{c}]^{SL_2} .$$

I.e., any SL_2-invariant can be written as a polynomial in $I_0, \ldots, I_7, J_1, \ldots, J_{10}$ and any syzygy can be written as a $\mathbb{C}[\mathbf{X}, \mathbf{Y}]$-linear combination of S_i and S_{ij}, $i, j = 1, \ldots 10$. The map also induces an isomorphism

$$\left(\mathbb{C}[\mathbf{X}] \oplus \mathbb{C}[\mathbf{X}]Y_1 \oplus \ldots \mathbb{C}[\mathbf{X}]Y_{10} \right) / \mathcal{S}_1 \cong \mathbb{C}[\mathbf{a}, \mathbf{b}, \mathbf{c}]^{SL_2} .$$

I.e., any SL_2-invariant I can be written as

$$I = p_0(I_0, \ldots, I_7) + p_1(I_0, \ldots, I_7)J_1 + \cdots + p_{10}(I_0, \ldots, I_7)J_{10} ,$$

and any syzygy among these is a $\mathbb{C}[\mathbf{X}]$-linear combination of $S_0, \ldots, S_{10}$.

We really want to study invariants of binary forms over the reals, but this makes no difference. A polynomial is invariant under the action of $SL_2(\mathbb{R})$ if and only if it vanishes under the induced action of the Lie algebra $\mathfrak{sl}_2(\mathbb{R})$. The one-to-one correspondence between homogeneous polynomials and symmetric tensors means that if $\mathcal{I}_\mathbf{d}$ denotes the space of homogeneous $SL_2(\mathbb{R})$-invariant polynomials of degree $\mathbf{d}$ (and order 0) then we have

$$\mathcal{I}_\mathbf{d} \subseteq S^\mathbf{d}\left(S^{2,2,3}(\mathbb{R}^2) \right) = S^\mathbf{d}\left(S^2(\mathbb{R}^2) \times S^2(\mathbb{R}^2) \times S^3(\mathbb{R}^2) \right) \subseteq \mathbb{R}[\mathbf{a}, \mathbf{b}, \mathbf{c}] ,$$

where $S^\mathbf{d}$ denotes the symmetric product. The invariant ring is $\mathbb{R}[\mathbf{a}, \mathbf{b}, \mathbf{c}]^{SL_2} = \bigoplus_\mathbf{d} \mathcal{I}_\mathbf{d}$. If we complexify we get

$$\mathcal{I}_\mathbf{d} \otimes_\mathbb{R} \mathbb{C} \subseteq S^\mathbf{d}\left(S^{2,2,3}(\mathbb{R}^2) \right) \otimes_\mathbb{R} \mathbb{C} = S^\mathbf{d}\left(S^{2,2,3}(\mathbb{C}^2) \right) \subseteq \mathbb{C}[\mathbf{a}] ,$$

and as $\mathfrak{sl}_2(\mathbb{C}) = \mathfrak{sl}_2(\mathbb{R}) \otimes_{\mathbb{R}} \mathbb{C}$ we see that the complex polynomials of multi-degree $\mathbf{d}$ that vanishes under the action of $\mathfrak{sl}_2(\mathbb{C})$ is $\mathcal{I}_{\mathbf{d}} \otimes_{\mathbb{R}} \mathbb{C}$. In other words there is a one to one correspondence between $SL_2(\mathbb{R})$-invariants and $SL_2(\mathbb{C})$-invariants.

Furthermore, we are in fact interested in GL_2-invariants, i.e., invariants with weight 0. None of the polynomial invariants has weight 0, but as $I_0 > 0$ we can obtain absolute GL_2-invariants by dividing with a suitable power of I_0. If we let ρ_i and δ_i be the weight of I_i and J_i respectively, and put

$$\widehat{I_i} = \frac{I_i}{I_0^{\rho_i/2}} \qquad \text{and} \qquad \widehat{J_i} = \frac{I_i}{I_0^{\delta_i/2}} \;, \tag{14}$$

then $\widehat{I_1}, \ldots, \widehat{I_7}$, and $\widehat{J_1}, \widehat{J_2}$ are rational $GL_2(\mathbb{R})$-invariants, but $\widehat{J_3}, \ldots, \widehat{J_{10}}$ are only invariant under the action of $GL_2^+(\mathbb{R})$. They changes sign if the linear transformation has a negative determinant, i.e., if the orientation is reversed. This gives the following corollary.

Corollary 9. *In Theorem 8 the field $\mathbb{C}$ can be replaced by $\mathbb{R}$. Furthermore, any rational $GL_2^+(\mathbb{R})$-invariant can be written as*

$$\frac{p_0 + p_1 \widehat{J_1} + \cdots + p_{10} \widehat{J_{10}}}{q_0 + q_1 \widehat{J_1} + \cdots + q_{10} \widehat{J_{10}}} \; ,$$

and any rational $GL_2(\mathbb{R})$-invariant can be written as

$$\frac{p_0 + p_1 \widehat{J_1} + p_2 \widehat{J_2}}{q_0 + q_1 \widehat{J_1} + q_2 \widehat{J_2}} + \frac{p_3 \widehat{J_3} + \cdots + p_{10} \widehat{J_{10}}}{q_3 \widehat{J_3} + \cdots + q_{10} \widehat{J_{10}}} \; ,$$

where p_i and q_i are polynomials in $\widehat{I_1}, \ldots, \widehat{I_7}$.

6   Implicit Surfaces

We now consider an implicitly defined surface $M = h^{-1}(0)$, where $h : \mathbb{R}^3 \to \mathbb{R}$. We first assume that $|\nabla h| = 1$ in some neighbourhood of M, so $h(x)$ is the *signed distance* from M to x for x in that neighbourhood. If $\mathrm{I\!I}$ and $\nabla \mathrm{I\!I}$ denotes the second fundamental form and its covariant derivative respectively, then $\mathrm{d}^2 h = -\mathrm{I\!I} \circ \pi$ and $\mathrm{d}^3 h = -\nabla \mathrm{I\!I} \circ \pi$ where $\pi : \mathbb{R}^3 \to T_p M$ is the orthogonal projection onto the tangent space. If we assume that ∇h is the third basis vector for $\mathbb{R}^3$, then $a_{ij} = \delta_{ij}$, $b_{ij} = h_{ij}$, and $c_{ijk} = h_{ijk}$ for $i, j, k = 1, 2$. Furthermore

$$\varepsilon^{ij} = \varepsilon^{ijk} h_k \quad \text{where} \quad \varepsilon^{ijk} = \begin{cases} 1 & \text{if } i, j, k \text{ is an even permutation of } 1, 2, 3 \\ -1 & \text{if } i, j, k \text{ is an odd permutation of } 1, 2, 3 \\ 0 & \text{otherwise.} \end{cases} \tag{15}$$

So to express the invariants in Table 1 in terms of the signed distance function we simply make the above substitutions. At first we have to sum from 1 to 2 only, but as

$\varepsilon^{3jk}h_k = \varepsilon^{i3k}h_k = 0$ we may sum from 1 to 3. The expression is now invariant so it holds for an arbitrary direction of ∇h.

We now consider an arbitrary C^3 function $f : \mathbb{R}^3 \to \mathbb{R}$ such that $M = f^{-1}(0)$ and $\lambda = |\nabla f| \neq 0$ in a neighbourhood of M. By differentiating the equation $dh(\nabla h) = |\nabla h|^2 = 1$ we see that ∇h is a null vector for the higher order derivatives of h, and using this fact we obtain

$$h_i = \frac{1}{\lambda}f_i ,\tag{16}$$

$$h_{ij} = \frac{1}{\lambda}f_{ij} - \frac{1}{\lambda^3}\left(f_{ik}f^k f_j + f_{jk}f^k f_i\right) + \frac{1}{\lambda^5}\left(f_{kl}f^k f^l\right)f_i f_j ,\tag{17}$$

$$h_{ijk} = \frac{1}{\lambda}f_{ijk} - \frac{1}{\lambda^3}\left(f_{il}f^l f_{jk} + f_{kl}f^l f_{ij} + f_{jl}f^l f_{ki}\right) + \text{terms with } f_i, f_j, f_k ,\tag{18}$$

where $f^l = \delta^{kl}f_k = f_l$. As $\varepsilon^{ijk}f_i f_k = \varepsilon^{ijk}f_j f_k = 0$ we can discard any term in (17) and (18) that contains f_i, f_j, or f_k as a factor. Summing up we have the following result:

Theorem 10. *Let $f : \mathbb{R}^3 \to \mathbb{R}$ be C^3 functions such that $\lambda = |\nabla f| \neq 0$ in a neighbourhood of the implicitly defined surface $M = f^{-1}(0)$. The invariants in Table 1 can be found by making the substitutions*

$$\varepsilon^{ij} \mapsto \frac{1}{\lambda}\varepsilon^{ijk}f_k , \qquad a_{ij} \mapsto \delta_{ij} ,$$

$$b_{ij} \mapsto \frac{1}{\lambda}f_{ij} , \qquad c_{ijk} \mapsto \frac{1}{\lambda}f_{ijk} - \frac{1}{\lambda^3}\left(f_{il}f^l f_{jk} + f_{kl}f^l f_{ij} + f_{jl}f^l f_{ki}\right) .$$

7 Applications

Coordinates on a surface where – at some point – the first fundamental form to first order is δ_{ij} and the second fundamental form to first order is diagonal are called *principal coordinates* at that point. E.g. if we in (1) have that the basis $\mathbf{r}_1, \mathbf{r}_2$ for the tangent plane is orthonormal and in the principal directions, then we have principal coordinates at $\mathbf{x}_0$. Now $a_{11} = a_{22} = 1$, $a_{12} = b_{12} = 0$ so $I_0 = 1$, $\hat{I}_i = I_i$, $\hat{J}_i = J_i$. If we put $b_{11} = L$, $b_{22} = N$, $c_{111} = P$, $c_{112} = Q$, $c_{122} = S$, and $c_{222} = T$ then we get the expressions in Table 2.

The equations $(*)$ is a system of linear equations in $P^2, PS, S^2, Q^2, QT, T^2$

$$\begin{bmatrix} 0 & -1 & -1 & 0 & 1 & 1 \\ 0 & -N & -L & 0 & N & L \\ 1 & 3 & 3 & 1 & 0 & 0 \\ N^3 & 3LN^2 & 3L^2N & L^3 & 0 & 0 \\ N & L+2N & 2L+N & L & 0 & 0 \\ N^2 & 2LN+N^2 & L^2+2LN & L^2 & 0 & 0 \end{bmatrix}\begin{bmatrix} P^2 \\ Q^2 \\ S^2 \\ T^2 \\ PS \\ QT \end{bmatrix} = \begin{bmatrix} I_3 \\ I_4 \\ I_5 \\ I_6 \\ J_1 \\ J_2 \end{bmatrix} ,\tag{19}$$

and the equations $(**)$ is a system of linear equations in PQ, PT, QS, ST

$$(L-N)\begin{bmatrix} 0 & -1 & 0 & 1 \\ 1 & 2 & 1 & 0 \\ N & L+N & L & 0 \\ N^2 & 2LN & L^2 & 0 \end{bmatrix}\begin{bmatrix} PQ \\ QS \\ ST \\ PT \end{bmatrix} = \begin{bmatrix} J_3 \\ J_4 \\ J_5 \\ J_6 \end{bmatrix} .\tag{20}$$

Table 2. The basic invariants in principal coordinates

$$I_1 = L + N,$$

$$I_2 = LN,$$

(*) $$I_3 = PS - S^2 - Q^2 + QT,$$

(*) $$I_4 = NPS - LS^2 - NQ^2 + LQT,$$

(*) $$I_5 = P^2 + 3Q^2 + 3S^2 + T^2,$$

(*) $$I_6 = N^3P^2 + 3LN^2Q^2 + 3L^2NS^2 + L^3T^2,$$

$$I_7 = 3Q^2S^2 - P^2T^2 - 4PS^3 - 4Q^3T + 6PQST,$$

(*) $$J_1 = NP^2 + (L+2N)Q^2 + (2L+N)S^2 + LT^2,$$

(*) $$J_2 = N^2P^2 + (2LN+N^2)Q^2 + (L^2+2LN)S^2 + L^2T^2,$$

(**) $$J_3 = (L-N)(PT - QS),$$

(**) $$J_4 = (L-N)(PQ + 2QS + ST),$$

(**) $$J_5 = (L-N)(NPQ + (L+N)QS + LST),$$

(**) $$J_6 = (L-N)(N^2PQ + 2LNQS + L^2ST),$$

$$J_7 = 2(PQ^3 - S^3T) + (P^2 + 3Q^2 - 3S^2 - T^2)PT + 3((Q+T)^2 - (P+S)^2)QS,$$

$$J_8 = 2(NPQ^3 - LS^3T) + (NP^2 + (L+2N)Q^2 - (2L+N)S^2 - LT^2)PT$$
$$+ ((L+2N)Q^2 + 2(2L+N)QT + 3LT^2 - NP^2 - 2(L+2N)PS - (2L+N)S^2)QS,$$

$$J_9 = 2(N^2Q^2 - N(2L+N)S^2)PQ + (N(NP^2 + (2L+N)Q^2) - L((2N+L)S^2 + LT^2))PT$$
$$- (3N^2P^2 - N(2L+N)Q^2 + L(L+2N)S^2 - 3L^2T^2)QS + (L(2N+L)Q^2 - L^2S^2)ST,$$

$$J_{10} = 2N^2(NQ^2 - 3LS^2)PQ + (N^3P^2 + 3LN^2Q^2 - 3L^2NS^2 - L^3T^2)PT$$
$$- 3(N^3P^2 - LN^2Q^2 + L^2NS^2 - L^3T^2)QS + 2L^2(3NQ^2 - LS^2)ST.$$

The determinant of the matrix is in both cases $(L - N)^7$, so if $L \neq N$, i.e, at a non-umbilical point, we can solve the equations, and the solutions are

$$\begin{bmatrix} PQ \\ QS \\ ST \\ PT \end{bmatrix} = \frac{1}{(L-N)^3} \begin{bmatrix} 0 & L^2 & -2L & 1 \\ 0 & -LN & L+N & -1 \\ 0 & N^2 & -2N & 1 \\ (L-N)^2 & -LN & L+N & -1 \end{bmatrix} \begin{bmatrix} J_3 \\ J_4 \\ J_5 \\ J_6 \end{bmatrix} . \qquad (21)$$

and

$$P^2 = \frac{L^3 I_5 - I_6 - 3L^2 J_1 + 3L J_2}{(L-N)^3} ,$$

$$Q^2 = \frac{-L^2 N I_5 + I_6 + (L+2N)L J_1 - (2L+N)J_2}{(L-N)^3} ,$$

$$S^2 = \frac{L N^2 I_5 - I_6 - (2L+N)N J_1 + (L+2N)J_2}{(L-N)^3} ,$$

$$T^2 = \frac{-N^3 I_5 + I_6 + 3N^2 J_1 - 3N J_2}{(L-N)^3} , \qquad (22)$$

$$PS = Q^2 + \frac{L I_3 - I_4}{L-N} ,$$

$$QT = S^2 - \frac{N I_3 - I_4}{L-N} .$$

Expressions similar to (22) were also found in [12], but there a different set of invariants was used namely:

$$\Lambda_1 = \frac{I_5}{I_0^3} , \qquad\qquad \Lambda_2 = \frac{2I_0 I_3 + I_5}{I_0^3} ,$$

$$\Lambda_3 = \frac{I_1 I_5 - I_0 J_1}{I_0^4} , \qquad\qquad \Lambda_4 = \frac{2I_0 I_1 I_3 + I_1 I_5 - 2I_0^2 I_4 - I_0 J_1}{I_0^4} ,$$

$$\Lambda_5 = \frac{I_1^2 I_5 - 2I_0 I_1 J_1 + I_0^2 J_2}{I_0^5} , \qquad\qquad \Lambda_6 = \frac{I_1^3 I_5 - I_0^3 I_6 - 3I_1^2 J_1 + 3I_0^2 I_1 J_2}{I_0^6} .$$

In principal coordinates the *principal curvatures* are simply $\kappa_1 = L$ and $\kappa_2 = N$, and the *principal directions* e_1 and e_2 are the coordinates directions. Furthermore, at a non umbilical point – where $\kappa_1 \neq \kappa_2$ – the directional derivatives of the principal curvatures are given by $\partial_{e_1} \kappa_1 = P, \partial_{e_2} \kappa_1 = Q, \partial_{e_1} \kappa_2 = S$, and $\partial_{e_2} \kappa_2 = T$, see [12].

We will now give a couple of examples to demonstrate how this can be used.

7.1 Fairing

Over the years there have been many suggestions of functions which should estimate the 'fairness' of a surface, see [13] for an extensive treatment. As a simple example we can take $|\nabla H|^2$, where $H = \frac{1}{2} I_1 / I_0$ is the mean curvature, just as $K = I_2 / I_0$ is the Gauss curvature. In principal coordinates we have

$$H(x^1, x^2) = \frac{1}{2} \left((L + x^1 P + x^2 Q) + (N + x^1 S + x^2 T) + \text{higher order terms} \right) ,$$

so at $(0,0)$ we have $\nabla H = \frac{1}{2}(P + S, Q + T)$ and hence

$$|\nabla H|^2 = \frac{P^2 + 2PS + S^2 + Q^2 + 2QT + T^2}{4}$$

If we now substitute (22) into the expression then we get

$$
\begin{aligned}
|\nabla H|^2 &= \frac{P^2 + 3Q^2 + 3S^2 + T^2}{4} + \frac{LI_3 - I_4}{2(L - N)} - \frac{NI_3 - I_4}{2(L - N)} \\
&= \frac{L^3 I_5 - I_6 - 3L^2 J_1 + 3L J_2}{4(L - N)^3} \\
&\quad + 3 \frac{-L^2 N I_5 + I_6 + (L + 2N) L J_1 - (2L + N) J_2}{4(L - N)^3} \\
&\quad + 3 \frac{L N^2 I_5 - I_6 - (2L + N) N J_1 + (L + 2N) J_2}{4(L - N)^3} \\
&\quad + \frac{-N^3 I_5 + I_6 + 3N^2 J_1 - 3N J_2}{4(L - N)^3} + \frac{I_3}{2} \\
&= \frac{(L^3 - 3L^2 N + 3LN^2 - N^3) I_5}{4(L - N)^3} + \frac{(-1 + 3 - 3 + 1) I_6}{4(L - N)^3} \\
&\quad + \frac{(-3L^2 + 3(L^2 + 2LN) - 3(2LN + N^2) + 3N^2) J_1}{4(L - N)^3} \\
&\quad + \frac{3L - (2L + N) + (L + 2N) - 3N) J_2}{4(L - N)^3} + \frac{I_3}{2} \\
&= \frac{I_5}{4} + \frac{I_3}{2} = \frac{\widehat{I_5}}{4} + \frac{\widehat{I_3}}{2} = \frac{2I_0 I_3 + I_5}{4I_0^3} \;.
\end{aligned}
$$

As $I_5 = P^2 + 3Q^2 + 3S^2 + T^2$ in principal coordinates we could have done the calculation faster. In any case, we have performed the calculation using special coordinates, but as both sides of the equality are invariant the equality holds in any parameterization. In a similar manner – see [12] – it can be shown that

$$|\nabla K|^2 = \frac{2I_0 I_2 I_3 + I_0 J_2}{I_0^4} \;,$$

$$\left(\nabla(|\kappa_1| + |\kappa_2|)\right)^2 = \begin{cases} \dfrac{2I_0 I_3 + I_5}{I_0^3} & \text{if } I_2 > 0 \;, \\[2mm] \dfrac{2I_0(4I_0 I_2 - I_1^2) I_3 + I_1^2 I_5 - 4I_0 I_1 J_1 + 4I_0^2 J_2}{I_0^3(I_1^2 - 4I_0 I_2)} & \text{if } I_2 < 0 \;, \end{cases}$$

$$\left(\nabla(\kappa_1^2 + \kappa_2^2)\right)^2 = 4 \frac{2I_0^2 I_2 I_3 + I_1^2 I_5 - 2I_0 I_1 J_1 + I_0^2 J_2}{I_0^5} \;,$$

$$\left(\partial_{e_1} \kappa_1\right)^2 + \left(\partial_{e_2} \kappa_2\right)^2 = \frac{(I_1^2 - I_0 I_2) I_5 - 3I_0 I_1 J_1 + 3I_0^2 J_2}{I_0^3(I_1^2 - 4I_2)} \;,$$

$$\frac{1}{\pi} \int_0^\pi \left(\frac{d\kappa_n}{ds}\right)^2 d\phi = \frac{6I_0 I_3 + 5I_5}{16I_0^3} \;.$$

7.2   Ridges and the Subparabolic Curve

A surface has two *focal surfaces* or *evolutes* given as the locus of the two principal centre of curvature. The focal surfaces will in general have cuspidal edges called *ribs* lying over curves, called *ridges*, in the original surface, see [14]. The parabolic curve - where the Gaussian curvature is zero – in the focal surfaces lies over a curve, called the *subparabolic curve*, in the original surface, see [14].

If we in the tangent plane use rectangular coordinates (x, y) such that the axes are in the principal directions, then we can write (2) as

$$z = \frac{1}{2}\left(Lx^2 + Ny^2\right) + \frac{1}{6}\left(Px^3 + 3Qx^2y + 3Sxy^2 + Ty^3\right) + \text{higher order terms} .$$

The unit normal is to first order $\mathbf{N} \approx (-Lx, -Ny, 1)$ and if $L \neq N$ then the principal curvatures are to first order $\kappa_1 \approx L + Px + Qy$ and $\kappa_2 \approx N + Sx + Ty$. The two sheets of the focal surface are to first order given by

$$\left(0, \frac{L-N}{L}y, \frac{1}{L} - \frac{P}{L^2}x - \frac{Q}{L^2}y\right) \quad \text{and} \quad \left(\frac{N-L}{N}x, 0\frac{1}{N} - \frac{S}{N^2}x - \frac{T}{N^2}y\right) .$$

The cross products of the partial derivatives are

$$\left(\frac{L-N}{L^3}P, 0, 0\right) \quad \text{and} \quad \left(0, \frac{N-L}{N^3}T, 0\right)$$

respectively. At a non umbilical point (21) shows that

$$\begin{aligned}(L-N)^3PT &= (L-N)^2J_3 - LNJ_4 + (L+N)J_5 - J_6 \\ &= \left((L+N)^2 - 4LN\right)J_3 - LNJ_4 + (L+N)J_5 - J_6 \\ &= \left(\widehat{I_1}^2 - 4\widehat{I_2}\right)\widehat{J_3} - \widehat{I_2}\widehat{J_4} + \widehat{I_1}\widehat{J_5} - \widehat{J_6} .\end{aligned}$$

We have assumed that $L \neq N$, but if $L = N$ then $J_3 = J_4 = J_5 = J_6 = 0$ so the equation holds in this case too. This give us the required invariant description of the ridges:

Theorem 11. *The ridges of a surface is the zero set of the invariant function*

$$\frac{\left(I_1^2 - 4I_0I_2\right)J_3 - I_2J_4 + I_1J_5 - I_0J_6}{I_0^{9/2}} .$$

We saw above that the ridges at non umbilical points are given by $\partial_{\mathbf{e}_1}\kappa_1 = 0$ or $\partial_{\mathbf{e}_2}\kappa_2 = 0$. Similar the subparabolic lines are given by $\partial_{\mathbf{e}_2}\kappa_1 = 0$ or $\partial_{\mathbf{e}_1}\kappa_2 = 0$, see [14]. In principal coordinates we get the equation $QS = 0$, and at a non umbilical point (21) shows that

$$(L-N)^3QS = -LNJ_4 + (L+N)J_5 - J_6 = -\widehat{I_2}\widehat{J_4} + \widehat{I_1}\widehat{J_5} - \widehat{J_6} .$$

Just as before this give us the invariant description of the subparabolic curve:

Theorem 12. *The subparabolic curve of a surface is the zero set of the invariant function*

$$\frac{-I_2J_4 + I_1J_5 - I_0J_6}{I_0^{9/2}} .$$

7.3 Darboux's Classification of Umbilical Points

At an umbilical point the first and second fundamental form are proportional and then 13 of the 18 basic invariants can be expressed as linear combination of the others. We are left with only 5 invariants I_0, I_3, I_5, I_7, and J_7, and a single syzygy $J_7^2 = Q(I_0, I_3, I_5, I_7)$. We have essentially the joint invariants of one quadratic and one cubic binary form.

The Darboux's classification depends on the pattern of the lines of curvatures around the umbilical point, which in turn depends on whether there are one or three real root lines of the cubic form $c = \nabla II$, and in the latter case whether the three root lines are contained in a right angle or not, see [14]. A root line of a cubic form is a direction where it vanishes, and there is at least one root line which we can assume it is the x-axis. I.e., we may assume that $P = 0$. Then $I_7 = 3Q^2S^2 - 4Q^3T$ and $c = (3Qx^2 + 3Sxy + Ty^2)y$. The quadratic factor has the discriminant $\frac{3}{4}(4QT - 3S^2) = -\frac{3}{4}I_7/Q^2$ so we have

$$I_7 < 0 \iff c \text{ has 3 distinct real root lines,}$$

$$I_7 > 0 \iff c \text{ has exactly 1 real root line.}$$

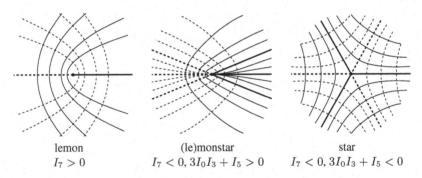

lemon	(le)monstar	star
$I_7 > 0$	$I_7 < 0, 3I_0I_3 + I_5 > 0$	$I_7 < 0, 3I_0I_3 + I_5 < 0$

Fig. 1. Curvature lines around an isolated umbilical point

In the case $I_7 < 0$ we can assume that the three real root directions are $\mathbf{v}_1 = (\alpha_1, \beta_1)$, $\mathbf{v}_2 = (\alpha_2, \beta_2)$, and $\mathbf{v}_3 = (\alpha_3, \beta_3) = (1, 0)$. The cubic form is then

$$c = \prod_{i=1}^{3}(\beta_i x - \alpha_i y) = (-\beta_1\beta_2 x^2 + (\alpha_1\beta_2 + \beta_1\alpha_2)xy - \alpha_1\alpha_2 y^2)y \ .$$

So $3Q = -\beta_1\beta_2$ and $T = -\alpha_1\alpha_2$. Hence

$$(\mathbf{v}_1 \cdot \mathbf{v}_2)(\mathbf{v}_1 \cdot \mathbf{v}_3)(\mathbf{v}_2 \cdot \mathbf{v}_3) = (\alpha_1\alpha_2 + \beta_1\beta_2)\alpha_1\alpha_2 = T^2 + 3QT = \frac{3I_0I_3 + I_5}{I_0^3} \ .$$

We can now see that if $I_7 < 0$, then we have

$$3I_0I_3 + I_5 > 0 \iff \text{The root lines of } c \text{ are contained in a right angle,}$$

$$3I_0I_3 + I_5 < 0 \iff \text{The root lines of } c \text{ aren't contained in a right angle.}$$

This gives the classification in Fig. 1, where we have sketched the three generic patterns possible for the lines of curvature around an isolated umbilical point.

Acknowledgment

I am indebted to my colleagues at the Department of Mathematics, Technical University of Denmark, for numerous valuable comments, especially to Agnes Heydtman who introduced me to the program 'Singular' and patiently answered many question about invariant theory.

References

1. Grace, J.H., Young, A.: The algebra of invariants. Cambridge University Press, Cambridge (1903)
2. Gurevich, G.B.: Foundation of the theory of algebraic invarians. P. Noordhoff, Groningen (1964)
3. Olver, P.J.: Classical invariant theory. Volume 44 of London Mathematical Society Student Texts. Cambridge University Press, Cambridge (1999)
4. Sturmfels, B.: Algorithms in Invariant Theory. Springer-Verlag, Wien, New York (1993) Text and Monographs in Symbolic Computation.
5. Turnbull, H.W.: The Theory of Determinants, Matrices, and Invariants. 3 edn. Dover Publications, New York (1960)
6. Weyl, H.: The Classical Groups, Their Invariants and Representations. Princeton University Press, Princeton, New Jersey (1946)
7. Kung, J.P.S., Rota, G.C.: The invariant theory of binary forms. Bulletin of the Amerian Mathematical Society **10** (1983) 27–85
8. Hilbert, D.: Über die Theorie der algebraischen Formen. Math. Ann. **26** (1890) 473–534
9. Fulton, W., Harris, J.: Representation Theory. A First Course. Volume 129 of Graduate Texts in Mathematics. Springer-Verlag (1991)
10. Greuel, G.M., Pfister, G., Schönemann, H.: SINGULAR 2.0. A Computer Algebra System for Polynomial Computations, Centre for Computer Algebra, University of Kaiserslautern (2001) http://www.singular.uni-kl.de/
11. Gravesen, J.: (Invariants) http://www.mat.dtu.dk/people/J.Gravesen/inv
12. Gravesen, J., Ungstrup, M.: Constructing invariant fairness measures for surfaces. Advances in Computational Mathematics (2001)
13. Nowacki, H., Kaklis, P.D., eds.: Creating Fair and Shape-Preserving Curves and Surfaces. B. G. Teubner, Stuttgart (1998) Papers from the International Workshop held in Kleinmachnow, September 14–17, 1997.
14. Porteous, I.R.: Geometric Differentiation for the intelligence of curves and surfaces. Cambridge University Press, Cambridge (1994)

Universal Rational Parametrizations and Spline Curves on Toric Surfaces

Rimvydas Krasauskas and Margarita Kazakevičiūtė

Vilnius University, Naugarduko 24, 2600 Vilnius, Lithuania,
{rimvydas.krasauskas,margarita.kazakeviciute}@mif.vu.lt
WWW home page: http://www.mif.vu.lt/cs2/cagl/

Abstract. Recently a constructive description of all rational parametrizations for toric surfaces was described in terms of the universal rational parametrizations (URP). We give an elementary introduction to this theory from the Geometric Modelling point of view: toric surfaces are defined via homogeneous coordinates; projections, singular cases, and non-canonical real structures are described; the URP theorem is explained. A theory of rational C^1 spline curves with certain interpolation properties on toric surfaces is developed. Applications for smooth blending of natural quadrics are sketched.

1 Introduction

Rational curves and surfaces are widely used in Geometric Modeling in the form of NURBS, which became a standard in industry about 25 years ago. Nevertheless, the full potential possibilities of NURBS were not completely realized because the geometry of rational parametrizations appeared too complicated and simple constructive methods were unavailable. Recently a constructive description of all rational parametrizations for a large important subclass of rational surfaces called almost toric surfaces was developed in terms of Universal Rational Parametrizations (URP).

The concept of a universal rational parametrization has been known under different names for about 10 years. URP first appeared as a generalized stereographic projection for the sphere and the hyperbolic paraboloid in the papers of Dietz, Hoschek and Jüttler [7, 8]. URP was extended to Dupin cyclides by Mäurer [17] under the name of generalized parameter representation, and to singular quadrics and the torus independently by Krasauskas [13], where the term *universal parametrization* appeared for the first time. Later it became clear that all these diverse examples are particular cases of almost toric surfaces: they are either canonical or non-canonical real parts of complex almost toric surfaces. For example, in recent papers [18, 10] several cases of cubic surfaces were considered, which under closer inspection appear to be toric or almost toric. Recently it was established by Cox, Krasauskas and Mustata [3] that all complex projective toric surfaces (in fact also higher dimensional toric varieties) and their general projections to lower dimensional projective spaces (i.e. almost toric surfaces) admit URP.

There are several reasons why (almost) toric surfaces are important in Geometric Modeling: many low degree rational surfaces (all quadrics and Dupin cyclides) are toric or almost toric, Bézier tensor-product and triangular surfaces are in fact parametrized by toric surfaces, and both can be generalized to multisided toric Bézier patches [15, 16].

This paper is devoted to an elementary introduction to toric surfaces and universal rational parametrizations from the Geometric Modeling point of view. In sect. 2 we give several important examples of URP and introduce a general notion of complex (almost) toric surface, including singular case. In sect. 3 the URP Theorem is formulated and explained. Section 4 is devoted to real structures on complex toric surfaces. The theory of rational C^1 spline curves on simplest toric surfaces is developed in sect. 5. Finally in sect. 6 a sketch of one application is given: we show how these splines on toric surfaces are used for smooth blending of natural quadrics.

2 Toric surfaces

A real d-dimensional projective space $\mathbb{R}P^d$ can be regarded as a quotient of $\mathbb{R}^{d+1} \setminus \{0,\ldots,0\}$ with respect to the multiplicative group $\mathbb{R}^* = \mathbb{R} \setminus 0$ action $\lambda \cdot (x_0,\ldots,x_d) = (\lambda x_0,\ldots,\lambda x_d)$. Thus points in $\mathbb{R}P^d$ are usually represented by homogeneous coordinates $[x_0,\ldots,x_d]$ unique up to a non-zero multiplier. Complex projective spaces $\mathbb{C}P^d$ are similarly defined: just change all appearances of $\mathbb{R}$ to $\mathbb{C}$ (complex numbers) in the previous definition. In particular, $\mathbb{C}^* = \mathbb{C} \setminus 0$ denotes the multiplicative group of non-zero complex numbers.

For simplicity, let us consider the complex case first. Rational curves and surfaces in a projective space $\mathbb{C}P^d$ can be treated as images of rational maps $\mathbb{C}^k \dashrightarrow \mathbb{C}P^d$, where $k = 1, 2$. Therefore, these maps are represented by collections of polynomials $F = (f_0,\ldots,f_d)$ in k variables. It is natural to cancel any common factors and to consider *irreducible collections*, i.e. $\gcd(f_0,\ldots,f_d) = 1$.

2.1 Examples of Universal Rational Parametrizations

Let $X \subset \mathbb{C}P^d$ be a surface (or a higher dimensional variety). We call an irreducible collection of polynomials $F = (f_0,\ldots,f_d)$ in k variables *rational parametrization* of X if the image of the corresponding rational map $\mathbb{C}^k \dashrightarrow \mathbb{C}P^d$ is contained in X. Note that this image may be smaller than X: for example, F can define a curve $\mathbb{C} \dashrightarrow X$ on a surface. Here we give two examples of surfaces with universal rational parametrizations. We use different letters s_i and t_j for variables in order to stress their relation to the associated lattice polygons (sect. 2.2).

Example 1. Consider a quadric surface Q given by the homogeneous equation $z_0 z_3 = z_1 z_2$ in the projective space $\mathbb{C}P^3$. One obvious rational parametrization of Q is given by the following formula in matrix form

$$P_Q : (s_0, s_1, t_0, t_1) \mapsto \begin{pmatrix} s_0 t_0 & s_1 t_0 \\ s_0 t_1 & s_1 t_1 \end{pmatrix} = \begin{pmatrix} z_0 & z_1 \\ z_2 & z_3 \end{pmatrix}. \tag{1}$$

Theorem 8 in sect. 3 implies that *all* rational parametrizations of Q are of the form $P_Q \circ F$ for some $F : \mathbb{C}^k \to \mathbb{C}^4$ represented by polynomials, that satisfy

$$F = (s_0, s_1, t_0, t_1), \quad \gcd(s_0, s_1) = \gcd(t_0, t_1) = 1. \tag{2}$$

Furthermore, although F is not unique, Theorem 8 describes the non-uniqueness precisely: if $P_Q \circ F = P_Q \circ F'$ for other equivalent $F' = (s_0', s_1', t_0', t_1')$ then

$$F' = (\lambda s_0, \lambda s_1, \lambda^{-1} t_0, \lambda^{-1} t_1) \tag{3}$$

for some nonzero scalar λ.

In the language of Theorem 8, we say that P_Q from (1) is a *universal rational parametrization* of the quadric Q. The key property of the quadric Q is that it derives from $\mathbb{C}P^1 \times \mathbb{C}P^1$. If s_0, s_1 are homogeneous coordinates on the first factor $\mathbb{C}P^1$ and t_0, t_1 are homogeneous coordinates on the second factor, then P_Q induces the Segre embedding $\mathbb{C}P^1 \times \mathbb{C}P^1 \longrightarrow \mathbb{C}P^3$ whose image is Q.

Example 2. Consider the Steiner surface St in $\mathbb{C}P^3$ defined in homogeneous coordinates by the equation $z_1^2 z_2^2 + z_2^2 z_3^2 + z_3^2 z_1^2 = z_0 z_1 z_2 z_3$.

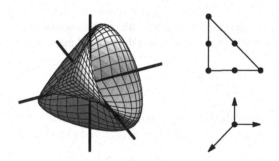

Fig. 1. The real Steiner surface with three double lines

Note that St is singular surface with three double lines (see Fig. 1)

$$z_1 = z_2 = 0, \quad z_2 = z_3 = 0, \quad z_3 = z_1 = 0. \tag{4}$$

One can easy verify that

$$P_{St}(t_0, t_1, t_2) = (t_0^2 + t_1^2 + t_2^2, t_0 t_1, t_1 t_2, t_2 t_0) \tag{5}$$

is a rational parametrization of St.

Now Theorem 8 tells us that *all* rational parametrizations H of St are of the form $H = P_{St} \circ F$ for some collection F satisfying

$$F = (t_0, t_1, t_2), \quad \gcd(t_0, t_1, t_2) = 1, \tag{6}$$

provided the image of H does not lie in the lines (4). Furthermore, Theorem 8 implies that F is unique up to a sign.

By Theorem 8, (5) is the universal rational parametrization of the Steiner surface St. In this case, the key property of St is that St derives from $\mathbb{C}P^2$ via the map $\mathbb{C}P^2 \to St$ induced by (5). This map is not an embedding but is birational (i.e., is generically one-to-one). Furthermore, the three lines (4) are the only places where this map fails to have an inverse.

Both $\mathbb{C}P^1 \times \mathbb{C}P^1$ and $\mathbb{C}P^2$ are examples of *smooth toric surfaces*, and the coordinates s_0, s_1, t_0, t_1 for $\mathbb{C}P^1 \times \mathbb{C}P^1$ and t_0, t_1, t_2 for $\mathbb{C}P^2$ are examples of *homogeneous coordinates* of toric surfaces. Hence we expect that there should be a toric generalization of these examples. For instance, we will see that the gcd conditions (2) and (6) are dictated by the data that determines the toric surface.

2.2   Complex Toric Surfaces

Consider the *lattice* $\mathbb{Z}^2$ of points with integer coordinates in the plane $\mathbb{R}^2$ and a convex *lattice polygon* $\Delta \subset \mathbb{R}^2$ (i.e. a convex polygon with vertices in the lattice). Let the edges of the polygon Δ be on the lines $\ell_i(x) = 0$, $i = 1, \ldots, r$, defined by the affine forms $\ell_i(x) = \langle n_i, x \rangle + a_i$, where $\langle , \rangle$ is a scalar product on $\mathbb{Z}^2$, the normals n_i of edges are inward oriented shortest vectors with integer coordinates.

With the lattice polygon Δ we associate the rational map

$$P_\Delta(u_1, \ldots, u_r) = (p_1, \ldots, p_{d+1}) = \sum_{m \in \Delta \cap \mathbb{Z}^2} q_m u_1^{\ell_1(m)} \cdots u_r^{\ell_r(m)}, \qquad (7)$$

where $q_m \in \mathbb{C}^{d+1}$ are called *homogeneous control points*. P_Δ defines a collection of polynomials $(p_1, \ldots, p_{d+1})$ in r variables $(u_1, \ldots, u_r)$ as a map $\mathbb{C}^r \to \mathbb{C}^{d+1}$.

Definition 3. *A toric surface X_Δ is a subset in $\mathbb{C}P^d$ which is parametrized by a rational map P_Δ with linearly independent control points q_m, $m \in \Delta \cap \mathbb{Z}^2$.*

It is well known that an implicit degree of X_Δ in $\mathbb{C}P^d$ is equal to the normalized area $\text{Area}(\Delta)$ of the polygon Δ, i.e. twice its usual Euclidean area (cf. [2, 4]). We can treat P_Δ as a rational map $\mathbb{C}^r \dashrightarrow \mathbb{C}P^d$ which is well defined (at least in the case of generic control points q_m) outside of the *exceptional set* $Z \subset \mathbb{C}^r$ given by the r monomial equations

$$\hat{u}_1 u_2 \cdots u_{r-1}\hat{u}_r = 0, \quad \hat{u}_1 \hat{u}_2 u_3 \cdots u_r = 0, \ldots \quad u_1 u_2 \cdots u_{r-2}\hat{u}_{r-1}\hat{u}_r = 0.$$

where $\hat{u}_i$ means that the parameter u_i is omitted [3]. We treat $u_1, \ldots, u_r$ as homogeneous coordinates and construct the *abstract* toric surface as follows. Consider a subgroup in $(\mathbb{C}^*)^r$

$$G = \{(\lambda_1, \ldots, \lambda_r) \in (\mathbb{C}^*)^r \mid \prod_{i=1}^r \lambda_i^{\langle n_i, m \rangle} = 1 \text{ for all } m \in \mathbb{Z}^2\} \qquad (8)$$

with the natural action $(\lambda_1, \ldots, \lambda_r) \cdot (u_1, \ldots, u_r) = (\lambda_1 u_1, \ldots, \lambda_r u_r)$ on $\mathbb{C}^r$. Notice that the exceptional set Z, the group G and the group action depends only on the collection of normals $\{n_1, \ldots, n_r\}$ which encode all the information about the *normal fan* Σ of the polygon Δ (see details in [2]).

Definition 4. *The* abstract toric surface *associated with the fan Σ is defined as the* quotient

$$X_\Sigma = (\mathbb{C}^r \setminus Z)/G. \tag{9}$$

The map P_Δ is homogeneous in the following sense: if $(\lambda_1, \ldots, \lambda_r) \in G$ then

$$P_\Delta(\lambda_1 u_1, \ldots, \lambda_r u_r) = \lambda_\Delta P_\Delta(u_1, \ldots, u_r), \quad \lambda_\Delta = \prod_{i=1}^{r} \lambda^{a_i}. \tag{10}$$

Therefore, the corresponding rational map $\mathbb{C}^r \setminus Z \dashrightarrow \mathbb{C}P^d$ factors into the composition $\mathbb{C}^r \setminus Z \to X_\Sigma \to X_\Delta \dashrightarrow \mathbb{C}P^d$. The first map $\Pi_\Delta : \mathbb{C}^r \setminus Z \to X_\Sigma$ is a well defined natural projection to the quotient (9). The second map is the embedding of X_Σ into $\mathbb{C}P^d$ (cf. [3]) with the image equal to the toric surface X_Δ. The third map $\pi : X_\Delta \dashrightarrow \mathbb{C}P^d$ is a central projection and has the same image $X \subset \mathbb{C}P^d$ as the whole composition.

Definition 5. We call X an *almost toric surface* if the map $\pi : X_\Delta \dashrightarrow \mathbb{C}P^d$ is *sufficiently nice*:
(i) π has no base points;
(ii) π is birational (is an isomorphism between dense open subsets).

Both conditions (i) and (ii) are satisfied when the control points $q_m \in \Delta \cap \mathbb{Z}$ are in general position. Unfortunately, their linear independence is possible only when the dimension of $\mathbb{C}P^d$ is high: $d + 1$ cannot be less than the number of lattice points in Δ. Therefore, usually in practice we deal with almost toric surfaces that are general projections of toric surfaces from higher dimensional spaces. The following proposition is useful for deciding if a given surface is almost toric.

Proposition 6. *A projection X of a toric surface X_Δ is almost toric if and only if their* implicit degrees coincide $\deg X = \deg X_\Delta$, *i.e.* $\deg X$ *is equal to the normalized area of the lattice polygon Δ.*

Proof. The proof follows directly from the degree formula for surfaces:

$$\deg(\pi) \deg X = \deg X_\Delta - \{\text{number of basepoints}\},$$

where $\deg(\pi)$ is the generic number of points in X_Δ which map to a point in X (see, e.g. [5]).

2.3 Singular Case and Desingularization

From the general theory of toric varieties (cf. [4], [9]) follows that an abstract toric surface X_Σ can be singular only at the vertices. These singular points correspond to the pairs of adjacent normals $\{n_i, n_{i+1}\}$ that do not form an integer basis for the lattice $\mathbb{Z}^2$.

The singular case is reduced to a smooth case by a desingularization procedure. The fan Σ is enriched to the regular fan $\tilde{\Sigma}$ (i.e. such that $X_{\tilde{\Sigma}}$ is smooth) by adding additional normals as shown in Fig. 2. Assume that $\tilde{\Sigma}$ has normals $\tilde{n}_i$, $i = 1, \ldots, s$, and let $\tilde{\ell}_i(x)$,

$i = 1, \ldots, s$, be affine forms that define supporting lines with these normals. Now we can rewrite the formula (7) in new variables as follows

$$P_{\Delta, \tilde{\Sigma}}(\tilde{u}_1, \ldots, \tilde{u}_s) = \sum_{m \in \Delta \cap \mathbb{Z}^2} q_m \tilde{u}_1^{\tilde{\ell}_1(m)} \cdots \tilde{u}_s^{\tilde{\ell}_s(m)}, \tag{11}$$

This rational map defines parametrization of the initial singular toric surface X by the abstract smooth toric surface $X_{\tilde{\Sigma}}$, which is called a desingularization of X. We will see in sect. 3 that $P_{\Delta, \tilde{\Sigma}}$ defines the URP for the singular surface X.

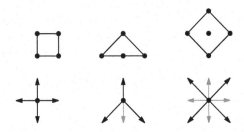

Fig. 2. Lattice polygons and their fans with desingularizations

Example 7. In the middle column of Fig. 2 we see a lattice triangle Δ with vertices $(-1, 0), (1, 0), (0, 1)$ and its fan of dark arrows below. Two of the three normals $(1, -1)$ and $(-1, -1)$ do not generate the entire lattice. In this case a regular fan $\tilde{\Sigma}$ has just one additional normal $(0, -1)$ (shown in grey), and we can write down the parametrization (11) with linearly independent control points

$$P_{\Delta, \tilde{\Sigma}} : (s_0, s_1, t_0, t_1) \mapsto \begin{pmatrix} s_0 \\ s_1 t_0^2 \; s_1 t_0 t_1 \; s_1 t_1^2 \end{pmatrix} = \begin{pmatrix} z_0 \\ z_1 \; z_2 \; z_3 \end{pmatrix}. \tag{12}$$

This is the parametrization of a singular quadric $C_2 : z_1 z_3 = z_2^2$ (i.e. z_0 is arbitrary) in $\mathbb{C}P^3$.

3 URP Theorem

Fix a ring of polynomials $R = \mathbb{C}[t_1, \ldots, t_k]$. We recall from sect. 2.1 that a rational parametrization of $X \subset \mathbb{C}P^d$ is an irreducible collection of polynomials $H = (h_1, \ldots, h_{d+1}) \in R^{d+1}$ such that the image of a map $\mathbb{C}^k \dashrightarrow \mathbb{C}P^d$ defined by H is contained in X.

For Theorem 8 below a toric analogy of irreducibility is needed. We say that $F = (f_1, \ldots, f_r) \in R^r$ is *Σ-irreducible* if $\gcd(f_{i_1}, \ldots, f_{i_s}) = 1$ whenever no vertex of Δ is incident with its edges defined by the lines $\ell_{i_1}, \ldots, \ell_{i_s}$.

Theorem 8. *Let P_Δ, X_Δ, π and a complex toric surface $X \subset \mathbb{C}P^d$ be defined as above, where X_Σ is smooth and $\pi : X_\Delta \to X$ is sufficiently nice, with an inverse defined on $U \subset X$ which we assume to be maximal. Then P_Δ is a **universal rational parametrization** of X in the following sense:*

1. *If a collection $F = (f_1, \ldots, f_r)$ is Σ-irreducible, then $P_\Delta \circ F$ is a rational parametrization of $X \subset \mathbb{C}P^d \in R^{d+1}$.*
2. *Conversely, given any rational parametrization $H \in R^{d+1}$ of X whose image meets $U \subset X$, there is a Σ-irreducible $F = (f_1, \ldots, f_r) \in R^r$ such that $H = P_\Delta \circ F$.*
3. *If F and F' are Σ-irreducible, then $P_\Delta \circ F = P_\Delta \circ F'$ as rational parametrizations if and only if $F' = \lambda \cdot F$ for some $\lambda \in G$, $\lambda_\Delta = 1$ (see (10)).*

Here, for simplicity, we presented only the smooth case of the URP Theorem, which is proved in general case of toric varieties of arbitrary dimensions in [3]. In a singular case we should change P_Δ to $P_{\Delta, \tilde{\Sigma}}$ defined in (11), and U to $U' = U \cup (\text{singular points})$.

Now we can check how Examples 1 and 2 of sect. 2.1 follow from Theorem 1. In case of Example 1, the lattice polygon Δ is the unit square in the plane with vertices $(0, 0)$, $(1, 0)$, $(1, 1)$, $(0, 1)$. The inner normals $(0, 1)$, $(-1, 0)$, $(0, -1)$, $(1, 0)$ correspond to variables t_0, s_0, t_1, s_1, A collection of polynomials $F = (s_0, s_1, t_0, t_1)$ is Σ-irreducible if $\gcd(s_0, s_1) = \gcd(t_0, t_1) = 1$ (this coincides with (2)), since only these pairs of edges are not adjacent. Also it is easy to calculate the group G

$$G = \{(\lambda_1, \lambda_1, \lambda_2, \lambda_2) \mid \lambda_1, \lambda_2 \in \mathbb{C}^*\} = \mathbb{C}^* \times \mathbb{C}^*, \qquad (13)$$

and $\lambda_\Delta = \lambda_1 \lambda_2$. Therefore, F is indeed defined up to multiplication of a group element $(\lambda_1, \lambda_1, \lambda_2, \lambda_2)$, where $\lambda_1 \lambda_2 = 1$.

In Example 2 the variables t_0, t_1, t_2 are associated with the inner normals $(0, 1)$, $(-1, -1)$, $(1, 0)$ of the lattice triangle Δ with vertices $(2, 0)$, $(0, 0)$, $(0, 2)$ (see Fig. 1(left)). The Steiner surface St is a projection of the corresponding toric surface X_Δ (this is a well-known Veronese surface)

$$\pi : X_\Delta \to St, \quad \begin{pmatrix} t_0^2 \\ t_0 t_1 \ t_0 t_2 \\ t_1^2 \ t_1 t_2 \ t_2^2 \end{pmatrix} \mapsto (t_0^2 + t_1^2 + t_1^2, t_0 t_1, t_1 t_2, t_0 t_2).$$

Since X_Δ has degree $\text{Area}(\Delta) = 4$ and its projection St has the same degree, from Proposition 1 follows that St is almost toric. Therefore, we can apply Theorem 1 to the projection P_{St} defined by (5). A collection of polynomials $F = (t_0, t_1, t_2)$ is Σ-irreducible if $\gcd(t_0, t_1, t_1) = 1$ (cf. (6)), since all pairs of edges are adjacent. It is easy to calculate the group: $G = \{(\lambda, \lambda, \lambda) \mid \lambda \in \mathbb{C}^*\} = \mathbb{C}^*$, and $\lambda_\Delta = \lambda^2$. Therefore, F is indeed defined up to multiplication of a group element $\lambda \in \mathbb{C}$, where $\lambda^2 = 1$, i.e. $\lambda = \pm 1$.

Example 9. Consider a cubic surface W given by the homogeneous equation $z_0 z_1^2 = z_2^2 z_3$ in the projective space $\mathbb{C}P^3$. The real part of W is called 'Whitney umbrella' (see Fig. 3). From Proposition 1 follows that W is alsmost toric, i.e. it is a sufficiently nice projection of a Hirzebruch surface $\mathbb{F}_1$ embedded in $\mathbb{C}P^4$, which is a toric surface associated with a lattice trapezoid Δ (shown in Fig. 3(right)) and given by the rational parametrization

$$P_\Delta : (s_0, s_1, t_0, t_1) \mapsto \begin{pmatrix} s_1 t_0 & s_1 t_1 \\ s_0 t_0^2 & s_0 t_0 t_1 & s_0 t_1^2 \end{pmatrix} = \begin{pmatrix} z_2 & z_1 \\ z_0 & z_4 & z_3 \end{pmatrix}. \tag{14}$$

Indeed, composing this map with a projection $\mathbb{C}P^4 \to \mathbb{C}P^3$ that forgets z_4 we get $P(s_0, s_1 t_0, t_1) = (s_0 t_0^2, s_1 t_1, s_1 t_0, s_0 t_1^2) = (z_0, z_1, z_2, z_3)$. According to Theorem 8, the parametrization P is a URP of the surface W.

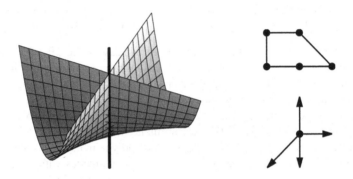

Fig. 3. The 'Whitney umbrella' surface

4   Real Structures on Complex Toric Surfaces

A real variety is a pair (X, c), where X is a complex variety and $c : X \to X$ is an anti-holomorphic involution, i.e. $\bar{c}$ is holomorphic and $c \circ c = $ id. The involution c is also called a *real structure* on X. The real part $\mathbb{R}_c X$ of X is just the fixed-point set $\{x \in X \mid c(x) = x\}$. The *canonical* real structure is defined by complex conjugation, so that the fixed-point set in this case coincides with all points with real coordinates.

There are three main types of quadric surfaces in $\mathbb{R}P^3$: double ruled, oval and singular. Consider first the complex quadric $Q : z_0 z_3 = z_1 z_2$ in $\mathbb{C}P^3$ as in Example 1.

4.1   Real Structures on Quadrics

Let $\mathbb{R}Q$ be the canonical real part of the complex quadric Q. Then it is a real double ruled quadric $x_0 x_3 = x_1 x_2$ in $\mathbb{R}P^3$. $\mathbb{R}Q$ can be similarly parametrized using (1) with all real variables. Then the group $\mathbb{R}^* \times \mathbb{R}^*$ acts on $\mathbb{R}^4$, and generates the real Segre map $\mathbb{R}P^1 \times \mathbb{R}P^1 \to \mathbb{R}P^3$ which defines an isomorphism $\mathbb{R}Q \cong \mathbb{R}P^1 \times \mathbb{R}P^1$.

Consider another involution $c : (z_0, z_1, z_2, z_3) \mapsto (\bar{z}_0, \bar{z}_2, \bar{z}_1, \bar{z}_3)$ in $\mathbb{C}P^3$. $\mathbb{R}P^3$ can be identified with the corresponding fixed point set $\mathbb{R}_c P^3 \subset \mathbb{C}P^3$

$$\mathbb{R}P^3 \to \mathbb{R}_c P^3 \subset \mathbb{C}P^3, \quad (x_0, x_1, x_2, x_3) \mapsto (x_0, x_1 + \mathrm{i}x_2, x_1 - \mathrm{i}x_2, x_3), \tag{15}$$

which takes a real oval quadric $S : x_0 x_3 = x_1^2 + x_2^2$ to the *non-standard real part* $\mathbb{R}_c Q$ of Q. The involution c naturally extends to the parametrization P_Q via $s_i \mapsto \bar{t}_i, i = 0, 1$. Calculating fixed points of (1) and using (15), we get a map

$$P_S : \mathbb{C}^2 \to \mathbb{R}^4, \quad (s_0, s_1) \mapsto (s_0 \bar{s}_0, \mathrm{Re}(\bar{s}_0 s_1), \mathrm{Im}(\bar{s}_0 s_1), s_1 \bar{s}_1). \tag{16}$$

The map P_S is homogeneous in the sense that $P_S(\lambda z_0, \lambda z_1) = |\lambda|^2 P_S(z_0, z_1)$ for every $\lambda \in \mathbb{C}$. This map is the parametrization of S by $\mathbb{C}^2 \setminus (0,0)$, where the group $G = \mathbb{C}^*$ acts by complex multiplication. In fact, we have arrived at the classical Riemann sphere construction $\mathbb{C}P^1 \to S \subset \mathbb{R}P^3$. Indeed, S is a sphere in the affine coordinates x, y, z if $x_0 = 1 + z$, $x_1 = x$, $x_2 = y$, $x_3 = 1 - z$. It is easy to check that the generalized stereographic projection [7] coincides with P_S.

4.2   Real Almost Toric Surfaces

Now we are ready to define the most general case of real surfaces with toric structures that will be used in applications. Further details about different real structures on toric surfaces (including structures with empty real parts) can be found in [6].

As earlier, we start with the lattice polygon Δ. In order to introduce a real structure, we fix an affine involution $c : \mathbb{Z}^2 \to \mathbb{Z}^2$ on the lattice that preserves the polygon Δ. In general, c permutes the lattice points of $\Delta \cap \mathbb{Z}$ and the edges of the polygon Δ. Hence the normals $\{n_1, \ldots, n_r\}$ are also permuted $- c(n_i) = n_{\sigma(i)}, i = 1, \ldots, r$.

Definition 10. *A real almost toric surface associated with a lattice polygon Δ with an involution c is the real part of a (complex) almost toric surface X parametrized by P_Δ (see (7)), where the anti-holomorphic involution acts on the domain $u_i \mapsto \bar{u}_{\sigma(i)}$, $i = 1, \ldots, r$, and the control points are compatible with c – that is, $\bar{q}_m = q_{c(m)}$, $m \in \Delta \cap \mathbb{Z}$.*

The latter condition ensures that the image of the real part $\mathbb{R}_c X_\Sigma$ is mapped to the canonical real part of $\mathbb{C}P^d$, i.e. the real almost toric surface $\mathbb{R}_c X$ is contained in the real projective space $\mathbb{R}P^d$.

Remark 11. In the case of canonical real structure Definition 10 gives real control points q_m. If we substitute variables u_i by the corresponding affine forms then we get exactly the definition of a toric Bézier patch $\mathcal{B}_\Delta(x) = P_\Delta(\ell_1(x), \ldots, \ell_r(x))$ as defined in [15, 16]. Therefore, all toric Bézier patches including tensor product Bézier surfaces and triangular Bézier patches with generic control points are almost toric and have a URP as formulated in Theorem 1.

4.3   The Torus Surface is Almost Toric!

Consider the particular case of a ring cyclide – the torus surface T in $\mathbb{R}P^3$

$$(x_1^2 + x_2^2 + x_3^2 + (a^2 - b^2)x_0^2)^2 = 4a x_0^2 (x_1^2 + x_2^2), \quad a > b > 0.$$

Via introducing an auxiliary variable x_4, T can be described as a central projection of a quartic surface T' given by the following two equations in $\mathbb{R}P^4$

$$x_1^2 + x_2^2 = x_4^2,$$

$$\left(\frac{a}{b}x_4 - \frac{a^2 - b^2}{b}x_0\right)^2 + \left(\frac{\sqrt{a^2 - b^2}}{b}x_3\right)^2 = x_4^2.$$

Since on the left side of these equations we have sum of squares, we can treat T' as a real part of the complex surface in $\mathbb{C}P^4$

$$\begin{cases} z_1 z_2 = z_4^2, \\ z_0 z_3 = z_4^2, \end{cases} \tag{17}$$

where the real structure is defined by the anti-holomorphic involution

$$c : (z_0, z_1, z_2, z_3, z_4) \mapsto (\bar{z}_3, \bar{z}_2, \bar{z}_1, \bar{z}_0, \bar{z}_4). \tag{18}$$

Now it is easy to check that this complex surface coincides with a toric surface $X_\Diamond$ associated with the lattice polygon $\Diamond$ (Fig. 2, right) and is parametrized by

$$P_\Diamond : \begin{pmatrix} s_1 & t_0 \\ & \\ t_1 & s_0 \end{pmatrix} \mapsto \begin{pmatrix} & s_0^2 t_1^2 & \\ s_0^2 t_0^2 & s_0 s_1 t_0 t_1 & s_1^2 t_1^2 \\ & s_1^2 t_0^2 & \end{pmatrix} = \begin{pmatrix} & z_0 & \\ z_1 & z_4 & z_2 \\ & z_3 & \end{pmatrix}. \tag{19}$$

Hence, the torus is a *real almost toric* surface according to Proposition 1. In order to calculate the universal parametrization of $\mathbb{R}_c X_\Diamond$ consider the normal fan $\Sigma = \Sigma(\Diamond)$ defined by 4 normals (solid arrows in Fig. 2). The desingularization $\widetilde{\Sigma}$ is defined by 8 normals (additional grey arrows in Fig. 2). Hence we extend the list of facet variables s_0, s_1, t_0, t_1 by additional variables u_0, u_1, v_0, v_1 and the following parametrization is generated

$$\widetilde{P}_\Diamond : \begin{pmatrix} s_1 u_1 t_0 \\ v_0 \quad v_1 \\ t_1 u_0 s_0 \end{pmatrix} \mapsto \begin{pmatrix} & s_0^2 t_1^2 u_0^2 v_0 v_1 & \\ s_0^2 t_0^2 u_0 u_1 v_1^2 & s_0 s_1 t_0 t_1 u_0 u_1 v_0 v_1 & s_1^2 t_1^2 u_0 u_1 v_0^2 \\ & s_1^2 t_0^2 u_1^2 v_0 v_1 & \end{pmatrix} \tag{20}$$

The real structure $(X_\Diamond, c)$ is defined by the anti-holomorphic involution (18) in $\mathbb{C}P^4$, which is associated with a central symmetry of $\Diamond$. The corresponding involution in the parameter space is $s_0, t_0, u_0, v_0 \mapsto \bar{s}_1, \bar{t}_1, \bar{u}_1, \bar{v}_1$. Denoting s_0, t_0, u_0, v_0 by s, t, u, v for simplicity we get the universal parametrization of $\mathbb{R}_c X_\Diamond$ in $\mathbb{R}P^4$

$$(s, t, u, v) \mapsto \left(|v|^2 \text{Re}(s\bar{t}u)^2, |v|^2 \text{Im}(s\bar{t}u)^2, |u|^2 \text{Re}(stv)^2, |u|^2 \text{Im}(stv)^2, |stuv|^2\right).$$

Here we identify $\mathbb{R}P^4$ with $\mathbb{R}_c P^4$ as follows (cf. (15))

$$(y_0, y_1, y_2, y_3, y_4) \mapsto (y_0 + iy_1, y_2 + iy_3, y_2 - iy_3, y_0 - iy_1, y_4).$$

The backward projection to $\mathbb{R}P^3$ gives exactly the formula in [13], where variables s, t, u, v were denoted by z_1, z_2, z_3, z_0 respectively.

More examples of real almost toric surfaces can be found in [14].

5   Spline Curves on Toric Surfaces

As we have seen (almost) toric surfaces have natural intrinsic coordinates which are called also homogeneous coordinates. In case of a projective plane these coordinates are familiar homogeneous coordinates used in plane projective geometry. Rational Bézier curves and surfaces traditionally are defined by affine control points and weights but also there is a parallel more flexible description via homogeneous control points. Our idea is to use toric homogeneous coordinates for modeling rational C^1 spline curves of minimal degree on toric surfaces. To begin, we consider an interpolation problem on the projective line $\mathbb{C}P^1$.

5.1   Interpolation on the Projective Line $\mathbb{C}P^1$

Let us fix homogeneous coordinates (z_0, z_1) on $\mathbb{C}P^1$. We say that a rational function in the form of a pair of complex polynomials $\mathbf{f}(\tau) = (f_0(\tau), f_1(\tau))$ has degree $(d_0|d_1)$ if $\deg f_0 = d_0$ and $\deg f_1 = d_1$.

Proposition 12. *For any integer numbers $d_0, d_1 \geq 0$ there exists a unique rational function of degree $(d_0|d_1)$ that interpolates $N = d_0 + d_1 + 1$ general points on $\mathbb{C}P^1$ at given real parameter values $\tau_0 < \tau_1 < \cdots < \tau_{N-1} \in \mathbb{R}$. Here N points are called non-generic if any $N - k$ of them can be interpolated by a rational function of degree $(d_0 - k|d_1 - k)$, $0 < k \leq \min(d_0, d_1)$, in the corresponding parameter values τ_i.*

Proof. We denote proportional vectors in $\mathbb{C}^2$ by $\mathbf{a} \sim \mathbf{b}$, the Hermitian scalar product by $\langle \mathbf{a}, \mathbf{b} \rangle := a_0 \bar{b}_0 + a_1 \bar{b}_1$, and introduce the operator $(a_0, a_1)^{\perp} = (-\bar{a}_1, \bar{a}_0)$. Let vectors $\mathbf{a}_0, \ldots, \mathbf{a}_{N-1} \in \mathbb{C}^2$ represent the given points on $\mathbb{C}P^1$. Then the interpolation conditions $\mathbf{f}(\tau_i) \sim \mathbf{a}_i$ (for non-zero $\mathbf{f}(\tau_i)$) are equivalent to the system of equations:

$$\langle \mathbf{f}(\tau_i), \mathbf{a}_i^{\perp} \rangle = 0, \quad i = 0, \ldots, d_0 + d_1. \tag{21}$$

Since there are $d_0 + d_1 + 1$ linear homogeneous equations and $d_0 + d_1 + 2$ unknown coefficients of polynomials $f_0(\tau)$, $f_1(\tau)$, the non-trivial solution exists. For any two solutions $\mathbf{f}(\tau)$ and $\mathbf{g}(\tau)$ we have $\langle \mathbf{f}(\tau_i), \mathbf{g}(\tau_i)^{\perp} \rangle = 0$ for all i. Hence the polynomial $\langle \mathbf{f}(\tau), \mathbf{g}(\tau)^{\perp} \rangle$ of degree $N - 1$ vanishes in N distinct points τ_i. Therefore, $\mathbf{f}(\tau) \sim \mathbf{g}(\tau)$ and $\mathbf{f}(\tau)$ is unique as a rational function. Also all $\mathbf{f}(\tau_i)$ are non-zero, since otherwise $f_0(\tau)$ and $f_1(\tau)$ should have common divisors, i.e. this contradicts the assumption that the points are general. ∎

Remark 13. Let points $a_0, \ldots, a_{m+n}$ be given in affine coordinates of the complex line $\mathbb{C} \subset \mathbb{C}P^1$. The explicit interpolation formula for parameter values $\tau_0, \ldots, \tau_{m+n}$ with a rational function of degree $(m|n)$ was known to A. Cauchy [1, p. 432] (here we use modern notations):

$$\frac{f_1(\tau)}{f_0(\tau)} = \frac{\sum_{[i]_m \subset I} \left(a_{i_0} \cdots a_{i_m} \prod_{j \in I \setminus [i]_m} \frac{\tau - \tau_j}{(\tau_{i_0} - \tau_j) \cdots (\tau_{i_m} - \tau_j)} \right)}{\sum_{[k]_{m-1} \subset I} \left(a_{k_0} \cdots a_{k_{m-1}} \prod_{j \in I \setminus [k]_{m-1}} \frac{(\tau_{k_0} - \tau) \cdots (\tau_{k_{m-1}} - \tau)}{(\tau_{k_0} - \tau_j) \cdots (\tau_{k_{m-1}} - \tau_j)} \right)}, \tag{22}$$

where $[i]_m = \{i_0 < \cdots < i_m\}$ and $[k]_{m-1} = \{k_0 < \cdots < k_{m-1}\}$ are ordered subsets of the set $I = \{0, \ldots, m+n\}$.

Several particular cases of Proposition 2 are well-known:

(1) Real coefficients, degree $(0|d)$: interpolation of $N = d+1$ points on $\mathbb{R}$ with a polynomial function of deg $= d$. The Cauchy formula (22) reduces to the La Grange interpolation polynomial.
(2) Real coefficients, degree $(d|d)$: interpolation of $N = 2d + 1$ points on $\mathbb{R}P^1$ with a rational function of deg $= d$.
(3) Complex coefficients, degree $(d|d)$: interpolation of $N = 2d + 1$ points on $\mathbb{C}P^1$ with a rational curve of deg $= d$ (or a curve of degree $2d$ on the sphere).

5.2 Splines on $\mathbb{C}P^1$ and $\mathbb{R}P^1$

We fix affine coordinate $z = z_1/z_0$ on $\mathbb{C}P^1$. Let us first consider a case of degree $(1|1)$ function on a single interval $[\tau_0, \tau_1]$ in the rational Bézier form:

$$f(\tau) = \frac{a_0 w_0(\tau_1 - \tau) + a_1 w_1(\tau - \tau_0)}{w_0(\tau_1 - \tau) + w_1(\tau - \tau_0)}, \tag{23}$$

where $a_0, a_1 \in \mathbb{C}$ can be considered as control points and $w_0, w_1 \in \mathbb{C}$ are non-zero weights. Thus, f interpolates two endpoints $a_i = f(\tau_i)$, $i = 0, 1$, with derivatives $v_i = f'(\tau_i)$, $i = 0, 1$,

$$v_0 = \left(\frac{a_1 - a_0}{\tau_1 - \tau_0}\right)\frac{w_1}{w_0}, \quad v_1 = \left(\frac{a_1 - a_0}{\tau_1 - \tau_0}\right)\frac{w_0}{w_1},$$

and we get a simple relation

$$v_0 v_1 = \left(\frac{a_1 - a_0}{\tau_1 - \tau_0}\right)^2. \tag{24}$$

Therefore, the function f is uniquely determined by any three of given four parameters a_0, a_1, v_0, v_1 satisfying Eq. (24). In the real case on $\mathbb{R}P^1$ the weights should be positive $w_0, w_1 > 0$ in order to get values in the interval with end points $f(\tau_0)$, $f(\tau_1)$. Also $v_0 v_1 > 0$, whenever $a_0 \neq a_1$, i.e. f is either increasing or decreasing. So we can expect only monotonic C^1 splines of degree $(1|1)$ on $\mathbb{R}P^1$.

C^1 *spline interpolation problem on* $\mathbb{C}P^1$. For given parameter values $\tau_0 < \cdots < \tau_{n-1} \in \mathbb{R}$ and points $a_0, \cdots, a_{n-1} \in \mathbb{C}$, we are going to find a C^1 rational spline of degree $(1|1)$ interpolating this data $f(\tau_i) = a_i$, $i = 0, \ldots, n - 1$. Thus f restricted on every interval $[\tau_i, \tau_{i+1}]$ must be a fractional linear function $f_{i,i+1}$ of type (23), such that derivatives on the endpoints coincide $f'_{i-1,i}(\tau_i) = f'_{i,i+1}(\tau_i) = v_i$. This results in the system of $n - 1$ equations for n unknown derivatives:

$$v_i v_{i+1} = d_i^2, \quad d_i = \frac{a_{i+1} - a_i}{\tau_{i+1} - \tau_i}, \quad i = 0, \ldots, n - 2. \tag{25}$$

For arbitrary choice of v_0 all other v_i are easily calculated:

$$v_i = \left(D_i^2 v_0\right)^{(-1)^i}, \quad D_i = d_0 d_1^{-1} d_2 \cdots d_{i-1}^{(-1)^i}, \quad i = 1, \ldots, n - 1. \tag{26}$$

We can close the spline f formally adding a point $a_n = a_0$, an 'interval' $\tau_{n-1} > \tau_0$, and the following equation:

$$v_{n-1}v_0 = d_{n-1}^2, \quad d_{n-1} = \frac{a_0 - a_{n-1}}{\tau_0 - \tau_{n-1}}. \tag{27}$$

In the case of odd n we can find a unique v_0 (up to a sign) from equation $v_0^2 = D_n^{-2}$. In the case of even n there are infinitely many solutions if $D_n = 1$ and there is no solutions otherwise.

The analogous C^1 interpolation problem with rational degree $(1|1)$ spline on $\mathbb{R}P^1$ may be solved in the same way as on $\mathbb{C}P^1$. As noted above, initial data should be monotonic, i.e. $a_0 < \cdots < a_{n-1} \in \mathbb{R}$, or $a_0 > \cdots > a_{n-1}$.

C^1 bi-arc spline on $\mathbb{C}P^1$ and $\mathbb{R}P^1$. Let points a_0, a_2 and derivatives v_0, v_2 are given. Then using (25) for $i = 0, 1$ we derive the equation

$$\frac{a_1 - a_0}{a_2 - a_1} = \pm\sqrt{\frac{v_0}{v_2}}$$

with the unknown a_1, which has two solutions on $\mathbb{C}P^1$ and only one in the real case ($v_0 v_1 > 0$ and the sign should be plus).

Hermite interpolation problem with degree $(1|2)$ rational curve on $\mathbb{C}P^1$ and $\mathbb{R}P^1$. For given points $a_0, a_2 \in \mathbb{C}$ and derivatives $v_0, v_1 \in \mathbb{C}$ we look for a rational curve of degree $(1|2)$ interpolating this data on interval $[0,1]$: $f(0) = a_0$, $f(1) = a_2$ $f'(0) = v_0$, $f'(1) = v_1$. We express f in a rational Bézier form

$$f(\tau) = \frac{a_0 w_0 (1 - \tau)^2 + a_1 w_1 2 (1 - \tau)\tau + a_2 w_2 \tau^2}{w_0(1 - \tau) + w_2\tau}. \tag{28}$$

where $w_1 = (w_0 + w_2)/2$. To solve this problem we need to solve the system of equations:

$$f'(\tau_0) = v_0, \quad f'(\tau_1) = v_1 \tag{29}$$

which gives control point $a_1 \in \mathbb{C}$ and the ratio of weights $w_0, w_2 \in \mathbb{C}$.

$$\frac{w_2}{w_0} = -\frac{a_2 - a_0 - v_0}{a_2 - a_0 - v_1}, \quad a_1 = a_0 + v_0\frac{a_2 - a_0 - v_1}{v_0 - v_1} \tag{30}$$

In the real case this system gives a bounded solution on the interval $[0, 1]$ if $w_2 w_0 > 0$. This is equivalent to a certain 'convexity' of initial data: $v_0 < a_2 - a_0 < v_1$ or $v_0 > a_2 - a_0 > v_1$.

It is easy to check that $(1|1)$ curves on the affine part $\mathbb{C}$ (with real coordinates $(\mathrm{Re}(z), \mathrm{Im}(z))$) of $\mathbb{C}P^1$ are circular arcs. Therefore, we have constructed on the plane $\mathbb{R}^2$ open and closed circular interpolating C^1 splines and special quartic curves with given derivatives at the endpoints. The image of these curves under generalized stereographic projection (cf. sect. 4.1)

$$(z_0, z_1) \mapsto \left(|z_0|^2 + |z_1|^2, 2\mathrm{Im}(z_0 \bar{z}_1), 2\mathrm{Re}(z_0 \bar{z}_1), |z_0|^2 - |z_1|^2\right)$$

are a circular spline and quartic curves on the unit sphere respectively (see Fig. 4 (left)).

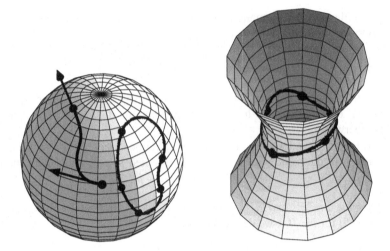

Fig. 4. C^1-splines on oval and double ruled quadrics

Remark 14. Quartic spherical rational curves are studied by Jüttler and Wang [11]. Our quartics on the sphere uniquely interpolate Hermite data but they are not spherical invariant. Nevertheless the construction can be used for other oval quadrics and other toric surfaces in nonsymmetric cases.

5.3   Splines on $\mathbb{R}P^1 \times \mathbb{R}P^1$

Since any curve on $\mathbb{R}P^1 \times \mathbb{R}P^1$ is encoded by two curves on a projective line, we can use results from the previous section. These C^1 splines have one disadvantage – they are monotonic in both directions.

For example, we can interpolate even number of points on the double ruled quadric $\mathbb{R}Q$ using the Segre map (1) and treating two pairs of variables (s_0, s_1) and (t_0, t_1) as homogeneous coordinates of two copies of $\mathbb{R}P^1$. Here monotonicity means that the spline curve cannot touch any of ruling lines as in Fig. 4 (right). For splines of more complicated shape splines of degree $(1|2)$ at least in one direction are required.

Remark 15. It is clear how to apply this construction to any tensor product surface of bidegree (p, q) with the resulting C^1 splines of degree $p + q$. This is the minimal possible degree of C^1-splines on such surfaces (only isoparametric curves can have lower degree). In traditional approach we draw at least quadratic C^1-spline curves on the square domain and then compose with the parametrization map. Hence, the resulting splines are of degree $2(p + q)$.

5.4   Splines on Hirzebruch Surfaces

A *Hirzebruch surface* $\mathbb{F}_k$ is an abstract toric surface associated with the fan $\Sigma_k = \{(1, 0), (0, 1), (-1, -k), (0, -1)\}$, $k = 1, 2, 3, \ldots$. Here we are mostly interested in a canonical real part $\mathbb{RF}_k$ of $\mathbb{F}_k$. According to Definition 4, $\mathbb{RF}_k$ is defined as a quotient of the parameter space $\mathbb{R}^4 \setminus Z$, $Z = \mathbb{R}^2 \times \{(0,0)\} \cup \{(0,0)\} \times \mathbb{R}^2$, by the action of the group $G = \mathbb{R}^* \times \mathbb{R}^*$:

$$(\lambda, \mu) \cdot (s_0, s_1, t_0, t_1) = (\lambda s_0, \lambda \mu^k s_1, \mu t_0, \mu t_1). \tag{31}$$

Hence the formula $(s_0, s_1, t_0, t_1) \mapsto (t_0, t_1)$ correctly defines a projection $\Phi : \mathbb{RF}_k \to \mathbb{R}P^1$.

Now we are ready to solve the following interpolation problem: for given points $a_i \in \mathbb{RF}_k$ and a given interpolation $f : \mathbb{R} \to \mathbb{R}P^1$ of their projections $f(\tau_i) = \Phi(a_i)$, find such curve $F : \mathbb{R} \to \mathbb{RF}_k$ that $\Phi \circ F = f$. For simplicity, let f have degree 1. This means that some linear polynomials $t_0(\tau)$, $t_1(\tau)$ are fixed, and we can find a pair of polynomials $(s_0(\tau), s_1(\tau))$ of degree $(d|d+k)$ that interpolate $N = 2d + k + 1$ points using Proposition 2.

We have seen in Example 9 that the cubic surface W is parametrized by $\mathbb{RF}_1$. Therefore, using degree $(1|2)$ functions in variables (s_0, s_1) we can interpolate any 4 points (monotonic in the direction of (t_0, t_1)) with cubic curves and also solve the Hermite interpolation problem.

Another interesting application is related to the quadratic cone C_2.

Example 16. From Example 7 we know that the desingularization of C_2 coincides with the Hirzebruch surface $\mathbb{F}_2$. Using polynomials (i.e. rational functions of degree $(0|2)$) in 'vertical variables' (s_0, s_1) and rational functions of degree $(1|1)$ in 'horizontal variables' (t_0, t_1) one can interpolate any three points on $\mathbb{R}C_2$. This leads to the theory of C^1-spline curves on $\mathbb{R}C_2$ similar to the splines of degree $(1|1)$ on $\mathbb{R}P^1$. In particular the Hermite interpolation problem can be solved using biarc conical C^1-splines on $\mathbb{R}C_2$.

5.5   Splines on a Hexagonal Toric Bézier patch

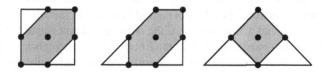

Fig. 5. Three cases of superscribed lattice polygons

We illustrate our approach to splines on general toric surfaces in the case of a hexagonal toric Bézier patch H (cf. [15, 16]) associated with lattice hexagon Δ (shaded in Fig. 5). Consider three different biquadratic parametrizations $\phi_i : \mathbb{R}P^1 \times \mathbb{R}P^1 \to H$, $i = 1, 2, 3$ defined by skipping different opposite pairs of coordinates in the URP. For example, $f_1 : (s_0, s_1, t_0, t_1) \mapsto P_\Delta(s_0, t_0, 1, s_1, t_1, 1)$ is associated with circumscribed square in Fig. 5 (left). Now we can draw C^1 quartic splines of any shape on the hexagonal patch choosing locally the most convenient parametrization. For example, in Fig. 6

we see a closed C^1 bi-arc spline of degree 8 on a hexagonal patch of depth 2 (i.e. associated with a hexagon shown on the right). This is the minimal possible degree of such spline.

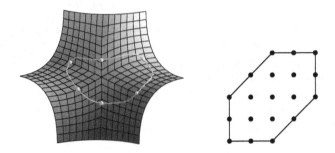

Fig. 6. A closed C^1 degree 8 spline on a hexagon of depth 2

Similarly we can consider six different parametrizations of Hirzebruch type of the hexagonal toric patch. These parametrizations correspond to the obvious six distinct circumscribed trapezoids (as it is shown in Fig. 5, middle). So we generate Hermite interpolation with quintic curves of any shape on the hexagonal patch of depth 1 choosing locally the most convenient parametrization.

6   Application: Blending Natural Quadrics

Spheres, circular cylinders and circular cones are perhaps the most popular surfaces in Geometric Modeling. These surfaces are called natural quadrics in order to distinguish them from more general quadrics (e.g. elliptic cones). One exceptional property of natural quadrics is that their offset surfaces are of the same type. Therefore, it is natural to use canal surfaces for smooth blending of natural quadrics. In fact, this blend is the traditional rolling ball blend. Unfortunately, if the radius of the ball is fixed, then its center curve is not rational in most cases. There have been several attempts to use a ball with rational radius when the corresponding canal surface can be rationally parametrized.

Here we propose a method that can help to improve this approach. We shall explain our idea for the case of blending of two circular cylinders C_1 : $x_2^2 + x_3^2 = r_1^2$ and C_2 : $x_1^2 + x_3^2 = r_2^2$, where $0 < r_1 < r_2$. (In [12] it is explained how two cylinders or cones in very general position can be transformed into a few canonical cases including this one.) Consider the space of all spheres in $\mathbb{R}^3$ as a 4-dimensional affine space $\mathbb{R}^4$ (or its projective version $\mathbb{R}P^4$), where the first three coordinates are the center point of a sphere and the last coordinate represents the radius of the sphere. The conditions that a sphere touches both cylinders C_1 and C_2 can be described by the following system of homogeneous equations in $\mathbb{R}P^4$

$$\begin{cases} x_2^2 + x_3^2 = (r_1 x_0 - x_4)^2, \\ x_1^2 + x_3^2 = (r_2 x_0 - x_4)^2. \end{cases}$$

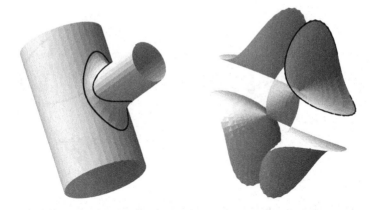

Fig. 7. Cylinder/cylinder blend and the curve $\pi(\gamma)$ on the surface $\pi(X)$

This system can be transformed to the equivalent system

$$\begin{cases} y_3^2 = y_0 y_4, \\ y_3^2 = y_1 y_2, \end{cases} \tag{32}$$

that describes a real toric surface $X \subset \mathbb{R}P^4$. Indeed, X is a canonical real part of $X_\diamond$, which is defined by the same equations (17).

Our task is to find a particular family of spheres touching both cylinders. Hence we need to find a curve γ on the surface X with some prescribed properties. A quartic curve for this case is described in [12]. It can be proved that four is the minimal possible degree for a curve on X with this topology. Let $\pi : \mathbb{R}^4 \rightarrow \mathbb{R}^3$ denote the orthogonal projection, that forgets x_4. In Fig. 7(right) we see a projection $\pi(X)$ (in fact a bisector surface of both cylinders) and a spine curve $\pi(\gamma)$ of the blending canal surface on $\pi(X)$. Unfortunately, as we can see in Fig. 7 (left) the blending ring has unwanted variation of the width.

To fix this problem, the natural approach is to use C^1 splines instead of the curve γ in order to improve the shape.

We chose a 'conical parametrization' of the toric surface X that correspond to the circumscribed triangle in Fig. 5 (right): variables s_0, t_1, u_1, v_0, v_1 are substituted by 1 in the universal parametrization $\widetilde{P}_\diamond$ (see (20)):

$$F(s_1, t_0, u_0) = \widetilde{P}_\diamond \begin{pmatrix} s_1 & 1 & t_0 \\ 1 & & 1 \\ 1 & u_0 & 1 \end{pmatrix} = \begin{pmatrix} & u_0^2 & \\ t_0^2 u_0 & s_1 t_0 u_0 & s_1^2 u_0 \\ & s_1^2 t_0^2 & \end{pmatrix} \tag{33}$$

Therefore, we can use C^1 splines on a Hirzebruch surface $\mathbb{R}\mathbb{F}_2$ as in Example 16, sect. 5.4. We fix degrees of polynomials $\deg s_1 = \deg t_0 = 1$ and $\deg u_0 = 2$. Hence, (s_1, t_0) are treated as 'horizontal variables and u_0 is 'vertical'. The particular choice of these polynomials defines the curve $\gamma(\tau)$, which has four symmetric arcs. Using Hermite interpolation with biarc C^1-splines one can modify $\gamma(\tau)$ in a symmetric fashion.

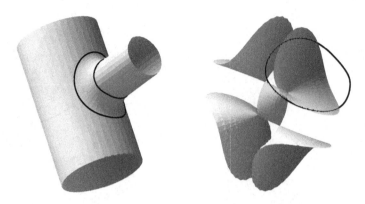

Fig. 8. New cylinder/cylinder blend and the spline curve $\pi(\gamma')$ on $\pi(X)$

The resulting curve $\gamma'(\tau)$ will have 8 quartic segments and will give us the desirable shape shown in Fig. 8.

Final remark. Other cases of blending for natural quadrics also lead to toric surfaces. For example, the case of an external sphere/cylinder blend defines a toric surface $\mathbb{R}_c X_\Diamond$ which has the non-canonical real structure (the same as in sect. 4). Since the required closed curve is homotopic to point on $\mathbb{R}_c X_\Diamond$, it can be shown that the minimal degree of such curve is 8. On the other hand, this curve can be approximated by a C^1 *quartic* spline!

7   Conclusions

In this paper we develop theory of toric surfaces in terms of toric homogeneous coordinates and have made an attempt to explain universal rational parametrizations (UPR) in quite elementary way. Theorem 1 (borrowed from [3]) describes the class of complex surfaces that admit URP as a large subclass of all rational surfaces: they are sufficiently nice projections of projective toric surfaces associated to lattice polygons. We extend this class further by considering both canonical and non-canonical real parts of such complex surfaces and call them *real almost toric surfaces*.

Our general idea is to develop design theory on abstract toric surfaces using homogeneous coordinates. This can reduce essentially the number of different cases that require consideration, since lattice polygons with the same normal fans (or desingularizations) have the same universal rational parametrizations. For example, all tensor product Bézier surfaces have the same URP as the product of two real projective lines $\mathbb{R}P^1 \times \mathbb{R}P^1$, the cone C_k over a rational curve of degree k has the same URP as the Hirzebruch surface $\mathbb{F}_k$ (sect. 5.4), toric Bézier patches of the different depths [16] have the same URP etc.

We develop, as an example, a theory of rational spline curves of minimal degree on toric surfaces. The key point is to use nice interpolation properties on the complex projective line (Proposition 2) and distinguished directions on simplest toric surfaces.

We conclude this paper with one blending construction between two cylinders, where quartic splines on a certain toric surface are essentially used. This approach can be extended to other blending cases between natural quadrics and is a subject of current research.

References

1. Cauchy, A.: Oeuvres Complètes, II-e série, tome III, Paris, Gauthier-Villars, 1897
2. Cox, D.: What is a toric variety?, in: Topics in Algebraic Geometry and Geometric Modeling, Contemporary Mathematics **334** (2003) 203–223
3. Cox, D., Krasauskas, R., Mustaţă, M.: Universal rational parametrizations and toric varieties, in: Topics in Algebraic Geometry and Geometric Modeling, Contemporary Mathematics **334** (2003) 241–265
4. Cox, D., Little, J., and O'Shea, D.: Using Algebraic Geometry, GTM, Springer, Berlin, 1998
5. D'Andrea, C., Khetan, A.: Implicitization of rational surfaces using toric varieties, 2004, arXiv:math.AG/0401403
6. Delaunay, C.: Real structures on smooth compact toric surfaces, in: Topics in Algebraic Geometry and Geometric Modeling, Contemporary Mathematics **334** (2003) 267–290
7. Dietz, R., Hoschek, J., and Jüttler, B.: An algebraic approach to curves and surfaces on the sphere and other quadrics, Computer Aided Geometric Design **10** (1993) 211–229
8. Dietz, R., Hoschek, J., and Jüttler, B.: Rational patches on quadric surfaces, Computer-Aided Design **27** (1995) 27–40
9. Fulton W.: *Introduction to Toric Varieties*, Princeton U. Press, Princeton, NJ, 1993
10. Jüttler, B., Rittenschober, K.: Using Line Congruences for Parameterizing Special Algebraic Surfaces, in Wilson, M. and Martin, R.R. (eds.): The Mathematics of Surfaces X, Lecture Notes in Computer Science, **2768**, Springer, Berlin, 2003, 223–243.
11. Jüttler, B., Wang, W.: The shape of spherical quartics, Computer Aided Geometric Design **20** (2003) 621–636
12. Kazakevičiūtė, M., Krasauskas, R.: Blending cylinders and cones using canal surfaces, in Nonlinear Analysis: Modelling and Control, Vilnius, IMI, **5** (2000) 77–89
13. Krasauskas, R.: Universal parameterizations of some rational surfaces, in A. Le Méhauté, C. Rabut and L.L. Schumaker (eds.): Curves and Surfaces with Applications in CAGD, Vanderbilt Univ. Press, Nashville, 1997, 231–238.
14. Krasauskas, R.: Shape of toric surfaces, in: Durikovic, R., Czanner, S. (eds.), Proceedings of the Spring Conference on Computer Graphics SCCG 2001, IEEE, 2002, 55–62
15. Krasauskas, R.: Toric surface patches, Advances in Computational Mathematics **17** (2002) 89–113
16. Krasauskas, R., Goldman, R.: Toric Bézier patches with depth, in: Topics in Algebraic Geometry and Geometric Modeling, Contemporary Mathematics **334** (2003) 65–91
17. Mäurer, C.: Generalized Parameter Representations of Tori, Dupin Cyclides and Supercyclides, in A. Le Méhauté, C. Rabut and L.L. Schumaker (eds.): *Curves and Surfaces with Applications in CAGD*, Vanderbilt Univ. Press, Nashville, 1997, 295–302.
18. Müller, R.: Universal parametrization and interpolation on cubic surfaces, Computer Aided Geometric Design **19** (2002) 479–502.

Panel Discussion

Contributors: Marc Daniel (MD), Tor Dokken (TD), Jens Gravesen (JG), Panagiotis Kaklis (PG), Rimvydas Krasauskas (RK), Ralph R. Martin (RM), Bernard Mourrain (BM), Josef Schicho (JS), Carlo Traverso (CT), Joab R. Winkler (JW)

Moderator: Bert Jüttler (BJ)

BJ: I have prepared a few questions to this audience for this panel discussion. Of course, these are just suggestions; any other comments or questions are welcome.

This workshop has brought together experts from different communities: Computer Aided Geometric Design, Symbolic Computation, and Algebraic Geometry.

Do you think that this combination has made sense? Did you benefit from the presentations and discussions at this workshop, also if they were not from your "native" field?

(Many:) Yes.

CT: During the programme of the workshop, new challenges for symbolic computation have been identified. We have seen examples where other approaches [from CAGD, B.J.] fail, and where symbolic techniques might help. It is unusual to see the failures of others in conferences, but it helps to identify new problems.

JW: This combination forces you to live in a wider world, not just in your native field.

MD: Currently, we are experiencing a great change in CAGD. While the traditional approaches relied almost exclusively on parametric representations, now implicitly defined curves and surfaces start to play a greater role. It is very interesting to exploit the advantages of both representations, and we will have to work to make the most out of it.

RM: In the past, symbolic methods were often regarded as being too slow to be of any practical use for geometric applications. This workshop has shown us that combining symbolic and numerical methods can be done in a useful way, to make practical algorithms — for example, being able to replace rationals with reals for most of the calculation, and only use more precise methods when specifically required.

JS: The workshop was a good opportunity to meet people who one does not meet usually at workshops and conferences in my field; I have enjoyed this very much.

JW: It has been stated by a leading mathematical authority that numerical analysis and computational linear algebra will become increasingly important because many problems only have a numerical solution, and thus a discrete (eg linear algebraic) rather than continuous (eg integral) equation is used to obtain the solution. The CAGD community must be aware of these numerical issues in order that reliable solutions to difficult problems be obtained.

TD: It seems that the CAD industry is currently not very interested in new ideas. Due to the concentration process CAD market, now very few vendors dominate the market, and they are trying to maximize their profits, not necessarily to improve their products. As another trend, high–end CAD systems are now being sold to everybody, since powerful computers are now available to everyone. Still, many things in CAD systems have to be improved, such as the robustness of intersection algorithms, and the data exchange. It is currently not clear to me whether the CAD industry is ready to address these issues.

BJ: It is hoped that the combination of knowledge from different fields helps to solve real problems and to gain new insights.

Which problems from Computer Aided Geometric Design, Symbolic Computation, or Algebraic Geometry could benefit from the use of methods or results originating in the other fields?

Are there any (obvious) (more or less) new questions in one of these fields, which are motivated by the other ones?

RM: Is there a way to parameterize an implicitly defined curve or surface by functions involving not only rational expressions, but also square roots, or even more general functions? More usefully, can algorithms be devised for doing this? Methods to decide if and when this is possible or not would be useful, but even more useful would be algorithms to generate the parameterization.

As another question, Sturm sequences and their multivariate generalizations can be used for counting the number of real zero-dimensional roots in a box. What if the solutions have higher dimension? Is an algorithm possible to count the number of (real) separate connected solution pieces of each degree which are contained in a box?

BM: The answer is yes! This is answered by Hermite's Theorem.

CT: No, the situation becomes more difficult if one is interested in real solutions.

MD: Is there still a future for NURBS surfaces? Is it still interesting to concentrate on these surfaces? Very often one starts from points, and it is not always straightforward to generate surfaces. Perhaps it would be better to continue working with the points, instead of bothering to generate surfaces.

BM: As an advantage, NURBS surfaces encode the geometry in a very compact way.

TD: NURBS surfaces are in the standards! However, NURBS are currently used only to represent relatively simple shapes. As a matter of fact, NURBS can represent much more complex shapes, but currently we do not know how to do this. We should try to develop techniques for fully exploiting the potential of NURBS representations. This is closely related to the parameterization problem, that is, to the problem of parameterizing implicitly defined surfaces.

JS: In Bernard Mourrain's talk, we have seen that it is difficult to correctly visualize surfaces, especially in the neighborhood singularities. What precisely is the mathematical information needed for visualizing surface singularities correctly?

TD: A new approach, which exploits exploit capabilities of recent hardware, is to avoid triangulation completely. Instead, the surface can be evaluated directly.

PK: I propose a "life–cycle" philosophy for CAD: During their life, from conceptual design via numerical analysis and simulation to detailed construction and manufacturing, curves and surfaces may need different representations, but the different representations should "talk to each other". What we need is research on different representations, also in order to explore the transformations between them. Obviously, different representations of curves and surfaces are differently well suited for certain applications in the product life cycle.

TD: Currently, a big problem is to built the results of numerical simulations into an existing CAD model. More precisely, the results of a simulation has to be reflected in the model. Currently, this is a very difficult problem in industry, and I expect that this will not be fully solved within the near future.

RM: Many geometric problems can be viewed as finding the solutions to a set of algebraic (or more general implicit) equations in a set of unknowns. These may represent geometry, constraints on the geometry, and so on.

Implicitization and parameterization can be looked at as being specific questions regarding particular sets of equations: if we have multiple equations, how can we reduce them to a single equation; if we have a single equation, how can we introduce extra variables to help us e.g. draw the geometry?

In the more general setting, we can now ask a more general question — how can we transform the set of equations into a new set of more or fewer, or just different, equations, in more or fewer or just different variables? When it is better to decrease the number of variables, at the expense of a more complicated representation? When is it better to increase the number of variables, in order to obtain a simpler description?

Clearly, Groebner bases are related to this question, but are not the complete answer, I believe, as the issue of parameterization is not really addressed by Groebner bases.

JS: In algebraic geometry, these are the concepts of projection vs. unprojection.

RK: Can we identify a class of surfaces which have a "good" implicit and parametric representation at the same time? As a good class of candidates one could look at the class of del Pezzo surfaces of degree 4, which we have seen in Josef Schicho's talk.

JG: I do not think that this will help. For instance, for practical problems from physics, the geometry of an object might already be given. In this situation, the restriction to such a special class of surfaces would make the life much more difficult.

TD: I agree that simple surfaces might not be flexible enough for everything. However, I have observed that designers simply like certain shapes.In recent years, CAD systems have not seen much development in user interfaces. Still, what we have are mostly 2D interfaces! For instance, 3D curves still mainly generated by intersections. This is not satisfying at all, and other techniques would be much better, since designers still have big problems to get their ideas into the CAD system, and sometimes they simply give up trying. So, the current CAD systems pose limits to the designers' creativity, instead of inspiring them. Perhaps, certain classes of surfaces, of CAD models can contribute to shape modeling. But how can we achieve this? How can we explain better that certain shapes, something like a library of shapes, are available?

RK: This is a question of good control handles!

TD: We need to interface of different technologies, perhaps virtual reality.
   As another issue, intersection algorithms are still difficult. I believe that a better theoretical basis for CAD will be needed in the future, since the growing demands for accuracy make problems like intersection even more difficult.

JG: Is there sensible way to restrict CAD systems to stable singularities? I feel that unstable singularities should be excluded beforehand, unless we are able to handle them in a reliable way.

JW: What is a stable singularity?

RM: Part of the issue is what is primary information, and what is derived information. An example of an unstable computation, given in the talk by Vibeke Skytt, is the attempt to compute intersections between tangential surfaces, e.g. a blending surface, and one of the base surfaces used to define it. Here the primary information is really the base surface, and the trimming curve on the base surface which defines where the blend meets it. The blend surface should be derived from this information, rather than trying to compute the intersection curve from the two surfaces.
   Some singularities can be avoided by being careful what operations the user interface allows the user to perform.

BM: I believe that, singularities cannot always be avoided. Instead, it is better to find the exact solution, and to deal with it directly.

JG: I agree.

TD: I have experienced that designers like singular shapes, and near-singular situations, such as surfaces meeting each other tangentially or almost tangentially. Thus, we will have to face these problems.

JG: This approach should be embedded in the underlying design philosophy, that is, in the user interface!

TD: According to my experience, trimmed surfaces often make problems.

RM: As a comment, the current version of STEP (the ISO standard for the exchange of CAD data) does not include the design history. The ISO certainly see this as a serious deficiency, and future versions will include such information. This will help to avoid some of the singularities and other problems caused by exporting data from one system and importing it into another with tighter tolerances. This symbolic information may also help systems to ascertain what the designer's intent was in singular cases.

TD: I agree, but converting design histories from one CAD system to another is much more difficult than converting models. This would imply to standardize not only the underlying representation, but even the available design tools? Already now, people simply cannot design certain shapes, due to the lack of available tools. A further standardization would make things worse.

JG: Here, open standards would do much better! It should be possible to include new methods, new types of surfaces or new classes of shapes into the standard.

TD: This may not be in the genuine interest of the CAD industry. Nevertheless, it will be important for other industries. In any case, this is a great challenge, this design history can sometimes be very, very difficult. For instance, if the design has partly be obtained by a numerical simulation.

As another remark, I feel that something like a "CAGD-hikers guide to algebraic geometry", and vice versa might be helpful. We use many similar concepts, but they come with different names.